Mathematics Applied to Engineering and Management

Mathematical Engineering, Manufacturing, and Management Sciences

Series Editor: Mangey Ram, Professor, Department of Mathematics; Computer Science and Engineering, Graphic Era Deemed to be University, Dehradun, India

The aim of this new book series is to publish the research studies and articles that contain the latest development and research applied to mathematics and its applications in the manufacturing and management sciences areas. Mathematical tools and techniques are the strength of engineering sciences. They form the common foundation of all novel disciplines as engineering evolves and develops. The series will include a comprehensive range of applied mathematics and its application in engineering areas such as optimization techniques, mathematical modeling and simulation, stochastic processes and systems engineering, safety-critical system performance, system safety, system security, high-assurance software architecture and design, mathematical modeling in environmental safety sciences, finite element methods, differential equations, and reliability engineering.

Sustainable Procurement in Supply Chain Operations

Edited by Sachin Mangla, Sunil Luthra, Suresh Jakar, Anil Kumar, and Nirpendra Rana

For more information on this series, please visit: https://www.crcpress.com/Mathematical-Engineering-Manufacturing-and-Management-Sciences/book-series/CRCMEMMS

Mathematics Applied to Engineering and Management

Edited by

Mangey Ram
S. B. Singh

CRC Press is an imprint of the
Taylor & Francis Group, an informa business

CRC Press
Taylor & Francis Group
6000 Broken Sound Parkway NW, Suite 300
Boca Raton, FL 33487-2742

First issued in paperback 2021

ISBN-13: 978-0-367-77930-6 (pbk)
ISBN-13: 978-0-8153-5804-6 (hbk)

Library of Congress Cataloging-in-Publication Data

Names: Ram, Mangey, editor. | Singh, S. B. (Engineer) editor.
Title: Mathematics applied to engineering and management / edited by Mangey Ram and S.B. Singh.
Description: Boca Raton : Taylor & Francis, a CRC title, part of the Taylor & Francis imprint, a member of the Taylor & Francis Group, the academic division of T&F Informa, plc, 2019. | Series: Mathematical engineering, manufacturing, and management sciences | Includes bibliographical references.
Identifiers: LCCN 2019015378 | ISBN 9780815358046 (hardback : acid-free paper) | ISBN 9781351123303 (e-book)
Subjects: LCSH: Engineering mathematics. | Industrial management--Mathematics.
Classification: LCC TA330 .M325 2019 | DDC 620.001/51--dc23
LC record available at https://lccn.loc.gov/2019015378

Visit the Taylor & Francis Web site at
http://www.taylorandfrancis.com

and the CRC Press Web site at
http://www.crcpress.com

Contents

Preface

There has been ever-increasing challenges, trends, and attention paid to applying mathematics to deal with engineering and managerial problems. Engineers and managers more often find that ultimately their problems are transformed into some of the mathematical forms, which itself brings mathematics into the concerned field of study. We further believe that researchers involved in problems dealing with practical engineering and management approaches may have to apply mathematics to find the solutions they're looking for. Moreover, decision-making by managers, practitioners, and engineers can be made more effective if the outcome is decided by the combination of management and engineering theories with the application of effective mathematical and statistical tools.

The fact of the matter is that mathematics is widely applied to solve the many problems of physical and the material world. Keeping these facts and the title of the book in mind, this volume contains applications of mathematical theories and models to problems such as warranty analysis, supervised machine learning, inventory systems, nonlinear mechanics, energy modeling, reliability evaluation, and hydromagnetic flow.

This book is useful to senior graduate, postgraduate students, and research scholars. Of course, it can also be used as a reference book by academicians, scientists, engineers, and managers working in different organizations. We wish to make it clear that this volume is small in an attempt to highlight some contemporary works in which mathematics is applied to engineering and managerial problems.

Mangey Ram
Graphic Era Deemed to be University
Dehradun, India

S. B. Singh
G. B. Pant University of Agriculture & Technology
Pantnagar, India

Acknowledgments

The Editors acknowledge CRC press for this opportunity and professional support. Also, we would like to thank all the chapter authors and reviewers for their availability for this work.

Editors

Dr. Mangey Ram received a PhD in Mathematics and a minor in Computer Science from G. B. Pant University of Agriculture and Technology, Pantnagar, India. He has been a faculty member for about 10 years and has taught several core courses in pure and applied mathematics at the undergraduate, postgraduate, and doctorate levels. He is currently a Professor at Graphic Era (Deemed to be University), Dehradun, India. Before joining Graphic Era, he was a Deputy Manager (Probationary Officer) with Syndicate Bank for a short period. He is Editor-in-Chief of the *International Journal of Mathematical, Engineering and Management Sciences* and the guest editor and member of the editorial board of various journals. He is a regular reviewer for international journals, including IEEE, Elsevier, Springer, Emerald, John Wiley, Taylor & Francis, and many others. He has published 150 plus research publications in IEEE, Taylor & Francis, Springer, Elsevier, Emerald, World Scientific, and many other national and international journals of repute and also presented his works at national and international conferences. His fields of research are reliability theory and applied mathematics. Dr. Ram is a Senior Member of the IEEE; life member of Operational Research Society of India; Society for Reliability Engineering, Quality and Operations Management in India; Indian Society of Industrial and Applied Mathematics; member of International Association of Engineers in Hong Kong; and Emerald Literati Network in the UK. He has been a member of the organizing committee of a number of international and national conferences, seminars, and workshops. He has been conferred with "Young Scientist Award" (2009) by the Uttarakhand State Council for Science and Technology, Dehradun. He has been awarded the "Best Faculty Award" (2011); "Research Excellence Award" (2015); and "Outstanding Researcher Award" (2018) for his significant contribution in academics and research at Graphic Era Deemed to be University, Dehradun, India.

Dr. S. B. Singh is a Professor in the Department of Mathematics, Statistics and Computer Science, G. B. Pant University of Agriculture and Technology, Pantnagar, India. He has more than two decades of teaching and research experience undergraduate and postgraduate students at different engineering colleges and universities. Professor Singh is a member of Indian Mathematical Society; Operations Research Society of India; and National Society for Prevention of Blindness in India. He is a regular reviewer of many books and international and national journals. He has been a member of the organizing committee of many international and national conferences and workshops. He is an editor of the *Journal of Reliability and Statistical Studies*. He has authored and coauthored eight books on different courses of applied/engineering mathematics. He has been conferred with four national awards. He has published his research works at national and international journals of repute. His area of research is reliability theory, fuzzy logic, and applied mathematics.

Contributors

Richard Arnold
School of Mathematics and Statistics
Victoria University of Wellington
Wellington, New Zealand

G. Arora
Lovely Professional University
Phagwara, India

M. Bashir
Lovely Professional University
Phagwara, India

Santosh Chaudhary
Department of Mathematics
Malaviya National Institute of Technology
Jaipur, India

Stefanka Chukova
School of Mathematics and Statistics
Victoria University of Wellington
Wellington, New Zealand

Alok Dhaundiyal
Szent Istvan University
Gödöllő, Hungary

Vladimir Dmitriev
Moscow Aviation Institute (SNRU)
Moscow, Russia

Afrooz Farhadi
Department of Industrial Engineering and Management
Sadjad University of Technology
Mashhad, Iran

Amos E. Gera
SCE College of Engineering
Ashdod, Israel

Muammel M. Hanon
Middle Technical University
Baghdad, Iraq

Yu Hayakawa
School of International Liberal Studies
Waseda University
Tokyo, Japan

Taha-Hossein Hejazi
Department of Industrial Engineering
Amirkabir University of Technology
Tehran, Iran

Bahareh Hekmatnia
Department of Industrial Engineering and Management
Sadjad University of Technology
Mashhad, Iran

KM Kanika
Department of Mathematics
Malaviya National Institute of Technology
Jaipur, India

Leonid Kondratenko
Central Research Institute of Machine Building Technology
Moscow, Russia

Amit Kumar
Lovely Professional University
Phagwara, India

Monika Manglik
University of Petroleum and Energy Studies
Dehradun, India

Sarah Marshall
Department of Mathematical Sciences
Auckland University of Technology
Auckland, New Zealand

Lubov Mironova
Moscow Aviation Institute (SNRU)
Moscow, Russia

Nisha Nautiyal
Department of Mathematics, Statics, and Computer Science
Govind Ballabh Pant University of Agriculture and Technology
Pantnagar, India

Bharatendra Rai
University of Massachusetts–Dartmouth
Dartmouth, Massachusetts

S. B. Singh
Department of Mathematics, Statistics, and Computer Science
Govind Ballabh Pant University of Agriculture and Technology
Pantnagar, India

1

Geometric and Geometric-Like Processes and Their Applications in Warranty Analysis

Richard Arnold, Stefanka Chukova, Yu Hayakawa, and Sarah Marshall

CONTENTS

1.1 Introduction

Stochastic processes, such as renewal processes, are often used to model the occurrence of recurrent events over time. Under the assumption of a renewal process, the time between events are modeled to be independent and identically distributed random variables. In many scenarios, this assumption is justifiable and reflects the modeled situation well. However, if trends over time are observed, this assumption does not hold. These types of trends can be observed in many practical problems in a variety of fields (e.g., engineering—a system's lifetime is stochastically decreasing because of aging or imperfect repairs, and at the same time the maintenance/repair time required to keep an aging system operational is stochastically increasing; in epidemiology—the number of infected cases is increasing at the start of an infectious disease outbreak and shows a decreasing trend at the later stages of the outbreak; in economics—the trends in the economic development of a country or a region show a periodic cycle in its gross national product, increasing at the early stage of the cycle and decreasing at the end of the cycle. All these examples have one common feature, their characterization involves a specific monotone trend over a substantial time interval).

To model these types of "simple" trends, a monotone (stochastically increasing or decreasing) behavior of the interevent times, a process called the geometric process (GP), was introduced by Yeh Lam [25]. The GP is a generalization of the renewal process, providing more modeling flexibility than the renewal process by allowing trends but still retaining simplicity, which preserves the tractability of its derived properties and its implementation. In addition, a variety of extensions of the GP, which we refer to in this chapter as geometric-like processes (GLPs), have been proposed to address an even wider range of scenarios and trends.

The aim of this chapter is to review GPs and GLPs and briefly outline some of their applications. We also provide some specific applications of GPs to warranty cost analysis. In Section 1.2 we provide an overview of GPs and GLPs. In Section 1.3 we discuss the application of GPs in warranty analysis and provide details of three alternating process models under two warranty strategies. We apply the models to real data in Section 1.4 and conclude this chapter in Section 1.5.

1.2 GPs and GLPs: An Overview and Applications

1.2.1 Geometric Process

The GP was introduced by Yeh Lam [25]. He points out that the well-studied "good as new" and "minimal" repair scenarios are not useful for modeling trends (e.g., decreasing) in successive survival times, and a new approach is needed to represent possible trends in survival/repair times of systems. The GP has many applications in reliability engineering. For a detailed review of the GP, refer to [21].

The Geometric Process (GP): Let $\{X_n, n=1,2,\ldots\}$ be a sequence of non-negative random variables. If they are independent and the distribution function of X_n, is given by $F_n(x)=F(a^{n-1}x)$ for $n=1,2,\ldots$, where a is a positive constant and $F_1(x)=F(x)$, then $\{X_n, n=1,2,\ldots\}$ is called a geometric process.

Equivalently, if a sequence of non-negative independent random variables $X_n, n=1,2,\ldots$ is such that the sequence $Z_n=a^{n-1}X_n, n=1,2,\ldots$ forms a renewal process (RP), then $X_n, n=1,2,\ldots$ is called a geometric process with parameter a. Letting $g(n)=a^{n-1}$, the distribution of X_n can be written as $F_n(x)=F(g(n)x)$. The function $g(n)$ can be thought of as providing the transformation required to convert a GP to a RP.

A GP is called a decreasing geometric process if $a>1$, and it is called an increasing geometric process if $0<a<1$. If $a=1$, the GP is a RP. The density of X_n is $f_n(x)=a^{n-1}f(a^{n-1}x)$, where $f_1(x)=f(x)$. If $E(X_1)=\lambda$ and $Var(X_1)=\sigma^2$, then

$$E(X_n)=\frac{\lambda}{a^{n-1}} \text{ and } \mathrm{Var}(X_n)=\frac{\sigma^2}{a^{2(n-1)}}.$$

The counting process defined by the GP is $N(x)=\sup\{n: S_n \le x\}, x\ge 0$ where $S_n=X_1+X_2+\ldots+X_n$, with $G_n(x)=P\{S_n\le x\}$. The expected number of failures at time x, $E(N(x))$, satisfies the following integral equation

$$E(N(x))=F(x)+\int_0^x E(N(a(x-y)))dF(y). \tag{1.1}$$

1.2.2 Classification of GLPs

Let $\{X_n, n=1,2,\ldots\}$ be a sequence of random variables with the distribution function of the nth random variable X_n given by $F_n(x)=P(X_n\le x)$. In the case of a RP, $\{X_n, n=1,2,\ldots\}$ are ***independent and identically distributed (iid)***, so $F_n(x)=F(x)$, where $F_1(x)=F(x)$ is the cumulative distribution function (cdf) of X_1. For GLP, the inter-event times are ***independent but not necessarily identically distributed***. Imposing a particular relationship between $F_n(x)$ and $F(x)$ identifies the type of the GLP.

In Figure 1.1, we present a simple taxonomy of GLPs and divide them into three main classes: basic, complex, and other. Many of the models we present here have the characteristics of accelerated life distributions—where the **form** of the distribution is always the same, but its time argument is transformed.

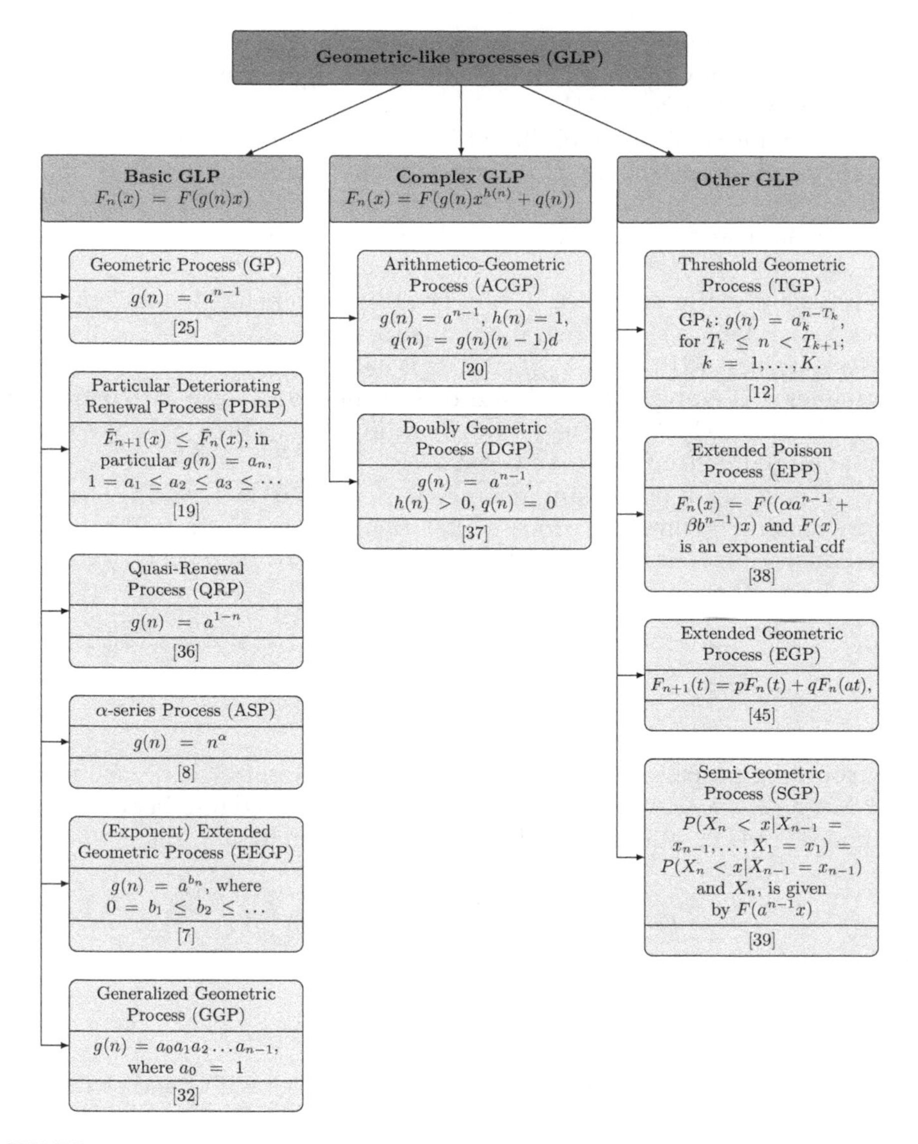

FIGURE 1.1
Taxonomy of geometric-like processes. (From John Braun, W. et al., *Nav. Res. Logisti.*, 52, 607–616, 2005; Maxim S. Finkelstein, M.S., *Microelectron. Reliab.*, 33, 41–44, 1993; Wu, S. and Wang, G., *IMA J. Manag. Math.*, 29, 229–245, 2018.)

- For the first class of GLPs, $F_n(x) = F(g(n)x)$, for some function $g(n)$. The function $g(n)$ could be thought of as providing the transformation required to convert a particular GP/GLP to a RP. We call this first class ***basic*** GLP.
- For the second class of GLP, the transformation required for conversion to RP has a more complex structure; thus, we refer to them

as ***complex*** GLPs. For these GLPs, $F_n(x) = F(g(n)x^{h(n)} + q(n))$, where $g(n)$, $h(n)$ and $q(n)$ are functions satisfying certain conditions that depend on the type of GLP. Again, the expression $g(n)x^{h(n)} + q(n)$ can be thought of as providing the transformation required to convert a particular complex GLP to a RP.

- The third class of GLP extends the GP in other ways, for example, by adding new assumptions about the inter-event times of the process. We refer to these processes as ***other*** GLPs.

Our classification scheme for GLPs is presented in Figure 1.1. The taxonomy we present here provides a starting point for further study of these processes, and the modeling of situations involving monotone trends (see [1] for further details).

1.2.3 Applications of GLPs

Next, we demonstrate the usefulness of GLPs by identifying some areas of their applications with related references. The application areas and selected references are summarized in Table 1.1.

TABLE 1.1

Applications of GLPs

Application	GLP	Selected Articles
Reliability and maintenance	ACGP	Repair-replacement [20]
	DGP	Failure times [37]
	EGP	Failure times [7]
	EGP	Maintenance [40,45,46]
	EPP	Corrective and preventive maintenance [38]
	GP	Failure times [3,11,24], Repair-replacement [14,15,22,25,30,33,41–44], Shock models [23,26,31]
	GGP	Preventive maintenance [32]
	QRP	Maintenance [34–36]
	TGP	Maintenance [21]
Warranty analysis	DGP	Number of claims [37]
	GP	Non-zero repair times [2,27]
	QRP	Instantaneous repair [4,18]
Other	GP	Coal mining disasters [3,11,24], Electricity prices [10], Drug-related arrests over time [13], Patient arrivals [24], vehicle arrivals [24]
	TGP	SARS epidemic [12]

Abbreviations: ACGP, arithmetico-geometric process; DGP, doubly geometric process; EPP, extended poisson process; EGP, extended geometric process; GGP, generalized geometric process; GP, geometric process; QRP, quasi-renewal process; SARS, severe acute respiratory syndrome; TGP, threshold geometric process.

1.3 Applications of GPs in Warranty Analysis

1.3.1 Basics of Warranty Analysis

A product warranty is an agreement offered by a producer to a consumer to repair or replace a faulty product or to partially or fully reimburse the consumer in the event of the product failure. The form of reimbursement as a result of (1) product failure or (2) dissatisfaction with the quality of the provided service is one of the most important characteristics of a warranty, see [5,6]. Warranties are written with coverage limitations, usually in time but also in other measures, such as usage. The two most common types of warranty coverage used in industry and discussed in the literature are:

1. **Nonrenewing warranty:** A newly sold item is covered by a warranty for some calendar time of duration, T, according to certain warranty agreements. The warranter assumes all, or a portion of the expenses associated with the failure of the product, from purchase or the start of its usage until the expiration of the warranty coverage.
2. **Renewing warranty:** The warranter agrees to repair or replace any failed item up to time, T (the length of the warranty period), from the time of purchase at no cost to the consumer. If the replacement or repair is completed within the coverage of the warranty, then the item is warranted anew for a further period of length, T.

Despite that warranties are so commonly used, the accurate pricing of warranties in many situations remains an unsolved problem. This may seem surprising because meeting warranty claim obligations can be costly for companies. Underestimating the true warranty costs results in losses for a company. On the other hand, overestimating them will lead to uncompetitive product prices. The data relevant to the modeling of warranty costs in industry are usually highly confidential because they are commercially sensitive. A good deal of warranty analysis, therefore, takes place in internal research divisions of large companies.

In most models for both maintenance and warranty, repairs are usually assumed to be instantaneous [9,28]. If the repair time is relatively small compared to the time to failure of the product and also if the down time of the product is not associated with high penalties, then the assumption of instantaneous repairs is reasonable. Under this scenario, and assuming that the repairs are perfect, the renewal reward process (see [29]) is an appropriate tool to model the total expected warranty costs.

When the repair time is not small compared with the lifetime of the item, the repair cost is related to its duration, or lengthy repair times incur a penalty cost, then the assumption of instantaneous repair may not be appropriate.

In these situations, incorporating repair times into the model can improve the estimation of warranty costs. Often, the repair cost is modeled as a linear function of repair time. This is a reasonable assumption when the repair activity is labor intensive, and the repair cost is dominated by hourly labor costs.

1.3.2 A Warranty Cost Model for Non-zero Repair Times

1.3.2.1 An Alternating Process (AP) Model

In this section, we describe a general model appropriate for evaluating the expected warranty cost under the assumption of non-zero warranty repair times. Our main goal is to use this model to evaluate the expected product warranty costs over a warranty period of duration, T, as well as the product life cycle of duration, L. The life cycle represents the period of time during which a product of the type under consideration is still of interest to or needed by a customer: and during which time it will be replaced by the customer if it fails, whether it is covered under a warranty. The notation that will be used in this section is summarized in Table 1.2.

The model can be described as follows. Consider an item, which initially operates (is "on") for a time, X_1, and then fails. It then undergoes repair (is "off") for a time, Y_1. After the completion of the repair, the item is again operative for a time, X_2, which is followed by a repair for a time, Y_2, and so on. We refer to this model as an alternating process (AP) model.

TABLE 1.2

Nomenclature for Alternating Process Models

X_i	ith operational ("on") time
X_i	ith repair ("off") time
Z_i	Length of ith cycle (operational time + repair time)
S_i	End time of ith cycle
$F_{X_i}(t)$	CDF of ith operational ("on") time
$F_{Y_i}(t)$	CDF of ith repair ("off") time
$H_i(t)$	CDF of ith cycle, Z_i
$G_1^i(t)$	CDF of end time of ith cycle
$f_{X_i}(t)$	PDF of ith operational ("on") time
$f_{Y_i}(t)$	PDF of ith repair ("off") time
C_i	Cost of the ith repair/claim
$N(t)$	Number of completed cycles by time, t
$m_1(t) = E(N(t))$	Expected number of completed cycles by time, t

Abbreviations: CDF, cumulative distribution function; PDF, probability density function.

The AP model is specified in the following way:

- The lifetime of a new item, denoted by $X = X_1$, is a positive continuous random variable with cumulative distribution function (CDF) $F_X(x) = F_{X_1}(x)$;
- After the ith failure, a repair action with non-zero duration, Y_i, takes place. Assume the repair times, Y_i, are positive continuous random variables with CDF $F_{Y_i}(y)$;
- After the ith repair, the item is restored to a functioning condition, with a new lifetime denoted by X_{i+1} with CDF $F_{X_{i+1}}(x)$.

For simplicity, we assume the existence of the corresponding probability density functions (PDF) $f_{X_i}(x)$ and $f_{Y_i}(y)$.

We suppose that all of the random variables in the two sequences $\{X_i\}_1^\infty$ and $\{Y_i\}_1^\infty$ are independent of each other. However, their distributions are not necessarily identical. By imposing different assumptions on the distribution of the sequences of "on" and "off" times, we will aim to evaluate the expected warranty costs under various warranty strategies.

Denote by $Z_i = X_i + Y_i$, the length of the ith cycle (i.e., the sum of the ith operational and ith repair times) with the CDF $H_i(t)$. Let $S_n = \sum_{i=1}^n (X_i + Y_i)$, with CDF $G_1^n(t)$. Further, let $N(t)$ denote the number of completed cycles by time, t. Then, the number of AP cycles completed by time, t, and its expected value are given respectively by

$$N(t) = \sup\{n : S_n \le t\} \text{ and } m_1(t) = E(N(t)).$$

Analogously to computing the renewal function [29], we can see that

$$m_1(t) = \sum_{k=1}^{\infty} P(S_k \le t) = \sum_{k=1}^{\infty} G_1^k(t),$$

where

$$G_i^k(t) = H_i * H_{i+1} * \cdots * H_{i+k-1}(t) \tag{1.2}$$

and $*$ denotes a convolution.

We will assume that the cost of the ith repair is a combination of a fixed cost and a term proportional to the (random) repair duration, $C_i = A + \delta Y_i$. Here A and δ are known non-negative constants, and Y_i is the length of the ith repair. If data on the length of the repair and associated costs are available, this relationship can be evaluated, for example, by using regression analysis.

1.3.2.2 Three AP Models

We focus on three variations of the AP model, which alter the definition of the sequences of $\{X_i\}_1^\infty$ and $\{Y_i\}_1^\infty$ to account for the aging and deterioration of the system, as well as for the length of the repair time required to bring the system back to an operational state. These processes are summarised as follows:

- Alternating renewal process (ARP)—no aging is observed [16,17]:
  - $\{X_i\}_1^\infty$ forms a RP
  - $\{Y_i\}_1^\infty$ forms a RP
- Generalized alternating renewal process (GARP)—aging causes increasing repair times [27]:
  - $\{X_i\}_1^\infty$ forms a RP
  - $\{Y_i\}_1^\infty$ forms an increasing GP with parameters $\{b, F_{Y_1(t)}\}, 0 < b \leq 1$.
- Alternating geometric process (AGP)—aging causes decreasing operational times and increasing repair times [2]:
  - $\{X_i\}_1^\infty$ forms a decreasing GP with parameters $\{a, F_{X_1(t)}\}, a \geq 1$.
  - $\{Y_i\}_1^\infty$ forms an increasing GP with parameters $\{b, F_{Y_1(t)}\}, 0 < b \leq 1$.

Observe that the ARP and GARP are special cases of the AGP. For instance, if $a = 1$ and $b = 1$, then an AGP is an ARP, and similarly if $b = 1$, then a GARP is an ARP. The expected values of the ith operational time, ith repair time, and ith warranty repair cost are summarized in Table 1.3 for the three models.

Because warranties are issued for a finite length of time, we need some finite time horizon theoretical results for the three alternating processes described. Table 1.4 provides a summary of the theoretical results needed

TABLE 1.3

Summary of Expected Values for the Alternating Processes ARP, GARP, and AGP

Result	ARP	GARP	AGP
$\{X_i\}_1^\infty$	RP	RP	Decreasing GP
$\{Y_i\}_1^\infty$	RP	Increasing GP	Increasing GP
$E(X_i)$	$E(X_1)$	$E(X_1)$	$E(X_1)/a^{i-1}$
$E(Y_i)$	$E(Y_1)$	$E(Y_1)/b^{i-1}$	$E(Y_1)/b^{i-1}$
$E(C_i)$	$A + \delta E(Y_1)$	$A + \delta E(Y_1)/b^{i-1}$	$A + \delta E(Y_1)/b^{i-1}$

Abbreviations: AGP, alternating geometric process; ARP, alternating renewal process; GARP, generalized alternating renewal process; GP, geometric process; RP, renewal process.

TABLE 1.4

Summary of Key Results for ARP, GARP, and AGP in a Finite Time Horizon

Result	ARP [16,29]	GARP [27]	AGP [2]
$P(\text{on at } t)$	$\bar{F}_X(t)+\int_0^t \bar{F}_X(t-y)dm_1(y)$	$\bar{F}_{X_1}(t)+\sum_{n=1}^{\infty}\int_0^t \bar{F}_{X_{n+1}}(t-z)dG_1^n(z)$ $=\bar{F}_X(t)+\int_0^t \bar{F}_X(t-z)dm_1(z)$	$\bar{F}_{X_1}(t)+\sum_{k=1}^{\infty}\int_0^t \bar{F}_{X_{k+1}}(t-s)dG_1^k(s)$
$E(Y_{N(T)+1} \mid \text{on at } T)$; for $T>0$	$E(Y)$	$\frac{E(Y_1)}{P(\text{on at } T)}\left\{\bar{F}_X(T)+\sum_{k=1}^{\infty}\frac{1}{b^k}\int_0^T \bar{F}_X(T-s)dG_1^k(s)\right\}$	$\frac{E(Y_1)}{P(\text{on at } T)}\left\{\bar{F}_{X_1}(T)+\sum_{k=1}^{\infty}\frac{1}{b^k}\int_0^T \bar{F}_{X_{k+1}}(T-s)dG_1^k(s)\right\}$
$P(S_{N(T)}+X_{N(T)+1}\le t \mid \text{on at } T)$; for $T\le t$	$\frac{\bar{F}_X(T)-\bar{F}_X(t)}{\bar{F}_X(T)+\int_0^T \bar{F}_X(T-u)dm_1(u)}+\frac{\int_0^T\left(\bar{F}_X(T-u)-\bar{F}_X(t-u)\right)dm_1(u)}{\bar{F}_X(T)+\int_0^T \bar{F}_X(T-u)dm_1(u)}$	$\frac{\bar{F}_X(T)-\bar{F}_X(t)}{\bar{F}_X(T)+\int_0^T \bar{F}_X(T-z)dm_1(z)}+\frac{\int_0^T\left(\bar{F}_X(T-z)-\bar{F}_X(t-z)\right)dm_1(z)}{\bar{F}_X(T)+\int_0^T \bar{F}_X(T-z)dm_1(z)}$	$\frac{\bar{F}_{X_1}(T)-\bar{F}_{X_1}(t)}{\bar{F}_{X_1}(T)+\sum_{k=1}^{\infty}\int_0^T \bar{F}_{X_{k+1}}(T-s)dG_1^k(s)}+\frac{\sum_{k=1}^{\infty}\int_0^T\left(\bar{F}_{X_{k+1}}(T-s)-\bar{F}_{X_{k+1}}(t-s)\right)dG_1^k(s)}{\bar{F}_{X_1}(T)+\sum_{k=1}^{\infty}\int_0^T \bar{F}_{X_{k+1}}(T-s)dG_1^k(s)}$
$P(S_{N(T)+1}+X_{N(T)+2}\le t \mid \text{off at } T)$ for $T\le t$	$\frac{1}{P(\text{off at } T)}\times\left(\int_0^T\int_{T-u}^{t-u}F_X(t-u-v)dF_Y(v)dF_X(u)+\int_0^T\int_0^{T-w}\int_{T-w-u}^{t-w-u}F_X(t-w-u-v)dF_Y(v)\,dF_X(u)dm_1(w)\right)$	$\frac{1}{P(\text{off at } T)}\times\left(\int_0^T\int_{T-u}^{t-u}F_X(t-u-v)dF_{Y_1}(v)dF_X(u)+\sum_{k=1}^{\infty}\int_0^T\int_0^{T-w}\int_{T-w-u}^{t-w-u}F_X(t-w-u-v)\,dF_{Y_{k+1}}(v)dF_X(u)dG_1^k(w)\right)$	$\frac{1}{P(\text{off at } T)}\times\left(\int_0^T\int_{T-u}^{t-u}F_{X_2}(t-u-v)dF_{Y_1}(v)dF_{X_1}(u)+\sum_{k=1}^{\infty}\int_0^T\int_0^{T-s}\int_{T-s-u}^{t-s-u}F_{X_{k+2}}(t-s-u-v)\,dF_{Y_{k+1}}(v)dF_{X_{k+1}}(u)dG_1^k(s)\right)$

Abbreviations: AGP, alternating geometric process; ARP, alternating renewal process; GARP, generalized alternating renewal process.

to evaluate the expected warranty costs under the three models. For further details regarding the ARP refer to [16], for the GARP refer to [27], and for the AGP refer to [2]. Note that some of the results related to the GARP and AGP are derived by extending results of [21]. For example, the probability that the system is "on" at time, t, extends Theorem 2.3.1 of [21].

1.3.3 Nonrenewing Free Repair Warranty Cost Analysis

In this section we consider a nonrenewing free repair warranty (NRFRW). As described in Section 1.3.1, under a nonrenewing warranty, a product is warranted for a fixed period of time, T, started immediately after the purchase. During the warranty period, all expenses are borne by the manufacturer. If the warranty expires while the item is under repair, the manufacturer is still liable for the completion of the repair, so that the warranty is effectively extended beyond, T, until the completion of the repair. This creates additional complexity when assessing warranty costs. In this section, we summarize the expected warranty costs over the warranty period, T, and life cycle, L, for the three models proposed in Section 1.3.2.

1.3.3.1 Expected Warranty Costs Over the Warranty Period $(0,T)$

The total cost to the manufacturer is the total cost of all repairs undertaken for failures that occur within the warranty coverage period $[0,T]$.

The total cost over the warranty period, $C(T)$ is given by

$$C(T)=\begin{cases}\sum_{i=1}^{N(T)} C_i, & \text{if "on" at time } T\\ \sum_{i=1}^{N(T)+1} C_i, & \text{if "off" at time } T.\end{cases}$$

Then it follows that

$$E\big(C(T)\big)=E\left(\sum_{i=1}^{N(T)+1} C_i\right)-E(C_{N(T)+1}\mid \text{on at } T)P\,(\text{on at } T) \tag{1.3}$$

where $E(C_i)$ is given in Table 1.3, $P\,(\text{on at } T)$ is given in Table 1.4, and $E\left(\sum_{i=1}^{N(T)+1} C_i\right)$ is given in Table 1.5. Note that expressions for $E\left(\sum_{i=1}^{N(T)+1} C_i\right)$ for an ARP follow from Wald's Equation (see, e.g., [16,29]), and for a GARP and AGP follow from Equation (2.4.2) of [21] (p. 45) (see also [2,27]).

TABLE 1.5

Summary of Key Results for ARP, GARP, and AGP Under NRFRW and RFRW Strategies

Process	NRFRW $E\left(\sum_{i=1}^{N(T)+1} C_i\right)$	RFRW $E(C(W_T))$
ARP	$(m_1(T)+1)E(C) =$ $(m_1(T)+1)(A+\delta E(Y_1))$	$\frac{F_X(T)}{1-F_X(T)}(A+\delta E(Y))$
GARP	$A(m_1(T)+1)+\delta E(Y_1)\frac{\{E(b^{-N(T)})-b\}}{1-b}$	$A\frac{F_X(T)}{1-F_X(T)}+\delta E(Y_1)b\frac{F_X(T)}{b-F_X(T)}$, if $F_X(T)<b$ and $F_X(T)<1$. $E(C(W_T))=\infty$ otherwise
AGP	$A(m_1(T)+1)+\delta E(Y_1)\frac{\{E(b^{-N(T)})-b\}}{1-b}$	$\sum_{k=1}^{\infty}\left(A+\frac{\delta E[Y_1]}{b^{k-1}}\right)\prod_{i=1}^{k}F_{X_i}(T)$ (divergent series)

Abbreviations: AGP, alternating geometric process; ARP, alternating renewal process; GARP, generalized alternating renewal process; NRFRW, nonrenewing free repair warranty; RFRW, renewing free repair warranty.

1.3.3.2 Expected Warranty Costs Over the Product Lifecycle $(0,L)$

Let L^* be a prespecified time during which a product is considered to be contemporary and competitive with similar products in the market. Let L be the time of the first off-warranty failure of the product after L^*. Then, we call $(0,L)$ a life cycle of the product. Let ξ_T represent the time between two consecutive purchases, that is,

$$\xi_T = \begin{cases} S_{N(T)} + X_{N(T)+1}, & \text{if "on" at time } T \\ S_{N(T)+1} + X_{N(T)+2}, & \text{if "off" at time } T \end{cases}$$

Then, the expected cost over $(0,L)$ is given by

$$E(C(L)) = (m^*_{\xi_T}(L)+1)E(C(T))$$

where $m^*_{\xi_T}(t)$ is a renewal function of the renewal process generated by ξ_T. The distribution of ξ_T can be represented via the respective conditional distributions of $S_{N(T)}+X_{N(T)+1}$ and $S_{N(T)+1}+X_{N(T)+2}$ (see Table 1.4).

1.3.4 RFRW and RRFRW(*n*) Cost Analysis

In this section, we consider a renewing free repair warranty (RFRW) under which, following a warranty repair, the item is warrantied anew for a period of length, T. If the warranty period ends during an operating period, the cost of the following repair is not incurred by the warranter, and the warranty coverage expires. Here, we will distinguish between warranty coverage, W_T, which is a random variable, and warranty period, which is a predetermined constant, T. We define W_T as the time from the purchase of the product until the expiry of the warranty coverage. We also consider a restricted renewing free repair warranty (RRFRW(n)), under which the number of repairs is limited to some predetermined number, n. We define W_T^n as the warranty coverage under an RRFRW(n).

1.3.4.1 Expected Warranty Costs Over $(0, W_T)$

Because of the mechanism of the renewing warranty, W_T is equal to:

$$W_T = \begin{cases} T, & \text{if } X_1 > T \\ T + \sum_{i=1}^{k} (X_i + Y_i), & \text{if } X_1 \le T, \cdots, X_k \le T, X_{k+1} > T, \text{for some } k. \end{cases}$$

Then, the warranty cost $C(W_T)$ over the warranty coverage is a random variable and its distribution is as follows:

$$C(W_T) = \begin{cases} 0, & \text{with probability} \quad 1 - F_{X_1}(T) \\ \sum_{i=1}^{k} C_i, & \text{with probability} \quad \left(1 - F_{X_{k+1}}(T)\right) \prod_{i=1}^{k} F_{X_i}(T), \text{for } k \ge 1 \end{cases} \tag{1.4}$$

The expected warranty cost, $E\left(C(W_T)\right)$, for the ARP, GARP, and AGP is provided in Table 1.5. Note that under an AGP model, it can be shown that the series $E\left(C(W_T)\right)$ is divergent, see [2]. Therefore, assigning a warranty period of length, T, for a product, with operational and repair times that form an AGP with parameters $\left\{a, F_{X_1}(T), b, F_{Y_1}(T)\right\}$ is not a viable business option. In these cases, a RRFRW(n) may be a suitable alternative. Under an RRFRW(n), the warranty coverage W_T^n can be represented as follows:

$$W_T^n = \begin{cases} T, & \text{if } X_1 > T \\ \sum_{i=1}^{k-1} (X_i + Y_i) + T, & \text{if } X_i \le T, i = 1, 2, \ldots, k-1;\ X_k > T, 2 \le k \le n \\ \sum_{i=1}^{n} (X_i + Y_i), & \text{if } X_i \le T, i = 1, 2, \ldots, n \end{cases}$$

Then, the warranty cost $C(W_T^n)$ over the warranty coverage is a random variable and its distribution is as follows:

$$C(W_T^n) = \begin{cases} 0, & \text{with probability} \quad 1 - F_{X_1}(T) \\ \sum_{i=1}^{k-1} C_i, & \text{with probability} \quad (1 - F_{X_k}(T)) \prod_{i=1}^{k-1} F_{X_i}(T), \text{ for } 2 \le k \le n \\ \sum_{i=1}^{n} C_i, & \text{with probability} \quad \prod_{i=1}^{n} F_{X_i}(T) \end{cases} \quad (1.5)$$

Under a RRFRW(n), for $T > 0$, the expected warranty coverage is

$$\begin{aligned} E(W_T^n) &= T(1 - F_{X_1}(T)) \\ &+ \sum_{k=2}^{n} \left(\left(\sum_{i=1}^{k-1} (E(X_i \mid X_i \le T) + E[Y_i]) + T \right) (1 - F_{X_k}(T)) \prod_{i=1}^{k-1} F_{X_i}(T) \right) \\ &+ \sum_{i=1}^{n} (E(X_i \mid X_i \le T) + E(Y_i)) \prod_{i=1}^{n} F_{X_i}(T), \end{aligned} \quad (1.6)$$

the expected warranty cost is

$$E\big(C(W_T^n)\big) = \sum_{k=1}^{n} \left(A + \frac{\delta E[Y_1]}{b^{k-1}} \right) \prod_{i=1}^{k} F_{X_i}(T), \quad (1.7)$$

and the expected number of cycles is

$$\begin{aligned} E\big(N(W_T^n)\big) &= \sum_{k=1}^{n-1} k\big(1 - F_{X_{k+1}}(T)\big) \prod_{i=1}^{k} F_{X_i}(T) + n \prod_{i=1}^{n} F_{X_i}(T) \\ &= \sum_{k=1}^{n} \prod_{i=1}^{k} F_{X_i}(T). \end{aligned}$$

For examples of $E(X_i \mid X_i \le T)$ for an AGP with specific distributions see [2].

1.3.4.2 Expected Warranty Costs Over Product Life Cycle $(0, L)$

Next, we consider the continuous positive random variable, ξ_T—the time between two consecutive purchases under a RFRW. By definition,

$$\xi_T = \begin{cases} X_1 & \text{if } X_1 > T \\ \sum_{i=1}^{n}(X_i + Y_i) + X_{n+1} & \text{if } X_1 \leq T, \cdots, X_n \leq T, X_{n+1} > T \text{ for some } n. \end{cases}$$

It can be shown that the cdf of ξ_T is given by

$$F_{\xi_T}(t) = \left(F_X(t) - F_X(T)\right) + \sum_{i=1}^{\infty} \int_0^{t-T} \left(F_X(t-s) - F_X(T)\right) dG_1^i(s),$$

where $G_1^i(s)$ is given by (1.2). Then, the expected costs over $(0, L)$, say $E\left(C(L)\right)$, are expressed in terms of ξ_T in the following way

$$E\left(C(L)\right) = \left(m^*_{\xi_T}(L) + 1\right) E\left(C(W_T)\right),$$

where $m^*_{\xi_T}(t)$ is the renewal function of the renewal process generated by ξ_T. Following a similar approach, the expected warranty cost over $(0, L)$ under an RRFRW(n) can be derived, see [2].

1.4 Data Analysis

In the previous sections we have presented a method for modeling warranty costs using ARP, GARP, and AGP. In this section we fit this model to claims from the warranty database of a large automotive manufacturer. The warranty database contains more than 200,000 claims from vehicles manufactured between 1998 and 2001. To fit an AP to these data, we need to extract the operational times, X_i, and repair times, Y_i, which are not explicitly included in the warranty database. The repair times have been modeled using a linear transformation of the labor costs (provided in the database). We assume that the age of the vehicle at the ith claim (provided in the database) is

$$S_{i-1} + X_i,$$

where $S_{i-1} = \sum_{j=1}^{i-1}(X_j + Y_j)$, and therefore, the operational times $X_i, i = 1, 2, \ldots, n$ can be identified. For further details refer to [2].

To demonstrate the process of fitting the three AP models to data, we have selected two vehicles with at least 9 claims, vehicles A and B, from this database. In Section 1.4.1, we fit the three AP models to vehicles A and B, and in Section 1.4.2, we use the fitted models to estimate the warranty costs.

1.4.1 Fitting APs to Warranty Claims Data

The key difference between the three AP models, introduced in Section 1.3.2.2, is whether $\{X_n\}$ and $\{Y_n\}$ form an RP or a GP. Therefore, in this section we demonstrate a method for assessing whether observed operational and repair times can be modeled by these two processes. This can be achieved by the completion of following three steps:

1. Hypothesis testing (H_0 = it is GP(RP); H_a = it is not GP(RP))
2. Comparing the fit of the GP and RP models
3. Identifying a parametric form for the RP and GP

We follow a procedure outlined by Lam [21, §4.2, pp. 101–104] to test if the operational times and repair times are consistent with an RP and GP and compare the fit of the models to the data. This procedure was used to assess the fit of a GARP to warranty data by [27] and an AGP to warranty data by [2]. The three steps are discussed in Sections 1.4.1.1 through 1.4.1.3, respectively.

1.4.1.1 Hypothesis Testing

The hypothesis tests outlined by [21, §4.2, pp. 101–104], were applied to operational times and the repair times for these two vehicles. For both operational and repair times, when testing for a GP, the null hypothesis is not rejected, and when testing for an RP, the null hypothesis is rejected by at least one of the tests for each vehicle. These results indicate that a GP is a suitable model for the operational and repair times for both vehicles. For detailed results of these tests refer to [2].

1.4.1.2 Comparing the Fit of the Models

Given a sample $Z_1, Z_2, \ldots, Z_n$, the mean (γ), variance (σ^2), and fitted values ($\hat{Z}_i$) of a RP and a GP can be estimated using the procedure described by [21]. Refer to [21] or [2] for further details. Note that we use the subscript $_R$ to indicate values that correspond to the RP. In our model the:

- RP for operational times has mean $E(X_1) = \hat{\gamma}_R = 1/\hat{\lambda}_R$
- RP for repair times has mean $E(Y_1) = \hat{\gamma}_R = 1/\hat{\mu}_R$

- GP for operational times has mean $E(X_1) = \hat{\gamma} = 1/\hat{\lambda}$ and ratio $\hat{a}$
- GP for repair times has mean $E(Y_1) = \hat{\gamma} = 1/\hat{\mu}$ and ratio $\hat{b}$.

The mean, variance and ratio of the GP and the mean and variance of the RP were computed for the operational and repair times (see [2] for details) Using these parameter estimates, the fitted values of the RPs and GPs for operational and repair times were computed. The fitted values of the RP and GP are compared with the data in Figure 1.2. Notice that for both vehicles, the GP provides a better fit to the operational times than an RP. However, for vehicle A, while the GP is better than the RP, the GP does not fit as well as it does for vehicle B, as shown in Figure 1.2a and b. For the repair times, the GP provides a better fit than the RP for both vehicles, as shown in Figure 1.2c and d. These results are consistent with the hypothesis test results.

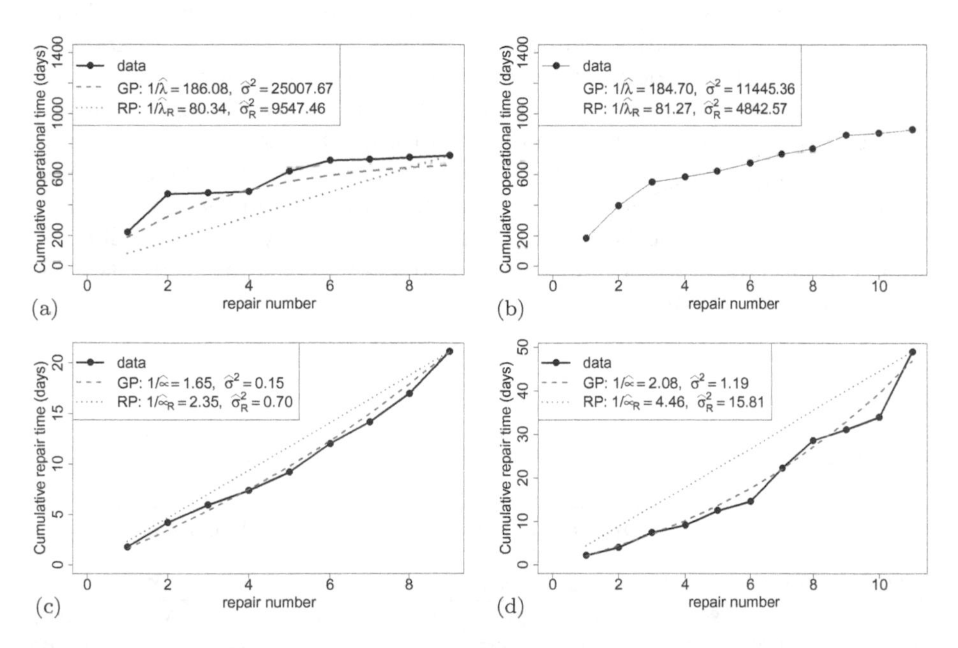

FIGURE 1.2
Fit of renewal and geometric processes to the cumulative operational and repair times: (a) vehicle A (operational), (b) vehicle B (operational), (c) vehicle A (repair), and (d) vehicle B (repair). AGP, alternating geometric process; ARP, alternating renewal process; GARP, generalized alternating renewal process.

1.4.1.3 Identifying a Parametric Form for the RP and GP

If observed data $\{Z_i, i = 1,2,...\}$ can be modeled by an RP, then $\{Z_i, i = 1,2,...\}$ are iid random variables. On the other hand, if the data can be modeled by a GP, then $\{Z_i\beta^{i-1}, i = 1,2,\ldots\}$ form a RP and are iid random variables. We fitted the exponential, gamma, and Weibull distributions to the following four sequences of random variables:

- RP model for operational times, $\{X_i, i = 1,2,...\}$;
- GP model for operational times, $\{X_i a^{i-1}, i = 1,2,\ldots\}$;
- RP model for repair times, $\{Y_i, i = 1,2,...\}$;
- GP model for repair times, $\{Y_i b^{i-1}, i = 1,2,\ldots\}$.

For all distributions, method of moments estimators are used to obtain the parameter estimates. Where needed, we add the subscripts on and off to the parameter names to distinguish between parameters for the operational (on) times and repair (off) times. Given the small sample sizes (9 and 11 claims for vehicles A and B, respectively), it is difficult to accurately fit the distributions to the four sequences of random variables listed for the each vehicle. For instance, none of the distributions provided a good fit for the operational times of vehicle A under the GP model. The Weibull distribution provided a reasonable fit to the data for most of the sequences, so we proceed using this distribution. Parameter estimates for the Weibull distribution are shown in Table 1.6. We emphasise that, given the small sample sizes for vehicles A and B (9 and 11 claims, respectively) and poor fit of the distributions, the following analysis is only for illustrative purposes and not for decision making.

TABLE 1.6

Parameters Values for the Weibull Distribution for the ARP, GARP, and AGP Used in the Simulation

		Operational Times				Repair Times			
Vehicle	**Process**	$E(X_1)$	$\hat{\theta}_{on}$	$\hat{k}_{on}$	$\hat{a}$	$E(Y_1)$	$\hat{\theta}_{off}$	$\hat{k}_{off}$	$\hat{b}$
A	ARP	80.3372	0.0138	0.8271	1.0000	2.3476	0.3808	3.0756	1.0000
A	GARP	80.3372	0.0138	0.8271	1.0000	1.6497	0.5557	4.8629	0.9213
A	AGP	186.0753	0.0051	1.1809	1.3579	1.6497	0.5557	4.8629	0.9213
B	ARP	81.2682	0.0116	1.1717	1.0000	4.4553	0.2152	1.1227	1.0000
B	GARP	81.2682	0.0116	1.1717	1.0000	2.0775	0.4266	1.9865	0.8799
B	AGP	184.7040	0.0048	1.7851	1.2261	2.0775	0.4266	1.9865	0.8799

Abbreviations: AGP, alternating geometric process; ARP, alternating renewal process; GARP, generalized alternating renewal process.

1.4.2 Estimation of Expected Warranty Cost Over (0,*T*)

In this section we estimate warranty cost for vehicles A and B using the AGP, GARP, and ARP models fitted in the previous section. The estimation of the warranty costs in this section will be somewhat poor because of the small sample sizes and the poor fit of the distributions in Section 1.4.1.3. However, to illustrate our approach, we continue with the analysis. The Weibull distribution was used to model the operational and repair times for the ARP, GARP, and AGP models (see Table 1.6). The subscripts "on" and "off" are used to indicate the parameters of the operational and repair time distributions respectively.

Preliminary analyses indicated that the estimated warranty cost was sensitive to the choice of cost parameters, A and δ. Therefore, to remove the effect of this modeling choice from the comparisons, we focus on two features of the processes over the warranty period: the number of claims (equivalent to $A=1$, $\delta=0$) and the cumulative repair time (equivalent to $A=0, \delta=1$). The estimated warranty cost under the three models is computed for times $T=$ {30, 50, 150, 250, ..., age at last claim} days using simulation and is the average over 10^5 simulation runs.

Figure 1.3a and b show the average number of repairs for various warranty periods for vehicles A and B, respectively. For vehicle A, the GAR and

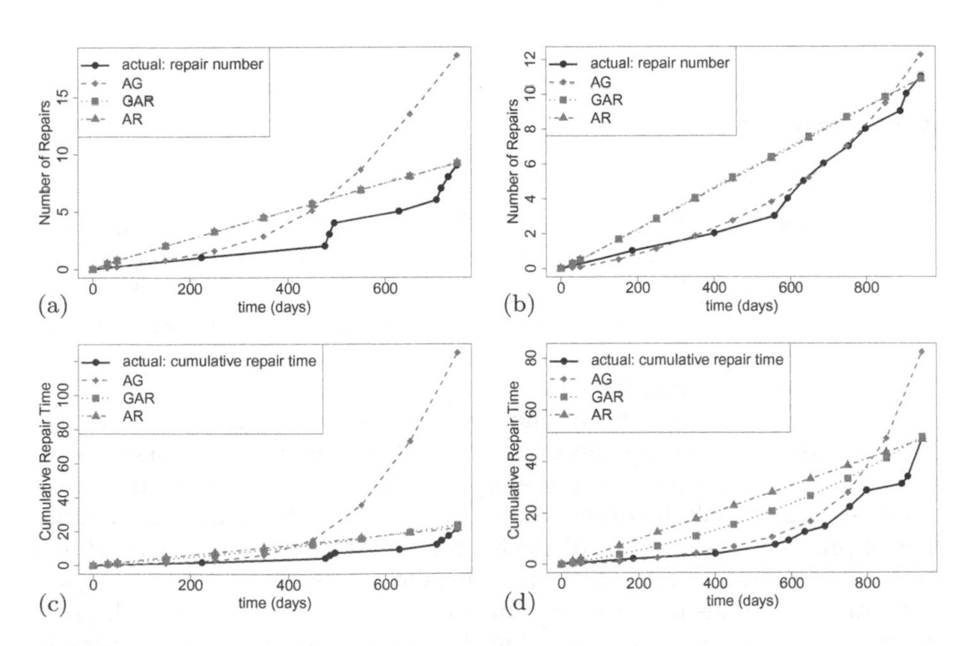

FIGURE 1.3
Estimated number of repairs and cumulative repair times under alternating renewal (AR), generalized alternating renewal (GAR), and alternating geometric (AG) processes (average more than 100,000 simulations): (a) vehicle A: repair count, (b) vehicle B: repair count, (c) vehicle A: repair time, and (d) vehicle B: repair time.

AR processes provide a better estimation of the number of repairs than the AG process. For vehicle B, the AG process provides a good estimation of the number of repairs. The poor fit of the AG process for vehicle A is not surprising given that the distributions fitted under a GP did not provide a good fit to the operational times (as discussed in Section 1.4.1.3). The curved nature of the repair number versus cumulative operational times (Figure 1.2a) suggests that an AG process may still be appropriate for vehicle A; however, additional data is needed to improve the estimation of the parametric form of the fitted models.

Figure 1.3c and d show the actual cumulative repair time and the estimated cumulative repair time from the simulation of the three models. For vehicle A, none of the models provide a satisfactory fit, and the AG model, in particular, hugely overestimates the cumulative repair time. For vehicle B, the AG model provides reasonable fit to the data, though does overestimate the repair times for larger warranty periods. One reason for this poor fit for vehicle A may be as follows. The cumulative repair time over the warranty period depends on the length of the repairs and the number of repairs that occur. The average number of repairs was overestimated as discussed and shown in Figure 1.3a. This may be compounded by the poor fit of the Weibull distribution to the repair times. Additional data may improve the parameter estimation and, thus, improve the fit of the models for vehicle A.

1.5 Conclusion

GPs and GLPs allow flexibility in modeling events over time where a monotone trend is present. The majority of applications of these types of processes have been in the field of maintenance and warranty analysis, where GPs have been used to model decreasing operational times and increasing repair times. We have summarized the key results relating to the use of GP in warranty analysis, including the ARP, GARP, and AGP.

The useful and successful application of these processes to real-life scenarios relies on reliable and accurate parameter estimation. In working with GLPs, the usual challenges apply—large numbers of parameters, small sample sizes, and tractability of the results. We have found that there are a number of practical analytical results available for APs built from GPs and that these provide insight into real data settings in our example.

From our findings it is our view that GPs and related processes will prove to be productive means of modeling data and maintenance and warranty settings.

References

1. Richard Arnold, Stefanka Chukova, Yu Hayakawa, and Sarah Marshall. Geometric-like processes: An overview and some applications. Manuscript submitted for publication, 2019.
2. Richard Arnold, Stefanka Chukova, Yu Hayakawa, and Sarah Marshall. Warranty cost analysis with an alternating geometric process. *Proceedings of the Institution of Mechanical Engineers, Part O: Journal of Risk and Reliability*, in press, January 2019. doi: 10.1177/1748006X18820379.
3. Halil Aydoğdu, Birdal Şenoğlu, and Mahmut Kara. Parameter estimation in geometric process with Weibull distribution. *Applied Mathematics and Computation*, 217(6):2657–2665, 2010.
4. Jun Bai and Hoang Pham. Repair-limit risk-free warranty policies with imperfect repair. *IEEE Transactions on Systems, Man, and Cybernetics-Part A: Systems and Humans*, 35(6):765–772, 2005.
5. Wallace R. Blischke and D. N. Prabhakar Murthy. *Warranty Cost Analysis*. Marcel Dekker, New York, 1993.
6. Wallace R. Blischke and D. N. Prabhakar Murthy. *Product Warranty Handbook*. Marcel Dekker, New York, 1996.
7. Laurent Bordes and Sophie Mercier. Extended geometric processes: Semiparametric estimation and application to reliability. *Journal of the Iranian Statistical Society*, 12(1):1–34, 2013.
8. W. John Braun, Wei Li, and Yiqiang Q. Zhao. Properties of the geometric and related processes. *Naval Research Logistics (NRL)*, 52(7):607–616, 2005.
9. Mark Brown and Frank Proschan. Imperfect repair. *Journal of Applied Probability*, 20:851–859, 1983.
10. Jennifer S. K. Chan, S. T. Boris Choy, and Connie P. Y. Lam. Modeling electricity price using a threshold conditional autoregressive geometric process jump model. *Communications in Statistics—Theory and Methods*, 43(10–12):2505–2515, 2014.
11. Jennifer S. K. Chan, Yeh Lam, and Doris Y. P. Leung. Statistical inference for geometric processes with gamma distributions. *Computational Statistics & Data Analysis*, 47(3):565–581, 2004.
12. Jennifer S. K. Chan, Philip L. H. Yu, Yeh Lam, and Alvin P. K. Ho. Modelling SARS data using threshold geometric process. *Statistics in Medicine*, 25(11):1826–1839, 2006.
13. Jennifer So Kuen Chan and Wai Yin Wan. Multivariate generalized Poisson geometric process model with scale mixtures of normal distributions. *Journal of Multivariate Analysis*, 127:72–87, 2014.
14. Jianwei Chen, Kim-Hung Li, and Yeh Lam. Bayesian computation for geometric process in maintenance problems. *Mathematics and Computers in Simulation*, 81(4):771–781, 2010.
15. Guo-Quang Cheng and L. Li. A geometric process repair model with inspections and its optimisation. *International Journal of Systems Science*, 43(9):1650–1655, 2012.

16. Stefanka Chukova and Yu Hayakawa. Warranty cost analysis: Non-zero repair time. *Applied Stochastic Models in Business and Industry*, 20(1):59–71, 2004.
17. Stefanka Chukova and Yu Hayakawa. Warranty cost analysis: Renewing warranty with non-zero repair time. *International Journal of Reliability, Quality and Safety Engineering*, 11(2):93–112, 2004.
18. Stefanka Chukova and Yu Hayakawa. Warranty cost analysis: Quasi-renewal inter-repair Times. *International Journal of Quality & Reliability Management*, 22(7):687–698, 2005.
19. Maxim S. Finkelstein. A scale model of general repair. *Microelectronics Reliability*, 33(1):41–44, 1993.
20. Leung Kit-Nam Francis. Optimal replacement policies determined using arithmetico-geometric processes. *Engineering Optimization*, 33(4):473–484, 2001.
21. Yeh Lam. *The Geometric Process and Its Applications*. World Scientific, Hackensack, NJ, 2007.
22. Yeh Lam. A geometric process maintenance model with preventive repair. *European Journal of Operational Research*, 182(2):806–819, 2007.
23. Yeh Lam and Yuan Lin Zhang. A shock model for the maintenance problem of a repairable system. *Computers & Operations Research*, 31(11):1807–1820, 2004.
24. Yeh Lam, Li-xing Zhu, Jennifer S. K. Chan, and Qun Liu. Analysis of data from a series of events by a geometric process model. *Acta Mathematicae Applicatae Sinica, English Series*, 20(2):263–282, 2004.
25. Yeh (Ye Lin) Lam. Geometric processes and replacement problem. *Acta Mathematicae Applicatae Sinica*, 4(4):366–377, 1988.
26. Xiaolin Liang, Yeh Lam, and Zehui Li. Optimal replacement policy for a general geometric process model with δ-shock. *International Journal of Systems Science*, 42(12):2021–2034, 2011.
27. Sarah Marshall, Richard Arnold, Stefanka Chukova, and Yu Hayakawa. Warranty cost analysis: Increasing warranty repair times. *Applied Stochastic Models in Business and Industry*, 34(4):544–561, 2018.
28. Hoang Pham and Hongzhou Wang. Imperfect maintenance. *European Journal of Operational Research*, 94(3):425–438, 1996.
29. Sheldon M. Ross. *Stochastic Processes*. John Wiley & Sons, New York, 2nd revised edition, 1996.
30. Wolfgang Stadje and Dror Zuckerman. Optimal strategies for some repair replacement models. *Advances in Applied Probability*, 22(3):641–656, 1990.
31. Ya-yong Tang and Yeh Lam. A δ-shock maintenance model for a deteriorating system. *Feature Cluster on Mathematical Finance and Risk Management*, 168(2):541–556, 2006.
32. Guan Jun Wang and Richard C. M. Yam. Generalized geometric process and its application in maintenance problems. *Applied Mathematical Modelling*, 49:554–567, 2017.
33. Guan Jun Wang and Yuan Lin Zhang. Geometric process model for a system with inspections and preventive repair. *Computers & Industrial Engineering*, 75:13–19, 2014.
34. Hongzhou Wang and Hoang Pham. Optimal age-dependent preventive maintenance policies with imperfect maintenance. *International Journal of Reliability, Quality and Safety Engineering*, 3(2):119–135, 1996.

35. Hongzhou Wang and Hoang Pham. Optimal maintenance policies for several imperfect repair models. *International Journal of Systems Science*, 27(6):543–549, 1996.
36. Hongzhou Wang and Hoang Pham. A quasi renewal process and its applications in imperfect maintenance. *International Journal of Systems Science*, 27(10):1055–1062, 1996.
37. Shaomin Wu. Doubly geometric processes and applications. *Journal of the Operational Research Society*, 69(1):66–77, 2018.
38. Shaomin Wu and Derek Clements-Croome. A novel repair model for imperfect maintenance. *IMA Journal of Management Mathematics*, 17:235–243, 2006.
39. Shaomin Wu and Guanjun Wang. The semi-geometric process and some properties. *IMA Journal of Management Mathematics*, 29(2):229–245, 2018.
40. Yuan Lin Zhang and Guan Jun Wang. An extended geometric process repair model with imperfect delayed repair under different objective functions. *Communications in Statistics—Theory and Methods*, 47(13):3204–3219, 2018.
41. Yuan Lin Zhang. A bivariate optimal replacement policy for a repairable system. *Journal of Applied Probability*, 31(4):1123–1127, 1994.
42. Yuan Lin Zhang. A geometric-process repair-model with good-as-new preventive repair. *IEEE Transactions on Reliability*, 51(2):223–228, 2002.
43. Yuan Lin Zhang. A geometrical process repair model for a repairable system with delayed repair. *Computers & Mathematics with Applications*, 55(8):1629–1643, 2008.
44. Yuan Lin Zhang and Guan Jun Wang. A geometric process repair model for a series repairable system with k dissimilar components. *Applied Mathematical Modelling*, 31(9):1997–2007, 2007.
45. Yuan Lin Zhang and Guan Jun Wang. An extended geometric process repair model for a cold standby repairable system with imperfect delayed repair. *International Journal of Systems Science: Operations & Logistics*, 3(3):163–175, 2016.
46. Yuan-Lin Zhang and Guan-Jun Wang. An extended geometric process repair model with delayed repair and slight failure type. *Communications in Statistics—Theory and Methods*, 46(1):427–437, 2017.

2

Supervised Machine Learning: Application Example Using Random Forest in R

Bharatendra Rai

CONTENTS

2.1 Introduction: What Is Machine Learning?

Machine learning has been gaining popularity along with recent interest in involving data science and big data among businesses and academics. Figure 2.1 shows that for last several years the popularity for searching the term "machine learning" is steadily going up.

In an interview sometime near 1990, Steve Jobs made use of a bicycle analogy for computers. There was a study done to assess the efficiency of locomotion for going from place A to place B for various species, including humans. Humans did not do well on that list. However, when the study was extended to humans with a bicycle, it was a total a game changer. Using the analogy of a bicycle, Jobs called computers as the "bicycle of the mind." About that time, Mac computers had a RAM of 1 megabyte and storage was just 40 megabytes. With rapid advances in technology, computers have become very powerful and that's one thing that made this rapid growth of machine learning possible. In machine learning, the word "machine" refers to computers and the learning takes place with data. Data can be classified mainly into two categories, structured data and unstructured data. Structured data are those that are organized into rows and columns. Data that are usually available in Excel or Comma Separated Values (CSV) files are usually structured data. Some examples of unstructured data include text (e.g., tweets or customer comments on Amazon),

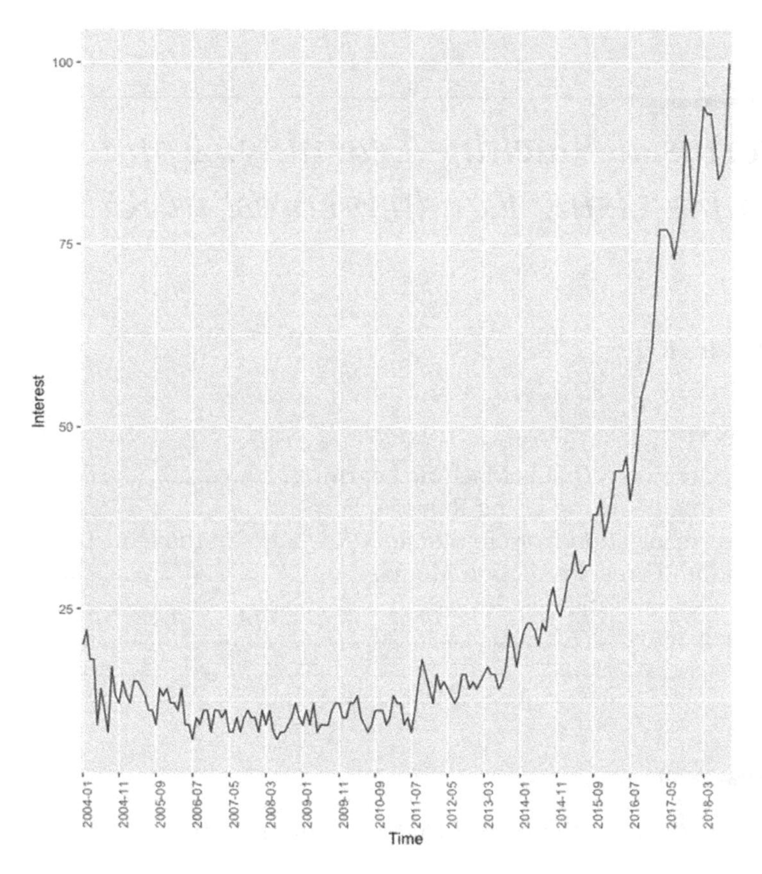

FIGURE 2.1
Search trend for term "machine learning" on Google.

image, and videos. Learning can be classified into supervised learning and unsupervised learning. In supervised learning, data has a response variable, whereas in unsupervised learning there is no response variable. There are several popular supervised machine-learning models such as linear regression, logistic regression, decision trees, naive Bayes, neural networks, support vector machines, and random forests. Within machine learning when big neural networks are used, it is called deep learning. An example that we commonly see in daily life includes face recognition on iPhone X or face recognition used by Facebook or Google for sorting pictures and images. Popular unsupervised-learning methods include recommender systems seen on Netflix or Amazon. When one watches a movie or a TV show or a documentary on Netflix, it will recommend what to watch next. Similarly if one searches for something on Amazon.com, it will suggest that those who look for item A also found item B useful. Another popular unsupervised learning-method is clustering.

This chapter illustrates use of a popular supervised machine-learning methods viz., random forest.

2.2 Machine Learning Using Random Forest

Predictive modeling in the presence of a large number of exploratory variables requires the use of methods that support feature selection. The random forest algorithm is a popular machine-learning method that automatically calculates variable importance measure as a by-product and has been successfully used by various researchers [1–2]. Variable importance measures that include mean decrease accuracy (MDA) and mean decrease Gini (MDG) provided by random forest has also been studied for stability. Studies involving simulations indicate that ranks based on MDA are unstable to small perturbations of the data set, whereas ranks based on MDG provide more stable results [3]. At the same time, in situations where there are strong within-predictor correlations, MDA rankings are found to be more stable than MDG [4]. It is also known that having too many features can not only slow down algorithms, but many machine-learning algorithms also exhibit a decrease in accuracy in such situations [5].

The Boruta algorithm that supports feature selection uses a wrapper approach build around random forest methodology [6]. An output of Boruta algorithm provides classification of explanatory variables or features into three categories, important, tentative, and unimportant variables or features. It also allows a rough fix for tentative variables, which can be used to fill missing decisions regarding importance or unimportance by simple comparison of the median attribute z-score with the median z-score of the most important shadow attribute. Many researchers have successfully applied this algorithm that provides several advantages in feature selection to support predictive modeling [7,8].

This chapter provides an application of Boruta algorithm for feature selection and then uses random forest algorithm for predictive modeling of housing data involving 79 explanatory features. These features describe various aspects people consider while buying a new house. The main objective is to develop a predictive model for the sale price of the house based on appropriate features.

2.3 Feature Selection from Exploratory Variables in the Housing Data

The data set used for this study consists of data on 1460 houses with 79 exploratory variables and property's sale price in dollars as target variable based on the location of Ames in Iowa, USA. This data set was made available through a competition on kaggle.com. There are 43 qualitative,

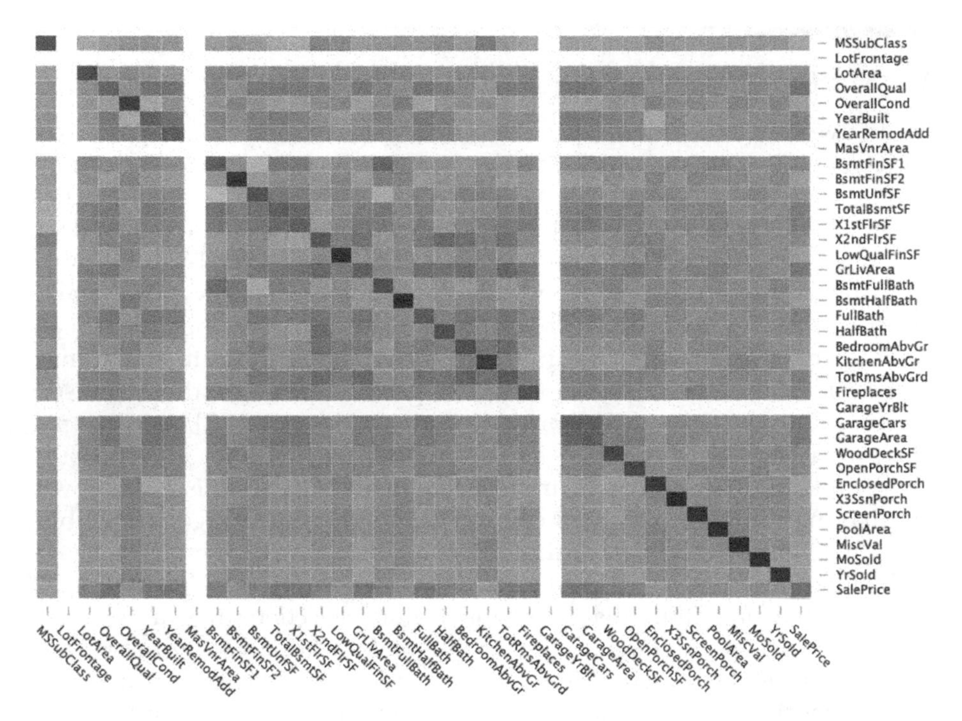

FIGURE 2.2
Correlation coefficient heatmap of quantitative variables.

31 quantitative, and 4 date-related variables out of 79 exploratory variables. A heatmap based on correlation coefficients of quantitative variables is shown in Figure 2.2.

There are 18 variables with missing data that range from a low of 8 to as high as 1406. The missing values for quantitative variables are replaced using average value of that variable, and missing values for qualitative variables are replaced by zero representing another level for that variable.

The Boruta package available in R software is used for feature selection in this study. It uses a wrapper algorithm and can work with any classification methodology that yields variable importance measure (VIM) as an output and by default uses random forest. This analysis performed 100 iterations in a total of about 7.13 minutes. The results yielded 49 attributes confirmed as important, 19 attributes confirmed as unimportant, and 11 tentative attributes as shown in Figure 2.3.

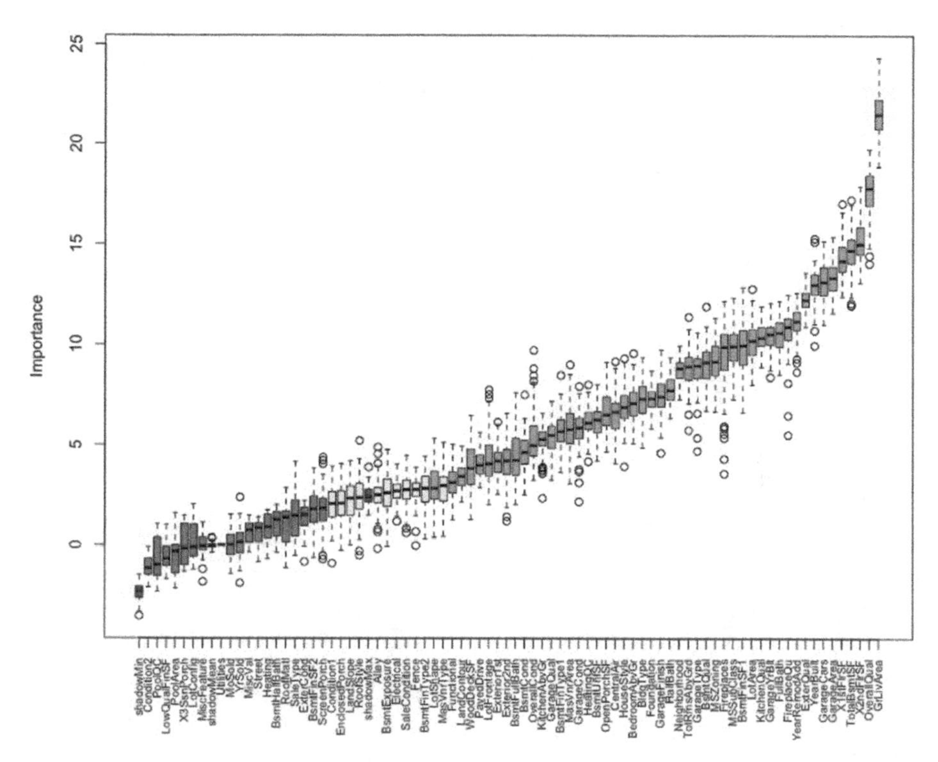

FIGURE 2.3
Important, tentative, and unimportant features based on Boruta analysis.

In Figure 2.3, box plots that are green represent features classified as important, boxplots that are yellow represent tentative features, and boxplots that are red represent unimportant features. Top three features based on the analysis are aboveground living area in square feet (GrLivArea), overall material and finish quality (OverallQual), and second-floor square footages (X2ndFlrSF) based on maximum importance values of 24.34, 19.74, and 17.91, respectively. The output from the analysis also provides mean, median, maximal and minimal importances, and number of hits normalized to number of importance source runs performed and the decision about feature importance. Figure 2.4 provides a plot of number of hits normalized to number of importance source runs performed versus mean importance for the feature categories.

Using a tentative rough fix, a final classification of 79 features is split into 57 important and 22 unimportant. Feature groupings based on Boruta analysis are summarized in Table 2.1.

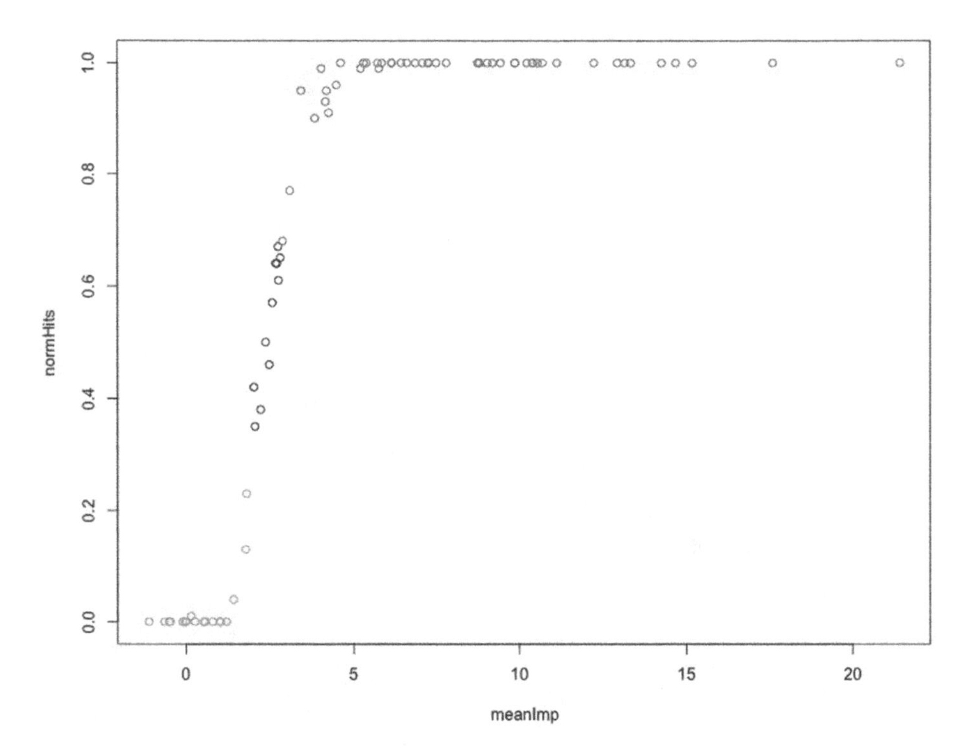

FIGURE 2.4
Number of hits normalized to number of importance source runs performed versus mean importance.

TABLE 2.1
Three Different Grouping of Features

Feature Groupings	Features Included
79 features	All 79 features included
49 features originally confirmed by Boruta analysis	MSSubClass + MSZoning + LotFrontage + LotArea + LotShape + LandContour + Neighborhood + BldgType + HouseStyle + OverallQual + OverallCond + YearBuilt + YearRemodAdd + Exterior1st + Exterior2nd + MasVnrArea + ExterQual + Foundation + BsmtQual + BsmtCond + BsmtFinType1 + BsmtFinSF1 + BsmtUnfSF + TotalBsmtSF + HeatingQC + CentralAir + X1stFlrSF + X2ndFlrSF + GrLivArea + BsmtFullBath + FullBath + HalfBath + BedroomAbvGr + KitchenAbvGr + KitchenQual + TotRmsAbvGrd + Functional + Fireplaces + FireplaceQu + GarageType + GarageYrBlt + GarageFinish + GarageCars + GarageArea + GarageQual + GarageCond + PavedDrive + WoodDeckSF + OpenPorchSF

(Continued)

TABLE 2.1 (*Continued*)

Three Different Grouping of Features

Feature Groupings	Features Included
57 features based on tentative rough fix	MSSubClass + MSZoning + LotFrontage + LotArea + Alley + LotShape + LandContour + Neighborhood + BldgType + HouseStyle + OverallQual + OverallCond + YearBuilt + YearRemodAdd + RoofStyle + Exterior1st + Exterior2nd + MasVnrType + MasVnrArea + ExterQual + Foundation + BsmtQual + BsmtCond + BsmtExposure + BsmtFinType1 + BsmtFinSF1 + BsmtFinType2 + BsmtUnfSF + TotalBsmtSF + HeatingQC + CentralAir + X1stFlrSF + X2ndFlrSF + GrLivArea + BsmtFullBath + FullBath + HalfBath + BedroomAbvGr + KitchenAbvGr + KitchenQual + TotRmsAbvGrd + Functional + Fireplaces + FireplaceQu + GarageType + GarageYrBlt + GarageFinish + GarageCars + GarageArea + GarageQual + GarageCond + PavedDrive + WoodDeckSF + OpenPorchSF + EnclosedPorch + Fence + SaleCondition

2.4 Random Forest Prediction Models

Random forests are extension of the idea of decision trees [9,10]. Unlike a single tree that is constructed in decision tree, multiple decision trees are constructed leading to a random forest. The output from all trees are combined to obtain a better model than what could be obtained from a single tree. Random forest models can be used for developing classification models when the response variable is a factor and can also be used for developing a prediction model when response variable is continuous as in this study. The model is developed using randomForest package available from R software. Random forest has two free parameters number of trees (ntree) and number of variables randomly sampled as candidates at each split (mtry). The default value for ntree is 500 trees in the random forest and default value for mtry is about $p/3$ for regression where p is the number of features. Coefficient of determination (r-square) and root mean square error (RMSE) are used for assessing the performance of the prediction model. Figure 2.5 shows error rate of a random forest model based on all 79 features with house sale price as the dependent variable.

It can be observed from Figure 2.5 that the error rate becomes flat after about 150 trees. This indicates that increasing the number of trees beyond the default value of 500 is unlikely to have a significant impact on the model accuracy. Therefore, for this study the default value for number of trees is kept constant at 500.

Data partitioning with 50:50, 60:40, 70:30, 80:20, and 90:10 splits into training and testing data sets, respectively, are used in the study. Random forest

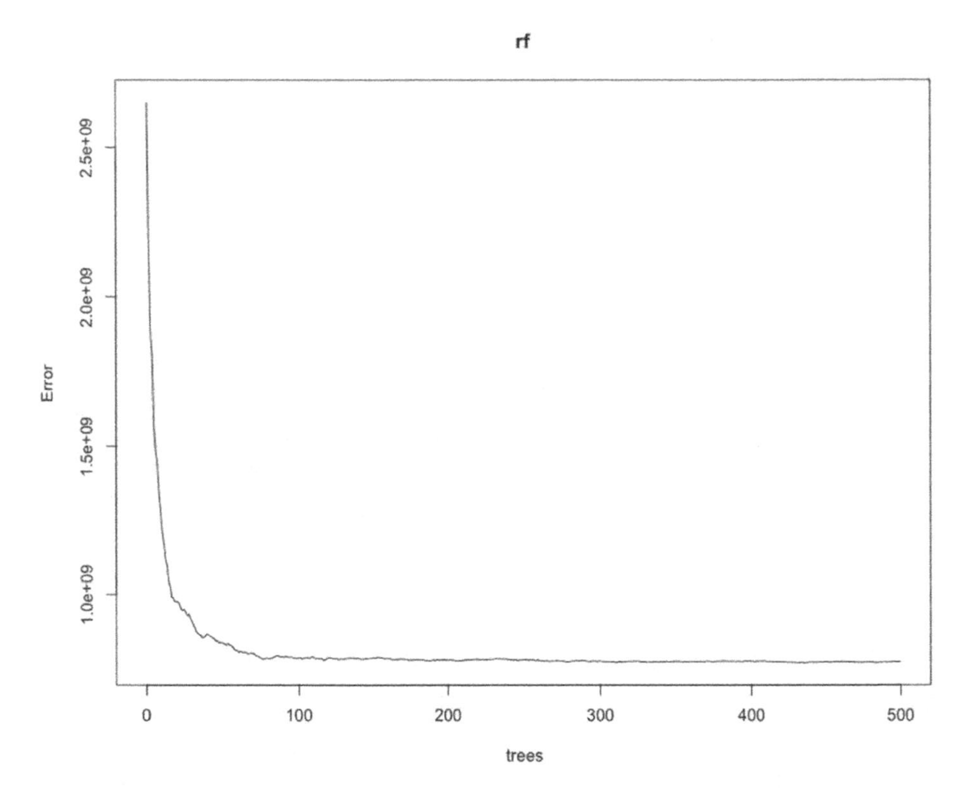

FIGURE 2.5
Error rate of a random forest model with all 79 features.

models are developed for three different feature groupings involving all 79 features, 49 originally confirmed features and 57 features confirmed with tentative rough fix. A random forest model is built using the training data set. To enable consistency in comparison of results across various training and testing data sets for each of the three feature groupings, a random seed with set.seed(123) is fixed for each data partitioning split. R-square and RMSE calculated using training data set and RMSE calculated using testing data sets are used for model assessment. Higher values of r-square and lower values of RMSE are desired. The results obtained for five different ratios of training and testing data splits and three different groupings of features are summarized in Table 2.2.

Table 2.2 shows that r-square values based on training data set are consistently higher when 49 features originally confirmed by the Boruta analysis are used, except when training and testing splits are 90:10. The highest r-square value of 87.98% is obtained with 60:40 split for 49 feature grouping. Similarly, RMSE based on training data set are consistently lower when 49 features originally confirmed by the Boruta analysis are used, except when the splits are 90:10. The lowest RMSE value of 27319.26 is obtained with 60:40

TABLE 2.2

Performance of Random Forest Models for Five Different Ratios of Training and Testing Data Splits and Three Different Grouping of Features

Split	Variables	Train Data R-Sq	Train Data RMSE	Test Data RMSE
50:50	79 (All)	85.83	29132.07	31603.16
50:50	49 confirmed	87.04	27860.26	31811.52
50:50	57 confirmed with rough fix	86.32	28619.47	31633.36
60:40	79 (All)	87.91	27398.04	34497.04
60:40	49 confirmed	87.98	27319.26	34130.56
60:40	57 confirmed with rough fix	87.55	27800.49	34644.65
70:30	79 (All)	86.93	29668.28	27015.77
70:30	49 confirmed	87.21	29353.45	26554.12
70:30	57 confirmed with rough fix	87.03	29554.35	26395.83
80:20	79 (All)	87.31	28522.43	29162.15
80:20	49 confirmed	87.57	28226.59	28648.07
80:20	57 confirmed with rough fix	87.41	28402.20	29154.09
90:10	79 (All)	87.64	28026.23	27663.77
90:10	49 confirmed	87.54	28144.27	25982.38
90:10	57 confirmed with rough fix	87.97	27652.51	27000.87

Abbreviations: r-square, coefficient of determination; RMSE, root mean square error.

split for 49 feature grouping. For 60:40 split, RMSE value is also lower for testing data when 49 features originally confirmed by Boruta analysis are used. These results indicate that use of unimportant and tentative variables in the random forest prediction model for house sale price do not help to improve model accuracy. The results also suggest that the data-partitioning ratio used for model development and assessment may also influence model accuracy. Although for the data set used in this study, the 60:40 split provides better model accuracy, and for a different data set, some other ratio may result in a better accuracy levels.

2.5 Discussion of the Results

Random forest models developed in the previous section suggested 49 confirmed features based on Boruta analysis and a 60:40 split provides improved model accuracy. Note that the number of trees in this random forest

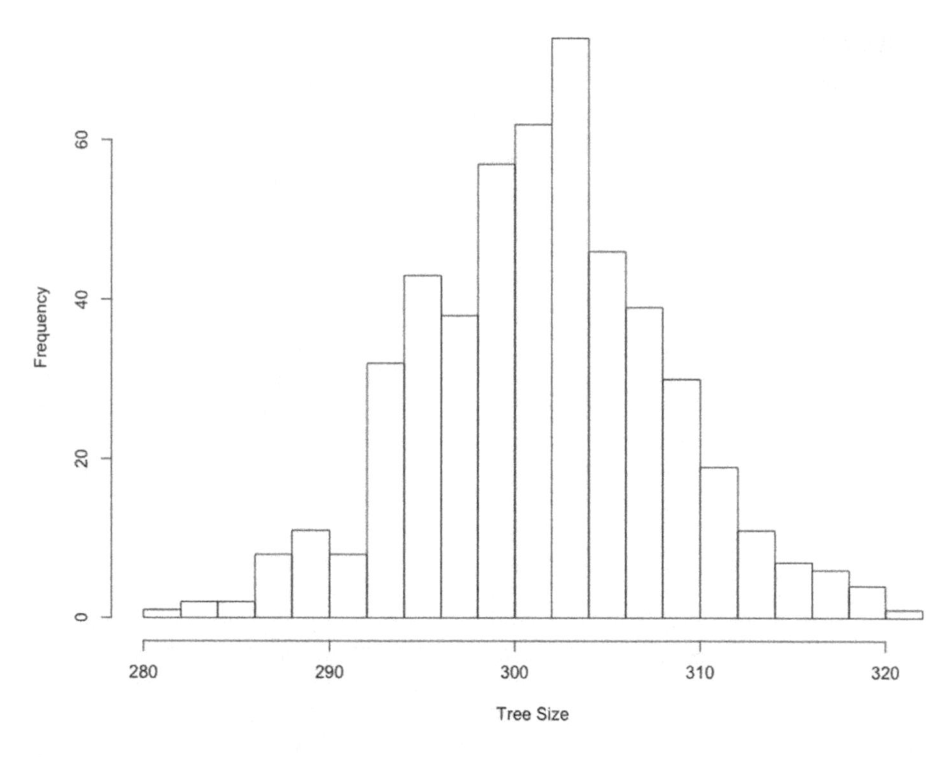

FIGURE 2.6
Histogram of tree size or number of nodes per tree in the random forest model based on 49 confirmed features.

model is 500. Figure 2.6 shows a histogram of tree size or number nodes in each of the 500 trees in the random forest model.

Figure 2.6 shows on an average there are about 300 nodes in a tree for the random forest model. The number nodes vary from about 280 to 325 per tree. The shape of the histogram is approximately symmetrical. Figure 2.7 provides the variable importance plot using random Forest package in R based on the random forest model from 49 confirmed features.

The variable importance plot in Figure 2.7 shows what impact each feature has if removed from the model. The importance is captured using percentage increase in mean square error (MSE) and increase in node purity. Removing GrLivArea from the random forest model has the highest impact on percentage increase in MSE. Similarly, dropping OverallQual has maximum impact on node purity. Note that this list of top 10 features is from among 49 confirmed features based on Boruta analysis. In addition, these 2 features were also in the top two list in importance for Boruta analysis. The performance of the random forest model is further assessed using training data set as shown in Figure 2.8.

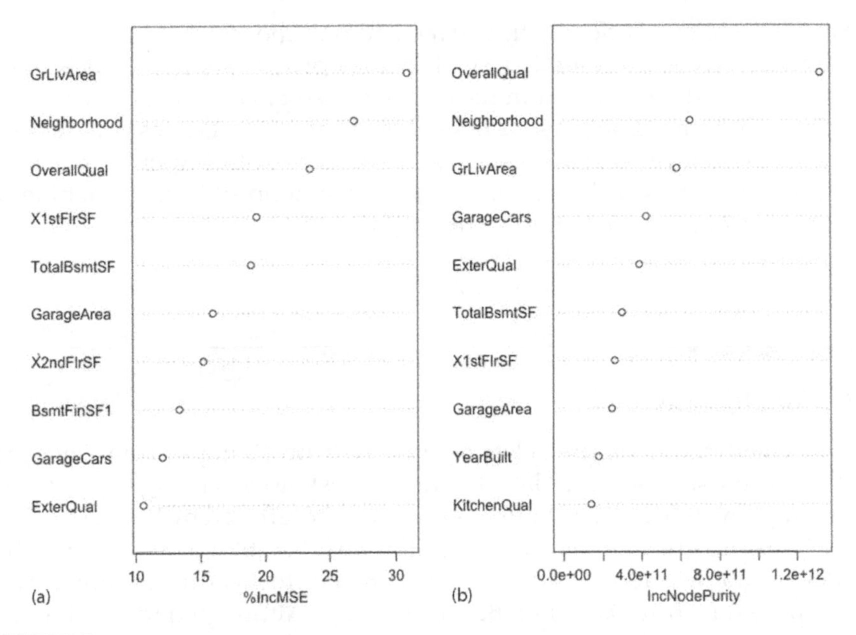

FIGURE 2.7
Top 10 variable importance plot based on the random forest model from 49 confirmed features. (a) Variable importance based on mean square error. (b) Variable importance based on node purity.

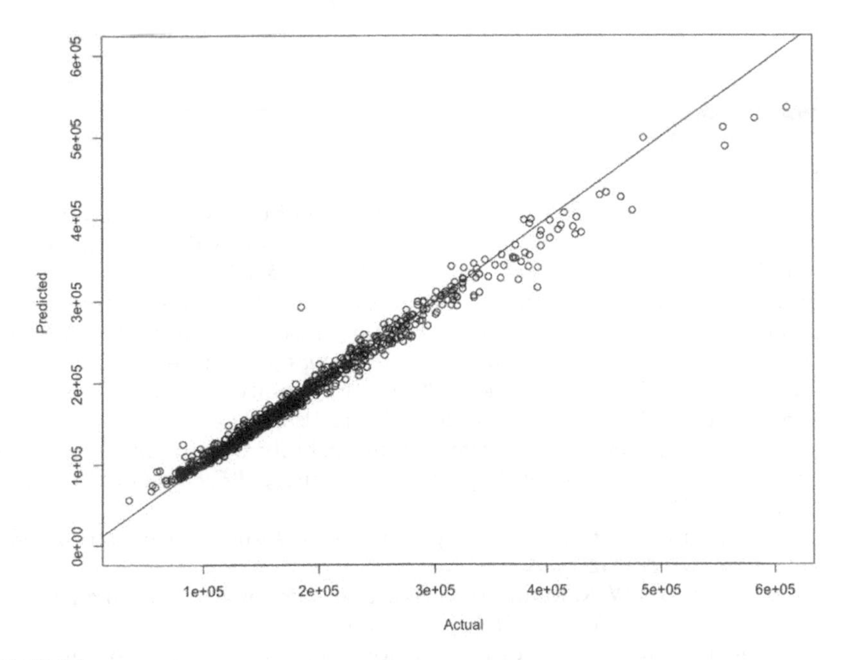

FIGURE 2.8
Actual versus predicted sales price based on the random forest model from 49 confirmed features.

Figure 2.8 shows a decent fit between actual and predicted house sales price based on the training data set. For sales price more than $500K, there seems to be underestimation in the sale of house prices. Such underestimation toward higher house prices is seen even more in the testing data set. This analysis and results also suggests need for exploring other machine-learning algorithms such as neural networks or support vector machines to further improve the prediction accuracy [11].

2.6 Conclusions

In this chapter, the feature selection approach involving Boruta algorithm is illustrated using housing data. Random forest models using three feature groupings involving all 79 features, 49 features confirmed by Boruta analysis and 57 features using tentative rough fix. Results obtained indicate better model accuracy in terms of r-square and RMSE for feature grouping with 49 confirmed features based on Boruta analysis. Although data partitioning with 60:40 split performed comparatively better than other four split ratios used, results also suggest scope for further improving the model accuracy by exploring other machine-learning approaches. This is especially true for house prices that are above $500K.

References

1. M. J. Hallett, J. J. Fan, X. G. Su, R. A. Levine, and M. E. Nunn, "Random forest and variable importance rankings for correlated survival data, with applications to tooth loss," *Statistical Modelling*, 14(6), 523–547, 2014.
2. A. L. Boulesteix, A. Bender, B. J. Lorenzo, and C. Strobl, "Random forest Gini importance favours SNPs with large minor allele frequency: Impact, sources and recommendations," *Briefings in Bioinformatics*, 13(3), 292–304, 2012.
3. M. L. Calle, and V. Urrea, "Letter to the editor: Stability of random forest importance measures," *Briefings in Bioinformatics*, 12(1), 86–89, 2011.
4. K. Nicodemus, "Letter to the editor: On the stability and ranking of predictors from random forest variable importance measures," *Briefings in Bioinformatics*, 12(4), 369–373, 2011.
5. R. Kohavi, and G. H. John, "Wrappers for feature subset selection," *Artificial Intelligence*, 97, 273–324, 1997.
6. M. B. Kursa, and W. R. Rudnicki, "Feature selection with the Baruta package," *Journal of Statistical Software*, 36(11), 1–13, 2010.
7. M. B. Kursa, "Robustness of random forest-based gene selection methods," *BMC Bioinformatics*, 15, 1–8, 2014.

8. Z. Yang, M. Jin, Z. Zhang, J. Lu, and K. Hao, "Classification based on feature extraction for hepatocellular carcinoma diagnosis using high-throughput DNA methylation sequencing data," *Procedia Computer Science*, 107, 412–417, 2017.
9. D. T. Larose, and C. D. Larose, *Discovering Knowledge in Data: An Introduction to Data Mining*. Hoboken, NJ: John Wiley & Sons, 2014.
10. G. Shmueli, N. R. Patel, and P. C. Bruce, *Data Mining for Business Intelligence: Concepts, Techniques, and Applications*. Hoboken, NJ: John Wiley & Sons, 2010.
11. B. K. Rai, "Classification, feature selection and prediction with neural-network Taguchi system," *International Journal of Industrial and Systems Engineering*, 4(6), 645–664, 2009.

3

Simulation Optimization Approach for Cost Reduction in Multi-echelon Inventory Systems Considering Cooperation among Retailers

Taha-Hossein Hejazi, Afrooz Farhadi, and Bahareh Hekmatnia

CONTENTS

3.1 Introduction

Cost optimization for supply chains has become crucial because of total profit maximization for economic reasons. Supply-chain management (SCM) should be studied with the goal of improving customer service, increasing variety of products, and reducing costs.

SCM is a study of direct or indirect members involved in customer satisfaction, the members of which can be classified as a supplier, producer, warehouse, retailer, transport, and customer (Gumus et al., 2010). In addition, supply-chain inventory management is an integrated approach to inventory planning and control throughout the organization's collaborative network, covering the original supplier to the last user on the network.

In this research, a model is proposed based on the multilevel inventory system that focuses on improving the amount of lost sales and costs. Research focuses on retailers who are directly related to customers and have interactions with other retailers with the goal of reducing lost sales and costs. In addition, the simulation approach has been used, and different operations of a multilevel supply chain have been simulated using the Arena simulation software and then the effect of different parameters of the supply chain structure on performance variables has been analyzed. Finally, optimization was performed to minimize costs and the optimal values of the decision variables were extracted from the optimization results and applied in the model. In the end, the results of simulation of the model with the optimal model are compared and suggestions are presented.

3.2 Overview

3.2.1 Definition

3.2.1.1 Supply Chain

In a supply chain (SC), the product flow starts with suppliers and manufactures, and then distributors deliver the final products to the customer groups to meet their demands (Afrouzy et al., 2016). The main reason for a company's focus on the SC is shortening the life cycle of products and changing demands of customers (threats) and development of information technology (opportunities).

An unmanaged SC is not absolutely stable. The problem often observed in unmanaged SCs is the bullwhip effect that creates fluctuations in the SC, while the main factors of bullwhip effect are changes in demand (Lee et al., 1997).

3.2.1.2 Supply Chain Management

SCM is a set of methods used for effective integration of suppliers, manufacturers, warehouses, and vendors. To minimize system costs and achieve appropriate services, goods are produced and distributed to a correct number at the right place and time.

SCM is a crucial approach to coordinate production, inventory, location, and transportation in a SC and attempts to achieve the best combination of responsiveness and efficiency in the market. The five main parts of SCM are as follows:

- Planning
- Resources
- Production
- Send
- Referral

3.2.1.3 Supply Chain Simulation

Simulation is the science and art of creating a model from a process or system for strategy evaluation, or in other words, simulation is a method to know the results of proposed ideas before they are implemented. Simulation is a set of methods and programs to show the actual system behavior, which is usually done by computer and appropriate software.

According to the bullwhip effect, a design can easily penetrate the entire SC. The effect of a poor design on the whole business is huge. A poor design leads to creating surplus cycles in production and the warehouse, poor production estimation, unbalanced capacities, poor customer service, uncertain production plans, high prices in inventory items, and even lost sale situations. Although, the methods of economic resource planning and SCM have great benefits to the industry, using them in academic research is very costly.

Simulation-event separation provides the possibility of performance evaluation before running a system. This will enable companies to analyze "what will happen-if ... " that will lead to better plan decision making. It also provides the possibility of comparing different functional solutions without interrupting the actual system, and ultimately the possibility of making decisions in a timely fashion. Most simulation tools are designed as interactive tools for use by human resources. Simulation tools are not like real-time decision-making tools that connect directly to the control system. These tools help humans make decisions properly by providing information. However, the designer must be able to interpret and modify the design (Kleijnen, 2005).

- Strategic and tactical applications of SC simulation
  Various types of inventory-management issues are meant by the term "multi-echelon inventory systems." In fact, the issue of

multi-echelon inventory systems in SCs can be considered as generalizations of classical inventory models.

The issues of the SC are divided into two groups of strategic or structural issues and operational or tactical issues. Strategic issues are discussed mainly in relation to how to deploy different components of a SC. Tactical issues are generally raised after structural issues and discuss the coordination policies among the different components of a SC. In the field of logistics and SCM, simulation-based models propose solutions to a wide range of issues at a strategic, operational, and tactical level. Specific examples of the issues that these models address are SC design and reconfiguration, inventory planning and management, production scheduling, and supplier selection (Tako and Robinson, 2012).

- Benefits of SC Simulation

  The benefits of supply chain simulations are as follows (Kleijnen, 2005):

  - Simulation helps to understand whole characteristics of the SC processes (by using graphics/animation).
  - Simulation can determine system dynamics and users could model unexpected events in certain domains by using probability distribution and determine the influence of these events on the SC.
  - Simulation should minimize the risk of changes in planning processes. Using the simulation "what will happen-if … " the user can test different strategies before changing the design.

  To study the simulation of SCM, the following trends:

  - Comprehending the supply chain processes (understanding the business process and industry indexes) and planning processes.
  - Design scenario (Often, modeling all details of the SC is not appropriate. The good idea is to focus on relevant sections of the problem).
  - Data collection.
  - Defining the final status.
  - Evaluating of supply chain policies/strategies.

- Simulation-modeling requirements

  In this section, we consider the simulation-modeling requirements for accurate analysis of SC issues. This information is derived from SC modeling researchers at IBM (formerly known as International Business Machines Corporation), which have been gathered in 10 years (Buckley and An, 2005).

a. *Data.* Simulation tools for business process general purposes usually do not support the details of SC data. Some critical data structures to model the SC are as follows:
   - Product definitions
   - Bill of materials (BOMs)
   - Customer demand of each product
   - Customer classes illustrating customer service requirements for different types of users
   - Demand forecasts
   - Initial inventory levels for products at a location
   - Storage-space definitions including storage size and associated costs
   - Reorder points for maintaining inventory levels
   - Lot sizes for inventory replenishment
   - Supply constraints restricting the number of products available from an external supplier over a period of time
   - Locations of customers, distribution centers, and manufacturing sites
   - Routes between locations and the transport time between them

b. *Processes.* The Supply Chain Operations Reference (SCOR) model (Council, 2008) suggests a starting point to develop a simulation model of SC. This model identifies five basic SCM processes: Plan, Source, Make, Deliver, and Return. For the forecasting and supply planning of many industry's customers, manufacturing, distribution, retail, transportation, and inventory policy are sufficient. It is important to notice each fundamental process with regard to the business functions. The plan process can be applied to a single business function or to a set of business functions. For example, a manufacturing function could plan its own activities according to the inputs it receives from other business functions in its SC. In other cases, planning can be applied to maximize overall SC value in business functions.

   To propose a model, it is possible to identify parameters of each business in terms of its fundamental processes. Overview of this parameterization follows:
   - *Customer*: It shows end customers that issue orders to other business functions. Customer functions enforce the fundamental processes plan, source, and return. Orders are generated on the basis of customer demand, which may be modeled as a sequence of specific customer orders. Customer

functions may send forecasts of future demand to other business functions.

- *Manufacturing*: It monitors the assembly, preservation, and finished inventory activities, as well as managing the essential processes of plan, source, make, deliver, and return. It should be noted that the function of manufacturing can supply another function; thus, it is not necessary to define a distinct function for the supplier model. Manufacturing function model information such as variety of products, cycle time, BOMs, manufacturing, and replenishment policies for components and finished products, reorder points, storage capacity, manufacturing resources, material-handling resources, and order-queuing policies.
- *Distribution*: It works with distribution centers and warehouses, which includes finished products inventory and material transportation. Distribution functions perform the fundamental processes plan, source, deliver and return. Typically, the distribution model includes inventory replenishment policies, reorder points, storage capacity, material handling resources and order queuing policies.
- *Retail*: It models retail stores, which includes finished products inventory and material-handling transportation. Retail stores manage the fundamental processes plan, source, deliver, and return. Inventory-replenishment policies, safety stock policies, reorder points, material-handling resources, storage capacity, and shelf space are considered in a retail model.
- *Transportation*: It designs types of transportation (e.g., trucks, planes, trains, and boats), cycle time between shipping locations, vehicle loading and transportation costs, and manages the basic planning, delivery, and return processes. A transportation model usually determines batching policies (by weight or volume), material handling, and transportation resources.
- *Inventory Planning*: It defines the planning of inventory target levels and implements the basic planning process. This function can be performed to determine the best inventory levels based on the customer serviceability, product lead time, and other considerations.
- *Forecasting*: It predicts product sales for future periods and processes. In addition, forecasting enforces the basic planning process. This business function could link to an optimization program.
- *Supply Planning*: It models BOMs and allocates distribution resources to future demand under capacity and supply

constraints. This business function could be linked to an optimization program.

In the SC, it is crucial to distinguish between execution and planning processes. The implementation of processes is driven by programs and policies generated by planning processes. Planning processes deal only with information and not physical goods.

c. *Entities*. Simulation modeling cases that enter and leave business processes are called entities or artifacts. Some specific entities are listed here:
   - *Request Orders*: Demonstrate customer or replenishment orders for physical goods. These entities carry order information from customers to manufacturing and distribution functions and from manufacturing and distribution functions to other manufacturing and distribution functions.
   - *Filled Orders*: Represent customer or replenishment orders for which physical goods have been provided.
   - *Shipments*: It carries filled orders from transportation functions to customers, manufacturing, and distribution functions.
   - *Forecasts*: It represents demand forecasts for customers and replenishment orders. Often, forecast information goes from forecast functions to supply planning, manufacturing, and distribution functions.
   - *Supply Plans*: They represent production and procurement plans generated by a supply planning function. Typically information is carried from supply planning functions to distribution and manufacturing functions by these entities.

d. *Resources*. General-purpose business processes, which are proposed from resource model simulators, are functional to simulate supply chain. Because cycle time and resource cost are important metrics in both business process and SC simulations, business process resource definitions can sometimes be reused for SC simulation. Other parameters needed to model the supply chain resources follow (Buckley and An, 2005):
   - *Storage Resources*: They represent the cost of model and capacity of the manufacturing, distribution, and transportation.
   - *Manufacturing Resources*: These resources represent cost and personnel capacity and equipment used to produce goods in the manufacturing part.
   - *Transportation Resources*: They determine cost of model and capacity of vehicles such as trucks, trains, and ships in transportation part.

- *Material Handling*: Resources the cost of model and capacity of personnel and equipment used for transfer of goods within manufacturing.

3.2.1.4 Multi-echelon Inventory Systems

Multi-echelon inventory systems issues are kind of tactical issues in the SC, in which analytical and simulation methods are generally applied to solve and model them. Tactical issues in the SC are divided into three categories: buyer-seller coordination issues, production-distribution coordination issues, and inventory-distribution coordination issues. Inventory-distribution coordination issues are also called inventory issues, and related systems are referred to as multi-echelon inventory systems (Thomas & Griffin, 1996).

Issues of multi-echelon inventory systems can be divided into two general groups of divergent and converging issues. The main difference between the issues of these two groups is actually the structure of the SC in which the issues of each group is divided into two categories of fixed interval order (FOI) and fixed order size (FOS) (Yao & Dresner, 2008). Also, according to the definition of each unit inventory per echelon, issues are divided into firm inventory and surface inventory. In the inventory of a firm, each unit in each echelon is only equivalent of its own unit inventory at each moment, but in the surface inventory, the inventory of each unit in each echelon includes its own unit inventory and all its units below.

In the multi-echelon inventory model, the most important argument is looking at the model seamlessly. In this model, strategic decisions and tactical decisions can be used (Rau et al., 2003). Strategic decisions seek to achieve the location's goals and determine the number of factories and allocate selected warehouses and assign retailers to selected warehouses. Tactical decisions determine the amount of safety stock and order in each warehouse.

3.2.2 Conclusion

The purpose of SCM is to meet customer's demand to guarantee the delivery of products of high-quality levels with low price and minimal delay. To achieve this goal, companies need to have a complete insight into the whole SC for their plans as well as plans of suppliers and customers. Today, companies must be agile to be able to implement and repair applications in real time and encounter unexpected events in the SC. These requirements require the use of separate events to analyze the whole SC. In addition, efficient SCM results from detailed precision of the material and information of capacity, which are available. Nowadays, companies are seeking to reduce inefficiencies in their business processes and redesigning business processes to achieve world-class business efficiencies.

Simulation helps companies inform their efficiency and dynamics of the SC. During the development of the SC simulation models, a more significant part is understanding the whole SC. Good understanding of commercial features (e.g., performance criteria and production for warehouse or according to order) is essential because each industry has its own business attributes and it would be better to focus on their problem and particular scenarios.

In this chapter, the advantages and data that are necessary to model and simulate the SC have been reviewed and in the following a review of studies will be discussed in the context of SC simulation.

3.3 Literature Review

In this section, most research, which have been published since 2000, have been reviewed. Articles are discussed in SCM and simulation, multi-echelon inventory system, and information sharing.

3.3.1 SCM and Simulation

Ingalls (1998) discussed the reason for applying simulation as the analysis methodology to evaluate SCs and its benefits and disadvantages. They compare simulation with other analysis methodologies such as optimization, business scenarios, and so on. Kritchanchail and MacCarthy (2000) reviewed the applicability of both approaches, specifically looking at discrete-event simulation, an event-driven approach, and system dynamics, which is a time-driven approach. In article by Terzi and Cavalieri (2004), more than 80 pieces of research were reviewed with the main purpose of ascertaining which general objectives simulation is generally called to solve, which paradigms and simulation tools are more acceptable, and deriving useful prescriptions both for practitioners and researchers on its applicability in decision-making processes within the SC context. The main contribution of papers by Kleijnen (2005) is twofold: (1) it classifies simulation for SCM, and (2) it surveyed several methodological issues. These different types of simulation are spreadsheet simulation, system dynamics, discrete-event simulation, and business games. Which simulation type should be applied depends on the type of managerial question to be answered by the model. Sensitivity analysis shows a shortlist of truly important factors in large simulation models with a hundred factors. Wang et al. (2008) characterize quality, budget, and demand as fuzzy variables. In a fuzzy vendor selection, expected value and a fuzzy vendor selection chance-constrained programming is modeled to maximize the total quality level. The two models have been proposed for selecting vendors in fuzzy environments. A genetic algorithm based on fuzzy simulations is applied to solve these two models. Numerical examples show the effectiveness of the algorithm. Schildbach and Morari (2016) proposed a novel approximation scheme using the scenario-based model to manage multi-echelon SCs. The suggested approach can handle SCs with stochastic planning uncertainties from various sources (e.g., demands, lead times, prices) and of a general nature (e.g., distributions, correlations). The main purpose of research is illustrated by a case study. It has been observed that the decisions of the method are reasonable and accurately keep the desired service level constraints in the long run. Finally, substantial cost savings are achieved compared to computationally tractable robust optimization approach deliveries that can be determined optimally in cooperation with up- and downstream members to achieve a minimum overall cost.

3.3.2 Multi-echelon Inventory System

The researches on multi-echelon inventory have been studied well during the last several decades.

Clark and Scarf (1960) made one of the most celebrated works in the field of multilevel inventory systems. By emphasizing the concept of "surface inventory," they analyzed the cost function of multilevel systems. Axsäter and Juntti (1996) studied worst-case results for the relative cost difference in serial and assembly multi-echelon inventory systems and applied simulation in distribution systems with stochastic demand. Diks and De Kok (1999) considered the problem of determining the control parameters of a divergent multi-echelon inventory system and minimized the expected holding and penalty costs and used a simulation example to illustrate validity of the proposed algorithm. Tee and Rossetti (2002) surveyed the behavior of a (R, Q) multi-echelon inventory model to predict the total system cost under a nonstationary Poisson demand process. Finally the analysis showed that mathematical tools are practical for multi-echelon inventory systems. A multi-echelon inventory consisting of a single supplier, a single producer, and a single retailer has been modeled by Rau et al. (2003). In his model, lead time is assumed to be negligible. Köchel and Nieländer (2005) applied the simulation optimization approach where a simulator is combined with an appropriate optimization tool to overcome different restrictive assumptions in multi-echelon inventory systems. Kiesmüller et al. (2004) consider the system that all stock points are controlled by continuous review (s, nQ) then assume stock policies with stochastic transportation times and compound renewal demand. Wan and Zhao (2009) presented a simulation model for the multi-echelon inventory optimization problems using Arena. They assumed five members in the multi-echelon inventory: a manufacturer, a distribution center, and three retailers. Their analysis focuses on the relationship between the fill rates of retailers and the average inventory level of the entire SC. They assumed that customer's demand, as well as the lead times, is uncertain. Seifert et al. (2012) studied coordination in a three-echelon SC and examined the impact of sub-SC coordination (subcoordination). Their analysis is based on the price-only contracts that are commonly used in practice. They consider the following cases: no coordination between any members of the SC (decentralized), coordination between only two members (sub-SC coordination), and coordination of the whole SC as a benchmark. Their analysis shows that both the supplier and the retailer would prefer to act alone rather than to coordinate with the manufacturer when sub-SC coordination is suggested. Ryu et al. (2013) presented a fractal-based approach for inventory management to minimize inventory costs and smooth material flows between SC members while responsively meeting customer demand. Within this framework, each member in the SC is defined as a self-similar structure, referred to as a fractal. The simulation results indicated the cost savings and also showed that trans-shipment between retailers reduces the total cost. Suraj et al. (2016) concentrated on the inventory in a three-stage model with two-echelon SC, considering two identical retailers with one distribution center and one manufacturer. They assume a continuous review in inventory replenishment policies and also a single product. In this model, the inventory is optimized by using arena simulation. In addition, it determines the reorder point and reorder quantity to optimize the inventory at both echelons in the SC.

3.3.3 Information Sharing

Considering articles and research on this topic, important elements and general ones that have been used in different fields in different publications have been identified to examine the effect of information sharing on SC dynamics. Huang and Iravani (2007) identified seven elements: SC configuration, decision-making level, information type, sharing style, dynamic performance index, SC modeling approach, and data analysis. Buzacott and Shanthikumar (1993) modeled various single-stage production-inventory systems. Using analytical models, Lee et al. (2000) proposed a model for a two-level SC. Their analysis shows that the value of demand information sharing can be quite high, particularly when demands are correlated over time. Ball et al. (2002) described their experience with performing and modeling supply chain information (SCI). They presented the integration architecture and the software components of their prototype implementation. Then a variety of information sharing methodologies had been discussed. They proposed a multi-echelon SC process model spanning multiple organizations, studied research on the benefits of intra-organizational knowledge sharing, and discussed performance scalability. Li et al. (2005) presented a detailed review of studies on the value of information sharing in SCs. By using the simulation technique, Chen et al. (2007) modeled different scenarios of information sharing and then analyzed the effects of information quality on multi-echelon SC performance. Also, this paper applied data envelopment analysis (DEA) to integrate multiple performance measures to guarantee the information sharing scenarios with enhanced performance. In research by Yang et al. (2011), the robustness of different SC strategies under various uncertain environments is studied by using simulation. Different techniques included Taguchi and multiple criteria decision-making methods (MCDMs) have been applied to solve the problem. The signal-to-noise (S/N) ratio is used to determine an overall evaluation among various SC information-sharing strategies. At last, the simulation results present that e-shopping has the most robust performance in uncertain environments. This study evaluates both the total inventory cost and the customer service-level as performance measures. Khan et al. (2016) made use of the model to study the impact of environmental and social costs on the information sharing in a two-level SC. The paper proposes a mathematical formulation for the reduction in buyer's unit price and the enhancement in the SC's annual profit by virtue of information sharing. Finally results showed that information-sharing results in better annual profit. This gain in profit depends on the buyer's unit price and the parameters for environmental and social cost. Srivathsan and Kamath (2017) focused on upstream inventory information sharing. They developed performance evaluation models of Supply Chain Networks (SCNs) that explicitly consider production capacity, inventory-related decisions, variability, transit delays, and inventory information sharing in a united manner. They applied a two-echelon SCN conjuration with two retail stores and two production facilities as a test bed. They modeled three levels of inventory information sharing in their study; the information shared ranges from the stock-out information at the lowest level of inventory and back-order level information at the highest level.

3.3.4 Conclusion

Considering the literature review, most research has applied simulation in multi-echelon inventory systems. Some research studied inventory information sharing at different levels, and these researchers assumed (S, s) ordering policy and do not optimize simulation results of the system. Some cases corresponded uncertainty with the SC and considered uncertainty in demand, lead time, lifetime of the product, and selecting vendors (Table 3.1).

3.4 Proposed Method

3.4.1 Problem Definition

In this section, a three-echelon SC model, which consists of manufacturer, distributor, and retailers, is shown in Figure 3.1.

In this study, the simulation model is a two-echelon SC, and Arena 14 has been applied to develop the model. The proposed model consists of one distribution center, three retailers, and customers and additional interrelationship among different retailers is considered, which means the retailers can share their own inventory to satisfy a customer order.

Simulated model developed in three different types:

1. A model that has three retailers without interrelationship.
2. A model that has three retailers with limited interrelationship.
3. A model that has three retailers with full interrelationship.

TABLE 3.1

Summary of the Information Provided in the Context of Multi-echelon Inventory and Information Sharing Issues

Article		Vieira (2004)	Wan and Zhao (2009)	Patil et al. (2011)	Barroso et al. (2013)	Ingalls (2014)	Khan et al. (2016)	Srivathsan and Kamath (2017)
Members		Manufacturer Supplier Distributer Retailer	Distributer Retailer	Distribute Retailer	Manufacturer Supplier Distributer Retailer	Manufacturer Supplier Distributer	Distributer Retailer	Manufacturer Retailer
Ordered policy	(S, s)	*	—	—	*	*	*	*
	(R, Q)	—	*	*	—	—	—	—
Number of products	One product	—	*	*	—	—	*	*
	Multi product	*	—	—	*	*	—	—
Production policy	Pull sys	*	—	—	*	*	—	*
	Push sys	—	—	—	—	—	—	—
Information sharing	Inventory	—	—	*	—	*	*	*
	Demand	—	—	—	*	—	—	—
	Supply	—	—	—	—	—	*	—
Analysis method	Optimization	—	*	—	—	*	*	—
	Simulation	*	*	*	*	*	—	*
Objective		Fixing the bullwhip effect	Reduce inventory levels and optimize order size and reorder point	Sales increase	Costs reduction	Costs reduction	Increase annual profit with a drop in buyer's price	Predict the performance measures at two lower levels of inventory

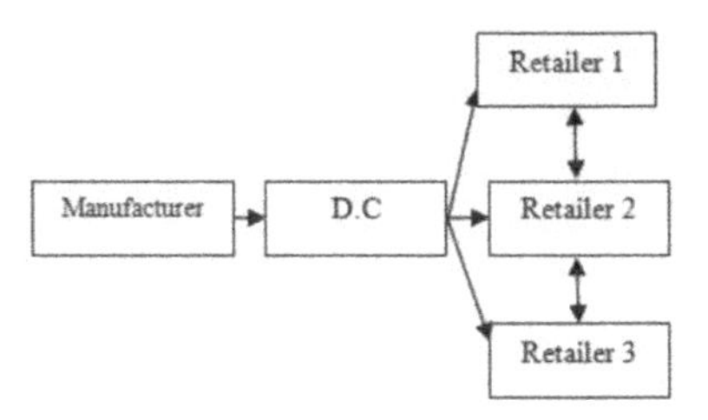

FIGURE 3.1
Supply chain system used in models.

System assumptions include:

1. Single product is considered throughout the whole chain.
2. Customer demand is uncertain and has Poisson distribution.
3. At the beginning of the simulation, all warehouses are full.
4. The lead time of the normal distribution is considered.
5. Inventory policy (S, s) is used.

The variables are as the following:

Size i: The customer demand at each retailer (i = 1, 2, and 3)
Maximum inventory i: Maximum inventory at each retailer (i = 1, 2, and 3)
Reorder point i: Reorder points at each retailer (i = 1, 2, and 3)
Lead time i: Lead time for each retailer for replenishment from distribution center.
Inspection interval of inventory: Inventory inspection interval of each retailer
Total demand: Total demand for retailer's sales and lost sales.

Three scenarios with the same parameters are simulated and initial values of the variables are shown in Table 3.2. The SC model is downstream and inventory policy (S, s) is used. Lead time is taken in hours, lead time among retailers is in minutes, and both have normal distribution. Demand is distributed in Poisson and occurs as an exponential process with the rate of 13, 14, and 15 in minutes.

To evaluate scenarios, the total cost of the SC is divided to the total cost of holding, the cost of lost sales and fixed and variable ordering cost unit is defined in terms of dollars.

The flowchart for the sales decision of retailers for the first scenario is shown in Figure 3.2. In this case, retailers are completely independent. The customer of each retailer is independently entered if the demand of customer is equal to the retailer inventory level. The demand of customer is fulfilled and registered as a sale; otherwise it will be recorded as lost sale.

TABLE 3.2

Initial Values of the Model

Parameters	Value
Demand for retailer 1	Pois(5)
Demand for retailer 2	Pois(4)
Demand for retailer 3	Pois(3)
Inventory level for retailer 1	300
Inventory level for retailer 2	450
Inventory level for retailer 3	600
Reorder point for retailer 1	30
Reorder point for retailer 2	40
Reorder point for retailer 3	50
Lead time for retailer 1	Norm [3.5,0.3]
Lead time for retailer 2	Norm [3.5,0.3]
Lead time for retailer 3	Norm [4.5,0.4]
Lead time from retailer 1 to retailer 2	Norm [13,0.2]
Lead time from retailer 1 to retailer 3	Norm [14,0.3]
Lead time from retailer 2 to 3	Norm [12,0.4]
Holding cost unit	$1
The cost lost sale per unit	$30
The variable ordering cost	$3
The fixed ordering cost	$32

Source: Patil, K. et al., Arena simulation model for multi echelon inventory system in supply chain management, In *Industrial Engineering and Engineering Management (IEEM), 2011 IEEE International Conference on*, pp. 1214–1217, IEEE, 2011.

The flowchart for the sales decision of retailers for the second scenario is shown in Figure 3.3. The customer entered if the demand of customer is equal to the inventory level of the first retail, the customer's demand is fulfilled and is registered as a sale; otherwise it will be recorded as lost sale and then the size of the second retailer's inventory levels should be checked. If the second retailer has sufficient inventory, the demand of customer will be fulfilled and is registered as a sale; otherwise it will be recorded as lost sale and then the size of the third retailer will be reviewed.

The flowchart for sales decision of retailers for the third scenario is shown in Figure 3.4. The customer entered if the demand of customer is equal to the inventory level of the first retail, and the customer's demand is fulfilled; otherwise the inventory levels of the second retailer should be checked. If the retailer has sufficient inventory, demand of customer will be fulfilled; otherwise inventory levels of the third retailer will be checked. If all three retailers do not have enough inventory, the amount of customer demand is recorded as lost sales.

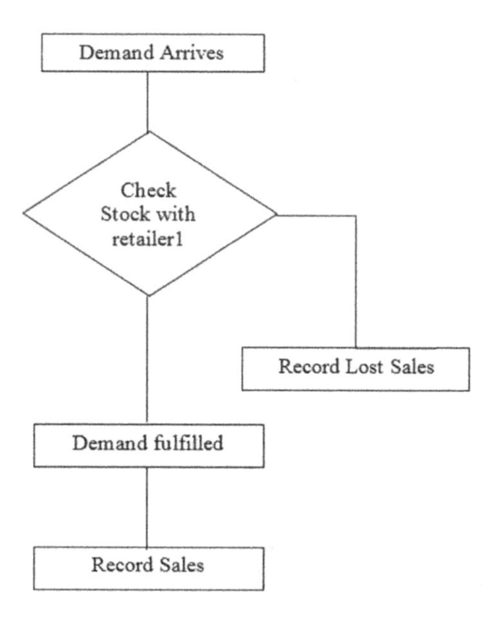

FIGURE 3.2
Retailers sales decision in first scenario flowchart.

3.4.2 Modeling by Arena

Each retailer's warehouse has upper and lower limits, and the inventory level represents the amount of retailer's inventory at the same moment.

1. *First scenario.* The three retailers do not have any interrelationship. If each retailer fails to fulfill the customer's demand, they will lose the customer and will incur the cost of lost sales. For example, an overview of the simulation for the first retailer is given in Figure 3.5.

   As shown in Figures 3.6 and 3.7, at first, customers will arrive at an average of 13 minutes with exponential distribution. Customer demand is distributed by Poisson. If the size of the customer demand was less than the retailer inventory level, the customer's request will be answered, and the retailer inventory level will be reduced to the amount of the demand that has been fulfilled. Otherwise, the amount of sales will be lost, and lost sale costs will be occurred in model.

   As seen in Figures 3.8 and 3.9 at the beginning of each day, every 16 hours, the inventory level is checked, and if the minimum inventory is lower than specific inventory, it will be ordered as the amount of difference between highest level of inventory and lowest inventory level at that moment (Figure 3.10). There are two types of

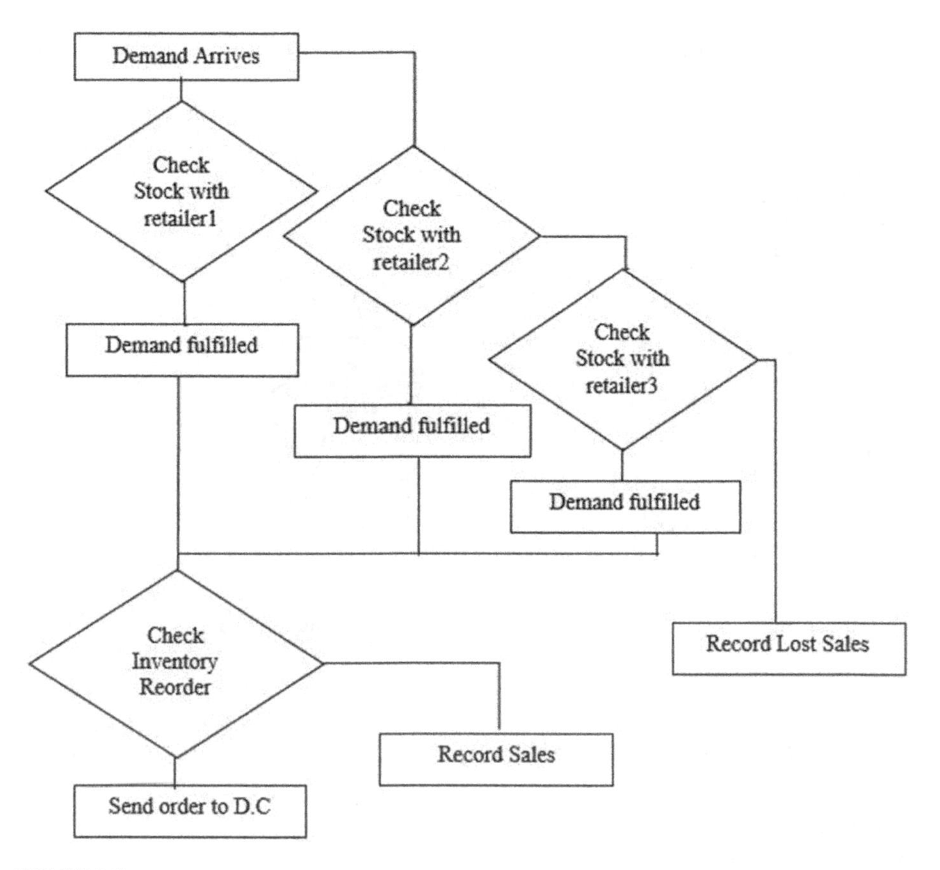

FIGURE 3.3
Retailers sales decision in second scenario flowchart.

ordering costs: Fixed cost and variable cost of order, which depends on the order quantity. After ordering, the time that it takes for the order to reach the retailer from the distribution center is distributed as normal in terms of hours. After a retailer receives the order, the inventory level will increase by the amount of order.

2. *Second scenario.* The three retailers have limited interrelationship. If any of the retailers cannot fulfill the customer demand, they will incur a lost sale cost; however, the next retailer will be introduced to the customer. An overview of the simulation for first and second retailers is given in Figure 3.11. All modules of this scenario are similar to the first scenario, and there is only time for the customer to go from the first retailer to second retailer and second retailer to the third.

3. *Third scenario.* The three retailers, in this case, have the most interrelationships. The retailers can share their own inventory to satisfy a customer order. Lost sales will emerge when none of the three

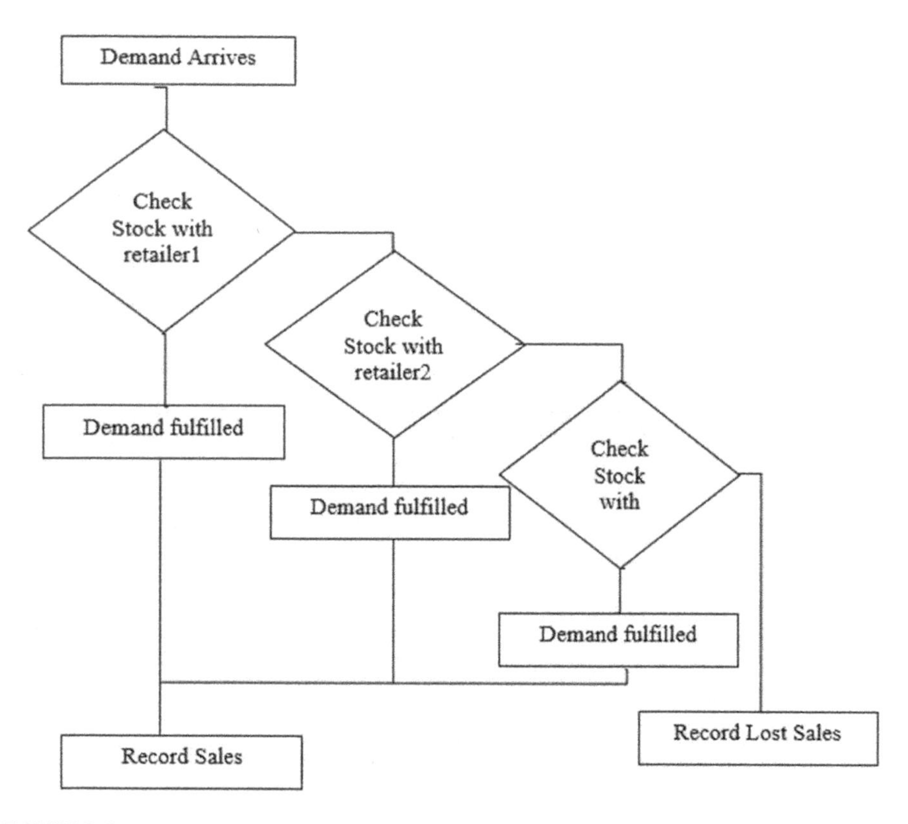

FIGURE 3.4
Retailers sales decision in third scenario flowchart.

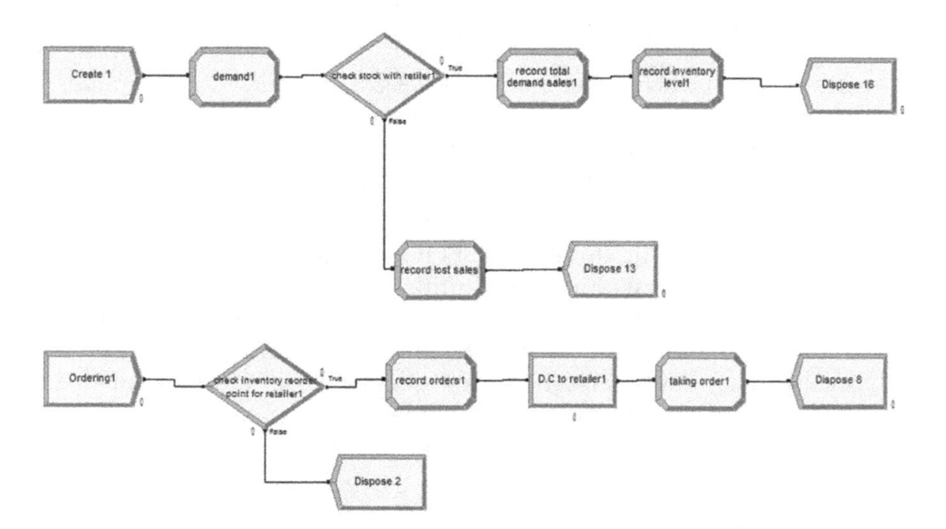

FIGURE 3.5
Overview of the simulation for first retailer in the first scenario.

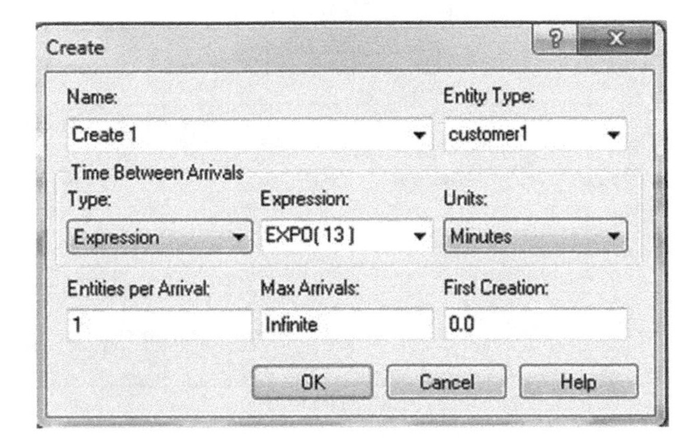

FIGURE 3.6
Demand of arrived customer.

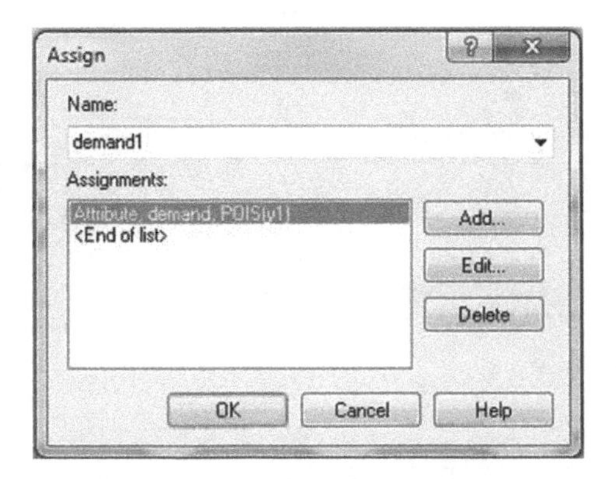

FIGURE 3.7
Time among two customer's entrance.

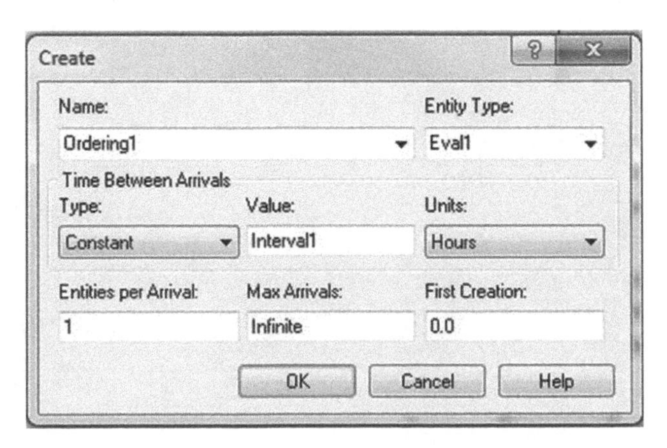

FIGURE 3.8
Inventory inspection interval.

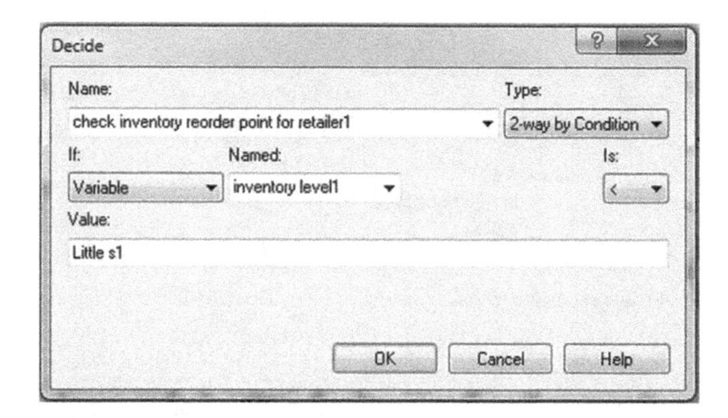

FIGURE 3.9
Ordering condition.

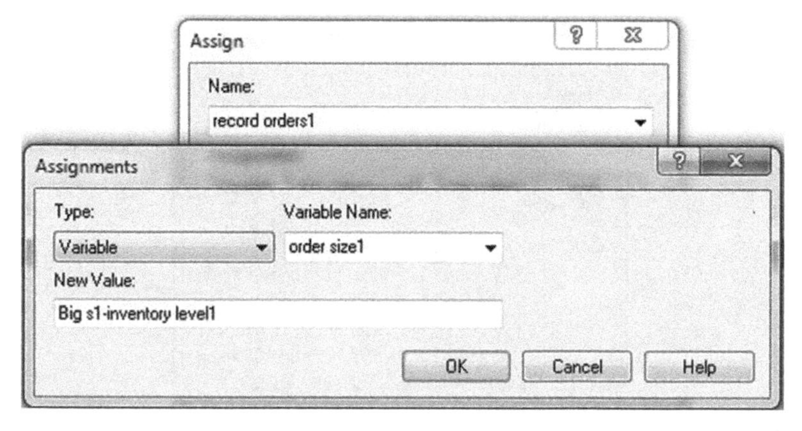

FIGURE 3.10
Quantity of order.

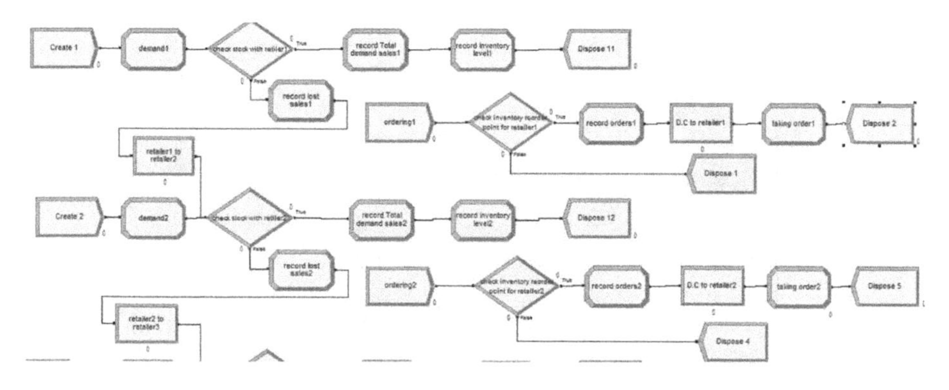

FIGURE 3.11
An overview of the simulation for first and second retailers in second scenario.

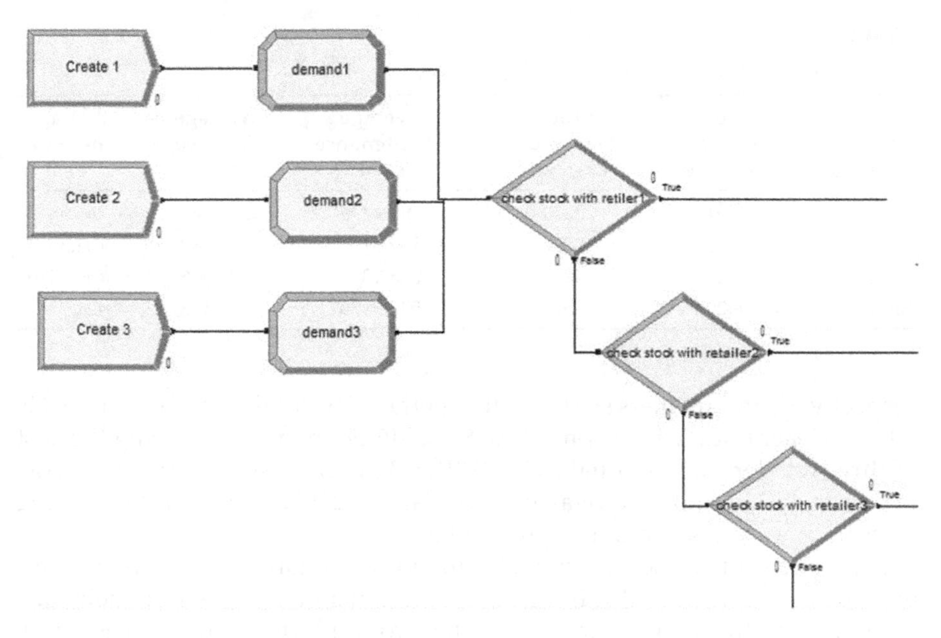

FIGURE 3.12
Customer entrance view in the third scenario.

retailers can fulfill customer's demand. In this scenario, less time is spent for the customer compared to the second scenario because if the first retailer inventory was not enough, the retailer would ask two other retailers, and each customer would be sent exactly to the same retailer. Another difference between this scenario and the previous scenarios is that every retailer does not have its own customer. The customer entrance view in the third scenario is given in Figure 3.12.

3.5 Experimental Results and Analyses

3.5.1 Results of the Initial System Implementation

The simulation was run for 10 replications with a 60-hour warm-up period and 250 hours of replication length, and each day is considered to be 16 hours (Patil et al., 2011).

The results for the first model (without interrelationship among retailers) are determined in Table 3.3. In this model, retailers are completely independent. As the results showed, retailers had sold 37%, 35%, and 28%,

TABLE 3.3

Simulation Results of the First Scenario

First Scenario	Total Demand of Sales	Total Demand of Lost Sales	Average Maintenance Cost	Average Ordered Cost	Average Lost Sales Cost
Retailer 1	3591.20	2224.80	91.4979	38.6540	266.98
Retailer 2	3416.10	869.60	180.31	32.5380	104.35
Retailer 3	2622	421.90	256.13	20.3596	50.6280
Total	9629.3	3516.3	527.9379	82.5516	421.958

respectively, and retailers sold 73% all together. The total cost of this model is $032.4475 per hour, which consists of $82.5516 per hour for ordering the cost of three retailers and is equal to 8%, $421.958 per hour shows lost sales costs of three retailers and is equivalent to 41%, and $527.9379 per hour for holding costs of three retailers and is equivalent 51%.

The results for the second model (limited interrelationship among retailers) are shown in Table 3.4. In this model, retailers will not share their own inventory. In this model, if the retailer fails to fulfill the customer's demand, profit of the customer will be lost and another retailer is introduced to the customer, but the customer will spend a long time in this system because the retailer may not have enough inventory and the customer can visit another retailer. With the results obtained, retailers had sold 40%, 35%, and 33%, respectively, and three retailers had sold 91% combined. The total cost of this scenario is $1362.9761 per hour, including $118.1532 per hour for ordering cost of three retailers and equals 9%, $740.13 per hour for lost sales costs of three retailers and is equivalent to 54%, and $504.6929 per hour for holding costs of three retailers and is equivalent 37%.

The results of the third model (full interrelationship among retailers) are shown in Table 3.5. In this model, the retailers could share their own inventory to fulfill customer orders. In this model, if the retailer fails to fulfill customer demand, he or she will contact two other retailers and those who have enough inventory customers will be sent to that retailer. In this case, the

TABLE 3.4

Simulation Results of the Second Scenario

Second Scenario	Total Demand of Sales	Total Demand of Lost Sales	Average Maintenance Cost	Average Ordered Cost	Average Lost Sales Cost
Retailer 1	3779.30	2128	95.3629	41.2480	255.36
Retailer 2	3850.50	2571.60	154.68	38.1740	308.59
Retailer 3	3844.20	1468.20	254.65	38.7312	176.18
Total	11474	6167.80	504.6929	118.1532	740.13

TABLE 3.5

Simulation Results of the Third Scenario

Third Scenario	Total Demand of Sales	Total Demand of Lost Sales	Average Maintenance Cost	Average Ordered Cost	Average Lost Sales Cost
Retailer 1	4783.60		5734.92	44.5920	
Retailer 2	4148.80	1077.50	154.74	43.7468	129.30
Retailer 3	3059.40		262.35	27.3580	
Total	11991.8	1077.50	474.4392	115.6968	129.30

customer spends less time. Also, lost sale emerges when none of the retailers have enough inventories to fulfill customer demand. From the results it can be concluded that a combination of three retailers and their interrelationship is beneficial. From the results obtained, retailers had sold 33%, 34%, and 33%, respectively, and all together retailers had sold 91%. The total cost of this model is $719.436 per hour, which is divided by $115.6968 per hour for ordering cost of three retailers and equals 16%, $129.30 per hour for lost sales costs of three retailers and equals 18% and $474.4392 per hour for holding costs of three retailers and equals 66%. Thus, this model satisfies most of the customer orders.

3.5.2 Optimization

In this section, the model has been optimized by Optquest. The minimum total amount of ordering costs for three retailers, total lost sales cost by three retailers, total holding costs for three retailers, and sum of all scenario's costs is obtained for three scenarios. For example, optimization steps for the first scenario are as follows:

First Step: Selection of decision variable: All variables of the model are selected. The variables in Figure 3.13 are integers and larger than zero, and the minimum and maximum values are set for them.

Second Step: Selection of objective function's components as in Figure 3.14; all three types of costs are selected for all three retailers.

Third Step: Selection of objective function as you can see in Figure 3.15; four objective functions are minimized:

1. Total holding costs for three retailers
2. Total lost sales costs for three retailers
3. Total amount of ordered costs for three retailers
4. Sum of all scenario's costs

User Specified Summary						
Included	Control	Element Type	Type	Low Bound	Suggested Value	High Bound
☑	Big s1	Variable	Integer	200	300	400
☑	Big s2	Variable	Integer	350	450	550
☑	Big s3	Variable	Integer	500	600	700
☑	Interval1	Variable	Integer	8	16	96
☑	Interval2	Variable	Integer	8	16	96
☑	Interval3	Variable	Integer	8	16	96
☐	inventory level1	Variable	Continuous	270	300	330
☐	inventory level2	Variable	Continuous	405	450	495
☐	inventory level3	Variable	Continuous	540	600	660
☑	Little s1	Variable	Integer	20	30	40
☑	Little s2	Variable	Integer	30	40	50
☑	Little s3	Variable	Integer	40	50	60

FIGURE 3.13
Selection of decision variables.

Responses User Specified

User Specified Summary			
Includ	Data Type	Response	Response Type
☑	Time Persistent	Holding cost1	DStat Average
☑	Time Persistent	Holding cost2	DStat Average
☑	Time Persistent	Holding cost3	DStat Average
☑	Output	lost cost1	Output Value
☑	Output	lost cost2	Output Value
☑	Output	lost cost3	Output Value
☑	Output	order cost1	Output Value
☑	Output	order cost2	Output Value
☑	Output	order cost3	Output Value
☐	Variable	Big s1	Variable Value

FIGURE 3.14
Selection of components in objective function.

Objectives Summary					
Select	Name	Linear	Goal	Description	Expression
☐	New Objective	NonLinear	Minimize		[Holding cost1]+[Holding cost2]+[Holding cost3]
☐	New Objective	NonLinear	Minimize		[lost cost1]+[lost cost2]+[lost cost3]
☐	New Objective	NonLinear	Minimize		[order cost1]+[order cost2]+[order cost3]
☐	New Objective	NonLinear	Minimize		[Holding cost1]+[Holding cost2]+[Holding cost3]+[lost cost1]+[lost cost2]+[lost cost3]+[order cost1]+[order cost2]+[order cost3]

FIGURE 3.15
Selection of objective function.

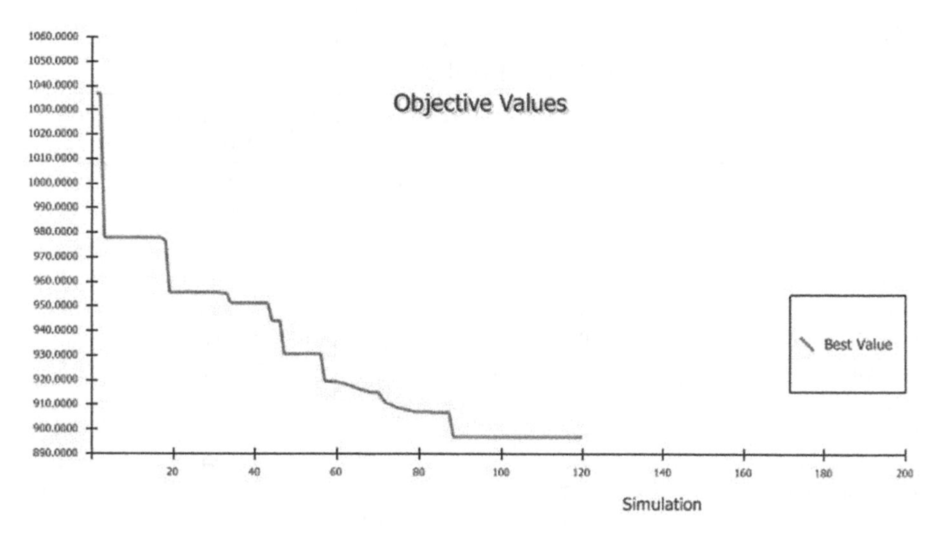

FIGURE 3.16
Optimization graph for the total cost of three retailers for first scenario.

Optimization graph for total cost of three retailers is shown in Figure 3.16. The optimal decision variables are reported in Section 3.4.3 to compare the scenarios.

3.5.3 Results of Optimization

The optimization results of four objective functions are represented in the previous section (Table 3.6). The results of the first scenario are shown in Table 3.7, for the second scenario in Table 3.8, and for the third scenario in Table 3.9. Specific values in the three tables represent the optimal value of variables corresponding to the objective function.

The optimal values of decision variables are derived from the optimization of the total cost and are presented in Table 3.10. The optimized system is simulated and the results are shown in Table 3.11, and the results of the initial system are shown in Table 3.12.

TABLE 3.6
Optimal Values of the Variables Corresponding to Objective Functions in Three Scenarios

Objective	First Scenario	Second Scenario	Third Scenario
Holding cost	153.64	81.21	58.30
Ordering cost	15.71	13.33	13.64
Lost sales cost	246.24	459.84	5.48
Total costs	897.05	1099.98	523.16

TABLE 3.7

The First Scenario Costs Optimization Results

First Scenario	Total Holding Cost	Total Ordering Cost	Total Lost Sales Cost	Total Costs
Big S1	200	201	397	267
Big S2	350	353	550	350
Big S3	500	694	500	500
Little S1	20	21	40	40
Little S2	50	50	50	31
Little S3	40	56	60	40
Interval1	95	95	11	8
Interval2	95	95	8	8
Interval3	96	96	8	96
Holding cost1	10.11	10.25	154.25	95.78
Holding cost2	36.37	36.94	241.65	145.41
Holding cost3	107.16	205.33	224.78	102.53
Total holding cost	153.64	252.52	620.68	343.72
Total ordering cost1	2.64	2.66	46.65	45.94
Total ordering cost2	4.44	4.48	39.03	35.67
Total ordering cost3	6.24	8.57	23.60	6.24
Total ordering cost	13.32	15.71	109.28	87.85
Total lost sales cost1	617.2	616.96	164.72	190.44
Total lost sales cost2	375.52	374.8	56.28	90
Total lost sales cost3	174.04	130.88	25.24	185.04
Total lost sales cost	1166.76	1122.64	246.24	465.48
Total costs	1333.72	1390.87	976.2	897.05

TABLE 3.8

The Second Scenario Costs Optimization Results

Second Scenario	Total Holding Cost	Total Ordering Cost	Total Lost Sales Cost	Total Costs
Big S1	200	200	287	235
Big S2	546	350	379	374
Big S3	503	500	532	503
Little S1	40	40	34	21
Little S2	31	30	39	36
Little S3	54	40	41	42
Interval1	92	96	8	16
Interval2	96	96	10	8
Interval3	87	96	10	11
Holding cost1	9.38	10.20	106.62	75.65
Holding cost2	46.24	28.60	140.50	128.69
Holding cost3	25.59	24.82	217.70	196.20

(Continued)

TABLE 3.8 (*Continued*)

The Second Scenario Costs Optimization Results

Second Scenario	Total Holding Cost	Total Ordering Cost	Total Lost Sales Cost	Total Costs
Total holding cost	**81.21**	63.62	464.82	400.54
Total ordering cost1	2.64	2.64	46.94	41.15
Total ordering cost2	6.80	4.44	47.63	54.54
Total ordering cost3	6.28	6.24	32.47	30.7
Total ordering cost	15.72	**13.32**	127.04	126.39
Total lost sales cost1	601.52	618	181.44	262.04
Total lost sales cost2	929	994.12	176.92	187.52
Total lost sales cost3	1047.36	1104.84	101.48	123.48
Total lost sales cost	2577.88	2716.96	**459.84**	573.04
Total costs	2674.81	2793.9	1051.7	**1099.98**

TABLE 3.9

The Third Scenario Costs Optimization Results

Third Scenario	Total Holding Cost	Total Ordering Cost	Total Lost Sales Cost	Total Costs
Big S1	200	200	300	205
Big S2	350	350	375	357
Big S3	501	500	502	501
Little S1	20	20	38	25
Little S2	50	50	46	37
Little S3	40	60	48	42
Interval1	84	96	10	8
Interval2	93	96	12	9
Interval3	76	96	8	19
Holding cost1	4.76	5.35	85.05	49.25
Holding cost2	12.71	28.50	135.04	120.96
Holding cost3	40.82	78.58	252.70	203.47
Total holding cost	58.30	112.43	472.79	373.68
Total ordering cost1	2.64	2.64	72.48	59.00
Total ordering cost2	4.51	4.51	42.51	50.66
Total ordering cost3	12.72	6.48	12.23	16.38
Total ordering cost	19.87	13.64	127.22	126.04
Total lost sales cost	1096.48	1151.64	5.48	23.44
Total costs	1174.65	1277.71	605.49	523.16

TABLE 3.10

Optimal Values of Decision variables from the Optimization of the Total Cost

Decision Variables	First Scenario	Second Scenario	Third Scenario
Big s1	267	235	205
Big s2	350	374	357
Big s3	500	503	501
Little s1	40	21	25
Little s2	31	36	37
Little s3	40	42	42
Interval1	8	16	8
Interval2	8	8	9
Interval3	96	11	19

TABLE 3.11

Optimal System Run Results

	First Scenario	Second Scenario	Third Scenario
Holding cost1	95.0510	73.6752	48.3288
Holding cost2	143.70	130.90	125.12
Holding cost3	105.77	194.07	221.43
Total holding costs	344.521	398.6452	394.8788
Ordering cost1	45.0000	39.6656	58.5296
Ordering cost2	35.5056	51.2092	51.7224
Ordering cost3	6.2440	33.5980	14.8556
Total ordering costs	86.7496	124.4728	125.1076
Lost sales cost1	206.08	268.40	
Lost sales cost2	95.8440	223.09	35.2920
Lost sales cost3	174.13	123.37	
Total lost sales costs	476.054	614.86	35.2920
Total costs	907.3246	1137.978	555.2784

3.5.4 Concluding Remarks

A simulation model had been developed with a multi-echelon inventory system in this research. The model has been optimized and finally, the third scenario was observed, which offers full interrelationships to retailers, reduces holding, and lost sales cost together. The second scenario that offers limited interrelationships to retailers reduces ordering cost. Overall, it depends on the retailer's objective; if they need the reduction of total cost (holding cost, ordering, and lost sales cost), then the model offers to follow the third scenario.

TABLE 3.12
Initial System Run Results

	First Scenario	Second Scenario	Third Scenario
Holding cost1	91.4979	95.3629	57.3492
Holding cost2	180.31	154.68	154.74
Holding cost3	256.13	254.65	262.35
Total holding costs	527.9379	504.6929	474.4392
Ordering cost1	38.6540	41.2480	44.5920
Ordering cost2	32.5380	38.1740	43.7468
Ordering cost3	20.3596	38.7312	27.3580
Total ordering costs	82.5516	118.1532	115.6968
Lost sales cost1	266.98	255.36	
Lost sales cost2	104.35	308.59	129.30
Lost sales cost3	50.6280	176.18	
Total lost sales costs	421.958	740.13	129.30
Total costs	1032.4475	1362.9761	719.436

3.6 Funding

This work was supported by the Iran's National Elites Foundation [15/96595].

References

Afrouzy, Z. A., Nasseri, S. H., & Mahdavi, I. (2016). A genetic algorithm for supply chain configuration with new product development. *Computers & Industrial Engineering, 101*, 440–454.

Axsäter, S., & Juntti, L. (1996). Comparison of echelon stock and installation stock policies for two-level inventory systems. *International Journal of Production Economics, 45*(1):303–310.

Ball, M. O., Ma, M., Raschid, L., & Zhao, Z. (2002). Supply chain infrastructures: System integration and information sharing. *ACM Sigmod Record, 31*(1): 61–66.

Barroso, A. P., Machado, V. H., & Machado, V. C. (2013). Demand information sharing impact on supply chain management under demand uncertainty. A simulation model. In *2013 IEEE International Conference on Industrial Engineering and Engineering Management* (pp. 924–928). IEEE.

Buckley, S., & An, C. (2005). Supply chain simulation. *Supply Chain Management on Demand*, 17–35.

Buzacott, J. A., & Shanthikumar, J. G. (1993). *Stochastic Models of Manufacturing Systems* (Vol. 4). Englewood Cliffs, NJ: Prentice Hall.

Chen, M. C., Yang, T., & Yen, C. T. (2007). Investigating the value of information sharing in multi-echelon supply chains. *Quality & Quantity, 41*(3): 497–511.

Clark, A. J., & Scarf, H. (1960). Optimal policies for a multi-echelon inventory problem. *Management Science, 6*(4), 475–490.

Council, S. C. (2008). Supply-chain operations reference-model. *Overview of SCOR Version, 5.*

Diks, E. B., & De Kok, A. G. (1999). Computational results for the control of a divergent N-echelon inventory system. *International Journal of Production Economics, 59*(1), 327–336.

Gumus, A. T., Guneri, A. F., & Ulengin, F. (2010). A new methodology for multi-echelon inventory management in stochastic and neuro-fuzzy environments. *International Journal of Production Economics, 128*(1), 248–260.

Huang, B., & Iravani, S. M. (2007). Optimal production and rationing decisions in supply chains with information sharing. *Operations Research Letters, 35*(5), 669–676.

Ingalls, R. G. (1998). The value of simulation in modeling supply chains. In *Proceedings of the 30th Conference on Winter Simulation* (pp. 1371–1376). IEEE Computer Society Press.

Ingalls, R. G. (2014). Introduction to supply chain simulation. In *Proceedings of the 2014 Winter Simulation Conference* (pp. 36–50). IEEE Press.

Khan, M., Hussain, M., & Saber, H. M. (2016). Information sharing in a sustainable supply chain. *International Journal of Production Economics, 181,* 208–214.

Kiesmüller, G. P., de Kok, T. G., Smits, S. R., & van Laarhoven, P. J. (2004). Evaluation of divergent N-echelon (s, nQ)-policies under compound renewal demand. *OR Spectrum, 26*(4), 547–577.

Kleijnen, J. P. (2005). Supply chain simulation tools and techniques: A survey. *International Journal of Simulation and Process Modelling, 1*(1–2), 82–89.

Köchel, P., & Nieländer, U. (2005). Simulation-based optimization of multi-echelon inventory systems. *International Journal of Production Economics, 93,* 505–513.

Kritchanchail, D., & MacCarthy, B. L. (2000). Discrete or continuous: Which is more appropriate for supply chain simulation modeling. In *Proceedings of the 2000 International Conference on Production Research* (pp. 101–108). Bangkok, Thailand.

Lee, H. L., Padmanabhan, V., &Whang, S. (1997). The bullwhip effect in supply chains. *MIT Sloan Management Review, 38*(3), 93.

Lee, H. L., So, K. C., & Tang, C. S. (2000). The value of information sharing in a two-level supply chain. *Management Science, 46*(5), 626–643.

Li, G., Yan, H., Wang, S., & Xia, Y. (2005). Comparative analysis on value of information sharing in supply chains. *Supply Chain Management: An International Journal, 10*(1), 34–46.

Patil, K., Jin, K., & Li, H. (2011, December). Arena simulation model for multi echelon inventory system in supply chain management. In *Industrial Engineering and Engineering Management (IEEM), 2011 IEEE International Conference on* (pp. 1214–1217). IEEE.

Rau, H., Wu, M. Y., & Wee, H. M. (2003). Integrated inventory model for deteriorating items under a multi-echelon supply chain environment. *International Journal of Production Economics, 86*(2), 155–168.

Ryu, K., Moon, I., Oh, S., & Jung, M. (2013). A fractal echelon approach for inventory management in supply chain networks. *International Journal of Production Economics, 143*(2), 316–326.

Schildbach, G., & Morari, M. (2016). Scenario-based model predictive control for multi-echelon supply chain management. *European Journal of Operational Research, 252*(2), 540–549.

Seifert, R. W., Zequeira, R. I., & Liao, S. (2012). A three-echelon supply chain with price-only contracts and sub-supply chain coordination. *International Journal of Production Economics, 138*(2), 345–353.

Srivathsan, S., & Kamath, M. (2017). Performance modeling of a two-echelon supply chain under different levels of upstream inventory information sharing. *Computers & Operations Research, 77*, 210–225.

Suraj, B. S., Sharma, S. K., & Routroy, S. (2016). Positioning of inventory in supply chain using simulation modeling. *IUP Journal of Supply Chain Management, 13*(2), 20.

Tako, A. A., & Robinson, S. (2012). The application of discrete event simulation and system dynamics in the logistics and supply chain context. *Decision Support Systems, 52*(4), 802–815.

Tee, Y. S., & Rossetti, M. D. (2002). A robustness study of a multi-echelon inventory model via simulation. *International Journal of Production Economics, 80*(3), 265–277.

Terzi, S., & Cavalieri, S. (2004). Simulation in the supply chain context: A survey. *Computers in Industry, 53*(1), 3–16.

Thomas, D. J., & Griffin, P. M. (1996). Coordinated supply chain management. *European Journal of Operational Research, 94*(1), 1–15.

Vieira, G. E. (2004, December). Ideas for modeling and simulation of supply chains with Arena. In *Simulation Conference, 2004. Proceedings of the 2004 Winter* (Vol. 2, pp. 1418–1427). IEEE.

Wan, J., & Zhao, C. (2009, December). Simulation research on multi-echelon inventory system in supply chain based on Arena. In *2009 First International Conference on Information Science and Engineering* (pp. 397–400). IEEE.

Wang, J., Zhao, R., & Tang, W. (2008). Fuzzy programming models for vendor selection problem in a supply chain. *Tsinghua Science & Technology, 13*(1), 106–111.

Yang, T., Wen, Y. F., & Wang, F. F. (2011). Evaluation of robustness of supply chain information-sharing strategies using a hybrid Taguchi and multiple criteria decision-making method. *International Journal of Production Economics, 134*(2), 458–466.

Yao, Y., & Dresner, M. (2008). The inventory value of information sharing, continuous replenishment, and vendor-managed inventory. *Transportation Research Part E: Logistics and Transportation Review, 44*(3), 361–378.

4

Applied Mathematic Technologies in Nonlinear Mechanics of Thin-Walled Constructions

Vladimir Dmitriev

CONTENTS

The chapter is written for engineers and science workers. Thin-walled structures—plates and shells of various form and geometry—are widely used in different branches of technics, such as space-air systems, machine-building, and building structures. This chapter briefly discusses the simplicity and availability of applied methods of mathematical modeling in nonlinear mechanics of thin-walled structures and the development of efficient calculation algorithms. For a more thorough study of applied methods of mathematical-modeling literature, sources are given in the end of the chapter.

4.1 Geometrical Relations

To investigate shell-deformation processes relations are mainly obtained from the three-dimensional theory of elasticity equations based on Kirchhoff-Love hypothesis (first-approximation model) and Timoshenko-Reissner hypothesis (second-approximation model) are used. Application area of Kirchhoff-Love equations is restricted by calculation of sufficiently thin shells at $R/h > (80 \div 100)$ (R is characteristic radius of curvature, h is shell thickness), made of traditional isotropic construction materials or composite materials with weak anisotropy of physicomechanical characteristics at smooth static and dynamic loads. Timoshenko-Reissner theory of shells (shear model) allows calculating relatively thick ($R/h > 10$) shells at nonsmooth loads, one-layer, and multilayer shells made of composite materials with the binder of a relatively small shear stiffness. Additionally, the Timoshenko-Reissner model describes more accurate dynamic processes connected with deformations waves propagation at loads of impact character.

Deformed state at a point is described by deformation tensor

$$T_\varepsilon = \begin{Vmatrix} \varepsilon_{11}\varepsilon_{22}\varepsilon_{33} \\ \varepsilon_{21}\varepsilon_{22}\varepsilon_{23} \\ \varepsilon_{31}\varepsilon_{32}\varepsilon_{33} \end{Vmatrix} = \left\| \varepsilon_{ij} \right\|, \tag{4.1}$$

where ε_{ii} are elongations along coordinate axes $\alpha_1,\alpha_2,\alpha_3$; $\varepsilon_{ij} = \gamma_{ij}/2$ are shears, and γ_{ij} are shear angles ($i, j = 1,2,3$). Deformation tensor can be decomposed into spherical tensor and strain deviator D_ε

$$T_\varepsilon = \frac{1}{3}\theta T_1 + D_\varepsilon, \tag{4.2}$$

where

$$D_\varepsilon = \left\| e_{ij} \right\|; \quad e_{ij} = \varepsilon_{ij} - \frac{1}{3}\theta\delta_{ij}, \tag{4.3}$$

and where $\theta = \varepsilon_{11} + \varepsilon_{22} + \varepsilon_{33}$ is volume deformation (relative volume change), $T_1 = \|\delta_{ij}\|$ is unit tensor, δ_{ij} is Kronecker symbol. Deformation intensity is defined by the formula

$$e_i = \frac{\sqrt{2}}{3}\sqrt{(\varepsilon_{11} - \varepsilon_{22})^2 + (\varepsilon_{22} - \varepsilon_{33})^2 + (\varepsilon_{33} - \varepsilon_{11})^2 + 6(\varepsilon_{12}^2 + \varepsilon_{13}^2 + \varepsilon_{23}^2)}. \tag{4.4}$$

Let us consider the connection of deformations and displacements in the framework of mean-bending theory for Kirchhoff-Love and Timoshenko-Reissner shells equations. Below we set the following symbols for deformation tensor components: $E_{ii} = \varepsilon_{ii}$, $E_{ij} = \gamma_{ij}$.

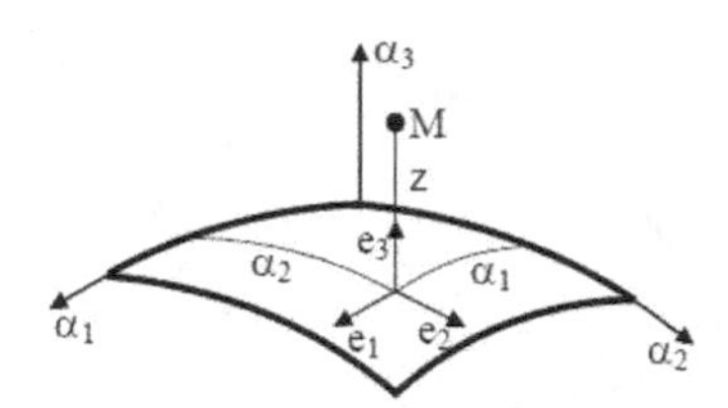

FIGURE 4.1
Unit axes of orthogonal trihedron e_1, e_2, e_3.

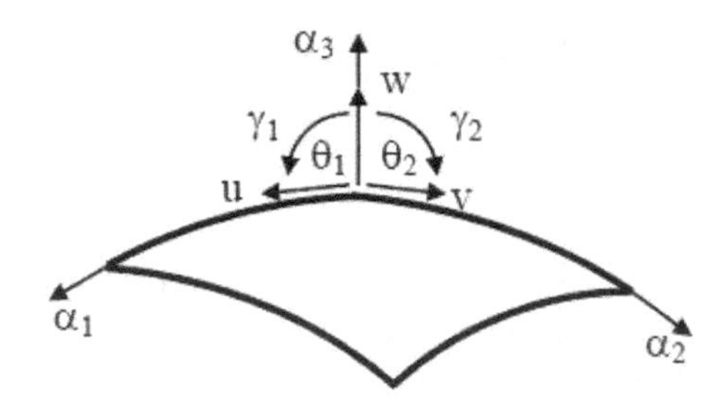

FIGURE 4.2
Positive directions for generalized displacements.

Equidistant surface point M displacement vector is determined by parameters

$$U = U(\alpha_1,\alpha_2,z);\ V = V(\alpha_1,\alpha_2,z);\ W = W(\alpha_1,\alpha_2,z), \tag{4.5}$$

where U, V, and W are point full displacement vector projections on the unit axes of orthogonal trihedron e_1, e_2, e_3 (Figure 4.1); α_1,α_2 are coordinates of the normal base on the coordinate surface $z = 0$, z is the distance in coordinate direction α_3.

If we take middle surface as coordinate one, then $h/2 \leq z \leq +h/2$. Let us introduce the following notations: $u = u(\alpha_1,\alpha_2)$, $v = v(\alpha_1,\alpha_2)$, $w = w(\alpha_1,\alpha_2)$ are displacements of coordinate surface $z = 0$ points; $\theta_1 = \theta_1(\alpha_1,\alpha_2)$, $\theta_2 = \theta_2(\alpha_1,\alpha_2)$ are angles of rotation in accordance with the hypothesis of "rigid" normal; and $\gamma_1 = \gamma_1(\alpha_1,\alpha_2)$, $\gamma_2 = \gamma_2(\alpha_1,\alpha_2)$ are full angles of normal rotation. Positive directions for generalized displacements u_k are shown on Figure 4.2; ($u_1 = u$, $u_2 = v$, $u_3 = w$, $u_4 = \gamma_1$, $u_5 = \gamma_2$; $k = 1,2,\ldots,5$).

Most widely spread are thin-walled structures in the form of shells of revolution, middle surface of which is formed by revolution of a flat smooth curve about a line lying in that same plane (Figure 4.3). So without losing generality, we consider this type of shell. As coordinate lines α_1 we take meridians, as coordinate lines α_2—parallels: $O_1M = R_1$ is curvature radius of

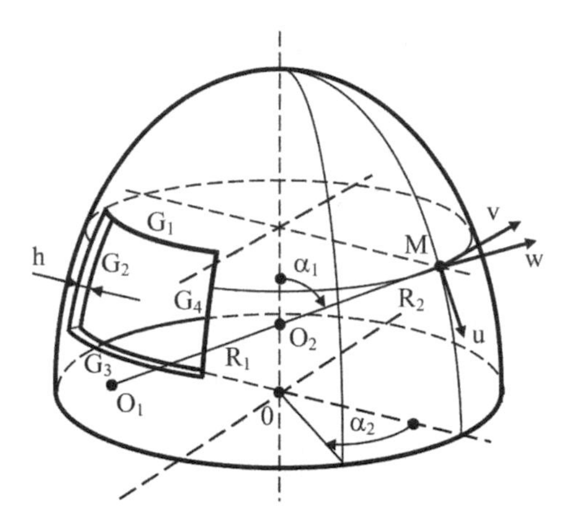

FIGURE 4.3
Shell of revolution construction.

shell meridian, and $O_2M = R_2$ is shell curvature radius along parallel circle of shell meridian. Curvature center $k_2 = 1/R_2$ is located in the point O_2 on surface symmetry axes 0z (Figure 4.3).

4.2 Kirchhoff-Love Theory of Shells Equations

Geometrically nonlinear expressions for shell coordinate surface tangential deformation parameters within the framework of mean-bending theory (quadratic theory) are written as [1,2]

$$E_{11} = \varepsilon_1 + \frac{1}{2}\theta_1^2; \quad E_{22} = \varepsilon_2 + \frac{1}{2}\theta_2^2; \quad E_{12} = \Omega_1 + \Omega_2 + \theta_1\theta_2, \tag{4.6}$$

where

$$\varepsilon_1 = \frac{1}{A_1}\frac{\partial u}{\partial \alpha_1} + k_1 w; \quad \varepsilon_2 = \frac{1}{A_2}\frac{\partial v}{\partial \alpha_2} + \psi u + k_2 w;$$

$$\Omega_1 = \frac{1}{A_1}\frac{\partial v}{\partial \alpha_1}; \quad \Omega_2 = \frac{1}{A_2}\frac{\partial u}{\partial \alpha_2} - \psi v; \tag{4.7}$$

$$\theta_1 = -\frac{1}{A_1}\frac{\partial w}{\partial \alpha_1} + k_1 u; \quad \theta_2 = -\frac{1}{A_2}\frac{\partial w}{\partial \alpha_2} + k_2 v; \quad \psi = \frac{1}{A_1 A_2}\frac{\partial A_2}{\partial \alpha_1},$$

and where A_1, k_1, A_2, k_2 are Lame parameters and principal curvatures of shell coordinate surface: $k_1 = 1/R_1$, $k_2 = 1/R_2$. Curvatures' changes K_{11}, K_{22} and torsion K_{12} of shell coordinate surface are determined by the expressions

$$K_{11} = \frac{1}{A_1}\frac{\partial \theta_1}{\partial \alpha_1}; \quad K_{22} = \frac{1}{A_2}\frac{\partial \theta_2}{\partial \alpha_2} + \psi\theta_1; \quad K_{12} = \tau_1 + k_1 \cdot \Omega_2 + \tau_2 + k_2 \cdot \Omega_1, \tag{4.8}$$

where

$$\tau_1 = \frac{1}{A_1}\frac{\partial \theta_2}{\partial \alpha_1}; \quad \tau_2 = \frac{1}{A_2}\frac{\partial \theta_1}{\partial \alpha_2} - \psi\theta_2. \tag{4.9}$$

4.3 Timoshenko-Reissner Theory of Shells Equations

Shell tangential deformation parameters are determined by formulas (Eqs. 4.6 and 4.7. Within the framework of the Timoshenko-Reissner model, full normal rotation angles are represented as a sum

$$\gamma_1 = E_{13} + \theta_1; \quad \gamma_2 = E_{23} + \theta_2, \tag{4.10}$$

where E_{13}, E_{23} are angles of transverse shear. Then components of transverse deformation of a shell will be determined as

$$E_{13} = \gamma_1 + \frac{1}{A_1}\frac{\partial w}{\partial \alpha_1} - k_1 u; \quad E_{23} = \gamma_2 + \frac{1}{A_2}\frac{\partial w}{\partial \alpha_2} - k_2 v. \tag{4.11}$$

Components of bending deformation have the form identical to Equations 4.8 and 4.9 [3,4]

$$K_{11} = \frac{1}{A_1}\frac{\partial \gamma_1}{\partial \alpha_1}; \; K_{22} = \frac{1}{A_2}\frac{\partial \gamma_2}{\partial \alpha_2} + \psi\gamma_1; \; K_{12} = \tau_1 + k_1 \cdot \Omega_2 + \tau_2 + k_2 \cdot \Omega_1, \tag{4.12}$$

where

$$\tau_1 = \frac{1}{A_1}\frac{\partial \gamma_2}{\partial \alpha_1}; \quad \tau_2 = \frac{1}{A_2}\frac{\partial \gamma_1}{\partial \alpha_2} - \psi\gamma_2. \tag{4.13}$$

For displacements and deformations, both in theories of Kirchhoff-Love and Timoshenko-Reissner linear law of distribution along shell thickness is taken [1–4]. For plates by reason of $k_1 = k_2 = 0$ the relations in Equations 4.6 through 4.13 are simplified correspondingly.

4.4 Physical Relations

Stress state in a point is characterized by symmetrical stress tensor

$$T_\sigma = \begin{Vmatrix} \sigma_{11}\sigma_{22}\sigma_{33} \\ \sigma_{21}\sigma_{22}\sigma_{23} \\ \sigma_{31}\sigma_{32}\sigma_{33} \end{Vmatrix} = \|\sigma_{ij}\|, \tag{4.14}$$

where σ_{ii} are normal stresses; and $\sigma_{12} = \sigma_{21}$, $\sigma_{13} = \sigma_{31}$, $\sigma_{23} = \sigma_{32}$ are tangential stresses. Below for tangential stresses we use notations $\tau_{ij} = \sigma_{ij}$. Stress tensor can be represented as a sum of spherical tensor and stress deviator, D_σ

$$T_\sigma = \sigma T_1 + D_\sigma, \tag{4.15}$$

where

$$D_\sigma = \|s_{ij}\|; \quad s_{ij} = \sigma_{ij} - \sigma\delta_{ij}, \tag{4.16}$$

and where $\sigma = (\sigma_{11} + \sigma_{22} + \sigma_{33})/3$ is mean stress. Stress intensity σ_i is determined by formula

$$\sigma_i = \frac{\sqrt{2}}{2}\sqrt{(\sigma_{11}-\sigma_{22})^2 + (\sigma_{22}-\sigma_{33})^2 + (\sigma_{33}-\sigma_{11})^2 + 6(\tau_{12}^2 + \tau_{13}^2 + \tau_{23}^2)}. \tag{4.17}$$

Internal forces and moments referred to shell coordinate surface are determined through stress tensor components according to formulas

$$T_{11} = \int_h \sigma_{11}(1+zk_2)dz; \quad T_{22} = \int_h \sigma_{22}(1+zk_1)dz; \quad T_{12} = \int_h \tau_{12}(1+zk_2)dz;$$

$$M_{11} = \int_h \sigma_{11}(1+zk_2)zdz; \quad M_{22} = \int_h \sigma_{22}(1+zk_1)zdz; \quad M_{12} = \int_h \tau_{12}(1+zk_2)zdz;$$

$$Q_{13} = \int_h \tau_{13}(1+zk_2)dz; \quad Q_{23} = \int_h \tau_{23}(1+zk_1)dz, \tag{4.18}$$

where $T_{ii} = T_{ii}(\alpha_1,\alpha_2)$, $T_{ij} = T_{ij}(\alpha_1,\alpha_2)$, and $Q_{i3} = Q_{i3}(\alpha_1,\alpha_2)$ are normal and shear forces, $M_{ii} = M_{ii}(\alpha_1,\alpha_2)$ and $M_{ij} = M_{ij}(\alpha_1,\alpha_2)$ are bending and torsional moments, $h = h(\alpha_1,\alpha_2)$ is shell thickness; and $(i, j = 1,2)$. Accepted positive directions for forces and moments are shown on Figure 4.4. Instead of shear forces, T_{12}, T_{21}, and torsional moments, M_{12}, M_{21}, in theory of shells we introduce statically equivalent symmetrical factors

$$\begin{gathered} S = T_{12} - k_2 M_{21} = T_{21} - k_1 M_{12}; \\ H = M_{12} = M_{21}. \end{gathered} \tag{4.19}$$

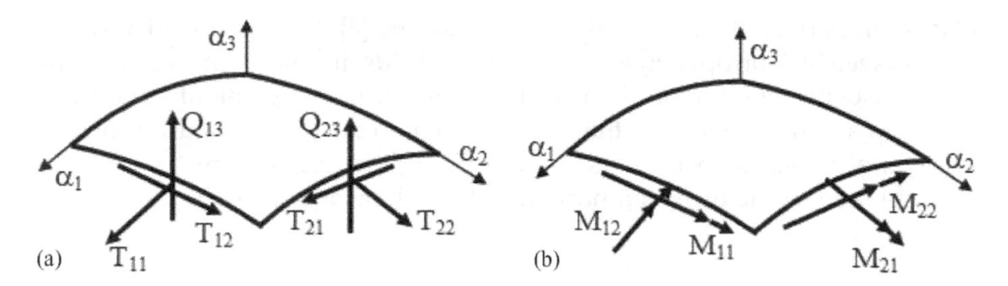

FIGURE 4.4
Positive directions for forces and moments. (a) is internal forces; (b) is internal moments

We shall restrict ourselves to the relations of Timoshenko-Reissner model as the most general case. We shall consider physical relations for a number of special cases of nonhomogeneous shells.

4.5 One-Layer Orthotropic Shells

For anisotropic material for which in each point there exist three orthogonal planes of elastic symmetry normal to corresponding coordinate directions $\alpha_1, \alpha_2, \alpha_3$ (for orthotropic material), relations of generalized Hook's law on the assumption $\sigma_{33} = 0$ can be written in the form [1–4]

$$\sigma_{11} = \frac{E_1}{1-\nu_{12}\nu_{21}}(E_{11} + \nu_{21}E_{22}); \quad \sigma_{22} = \frac{E_2}{1-\nu_{21}\nu_{12}}(E_{22} + \nu_{12}E_{11});$$
$$\tau_{12} = G_{12}E_{12}; \quad \tau_{13} = G_{13}E_{13}; \quad \tau_{23} = G_{23}E_{23}, \tag{4.20}$$

where E_1, E_2 are Young's moduli in coordinate directions, α_1, α_2; G_{12}, G_{13}, G_{23} are shear moduli, and ν_{12}, ν_{21} are Poisson's coefficients, the first index points the direction of the acting stress, the second one in the direction of the resulting transverse deformation. Here

$$E_1\nu_{21} = E_2\nu_{12}. \tag{4.21}$$

Physicomechanical characteristics of shell material—E_1, E_2, ν_{12}, ν_{21}, G_{12}, G_{13}, G_{23}—in the general case are functions of coordinates α_1, α_2, z. Most contemporary composite materials (e.g., fiber composites) for the reason of their production technology peculiarities are transversely isotropic in transverse plane. In transversely isotropic body, all the directions in the plane of isotropy and the direction normal to this plane are principal directions of elasticity. For such body principal axes of deformed state coincide with principal directions of stress state, if one of the principal axes of stress state is normal to anisotropy plane. The number of independent coefficients characterizing elastic properties of such a body is equal to five [4]. For the considered case of transversal isotropy: E_2 is elasticity modulus in the plane of isotropy, E_1 is elasticity modulus in the direction normal to the plane of anisotropy, $G_{13} = G_{12}$ are shear moduli in the planes normal to the plane of anisotropy, and ν_{12},ν_{21} are Poisson's coefficients characterizing compression in the plane of isotropy and in the direction normal to this plane at tension in the plane of isotropy. Here

$$G_{23} = \frac{E_2}{2(1+\nu_{23})}. \tag{4.22}$$

Formulas for forces and moments are obtained by integrating Equation 4.18 over shell thickness and neglecting terms of the order h/R. For the case of one-layer orthotropic shell within the framework of Timoshenko model one can obtain the following expressions for forces and moments

$$\begin{gathered} T_{11} = B_{11}E_{11} + B_{12}E_{22}; \quad M_{11} = D_{11}K_{11} + D_{12}K_{22}; \quad Q_{13} = k^2 B_{13}E_{13}; \\ T_{22} = B_{22}E_{22} + B_{21}E_{11}; \quad M_{22} = D_{22}K_{22} + D_{21}K_{11}; \quad Q_{23} = k^2 B_{23}E_{23}; \\ S = B_{33}E_{12}; \quad H = D_{33}K_{12}, \end{gathered} \tag{4.23}$$

where

$$\begin{gathered} B_{11} = \frac{E_1 h}{1-\nu_{12}\nu_{21}}; \quad B_{22} = \frac{E_2 h}{1-\nu_{12}\nu_{21}}; \quad B_{12} = \nu_{21}B_{11}; \\ D_{11} = \frac{E_1 h^3}{12(1-\nu_{12}\nu_{21})}; \quad D_{22} = \frac{E_2 h^3}{12(1-\nu_{12}\nu_{21})}; \quad D_{12} = \nu_{21}D_{11}; \\ B_{33} = G_{12}h; \quad B_{13} = G_{13}h; \quad B_{23} = G_{23}h; \quad D_{33} = \frac{G_{12}h^3}{12}, \end{gathered} \tag{4.24}$$

and where $k^2 = 5/6$ is shear coefficient taking into account parabolic law of transverse tangential stresses distribution across thickness. For Kirchhoff-Love model: $Q_{13} = Q_{23} = 0$.

4.6 Multilayer Shells of Composite Materials

For multilayer shell consisting of N anisotropic rigidly connected layers of various thickness, we suppose that the layers deform without relative sliding and detachment so that for the package as a whole we can take Timoshenko-Reissner hypothesis (Figure 4.5). Index "i" is used for layers numeration as well as for identification of calculated parameters and physicomechanical layer characteristics ($i = 1,2,3,\ldots,N$). As coordinate surfaces, we take a middle surface of a shell of any layer or one of layer contact surfaces.

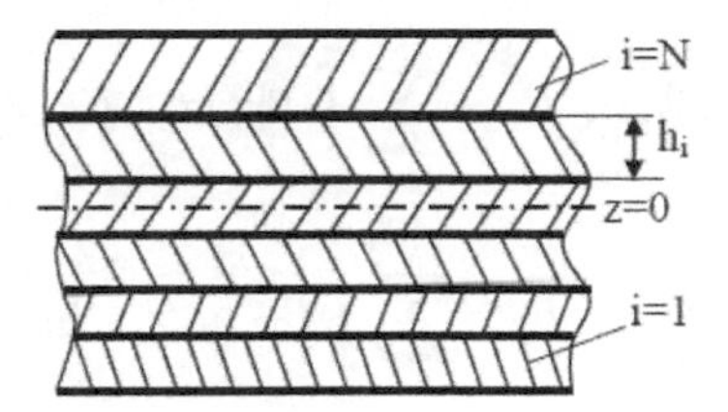

FIGURE 4.5
Multilayer shell construction with anisotropic rigidly connected layers.

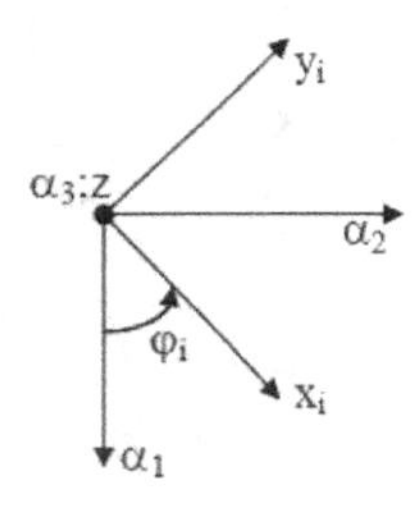

FIGURE 4.6
Principal directions of material elasticity x_i, y_i of the ith layer and coordinate directions α_1, α_2 for shell of revolution.

Principal directions of material elasticity x_i, y_i for the ith layer can be oriented relative to coordinate directions, α_1,α_2, at some angle φ_i (Figure 4.6).

Force factors in a multilayer shell can be expressed through coordinate surface-deformation components by the formulas [4]

$$\begin{aligned}
T_{11} &= C_{11}E_{11} + C_{12}E_{22} + C_{16}E_{12} + B_{11}K_{11} + B_{12}K_{22} + B_{16}K_{12};\\
T_{22} &= C_{22}E_{22} + C_{21}E_{11} + C_{26}E_{21} + B_{22}K_{22} + B_{21}K_{11} + B_{26}K_{21};\\
S &= C_{16}E_{11} + C_{26}E_{22} + C_{66}E_{12} + B_{16}K_{11} + B_{26}K_{22} + B_{66}K_{12};\\
M_{11} &= B_{11}E_{11} + B_{12}E_{22} + B_{16}E_{12} + D_{11}K_{11} + D_{12}K_{22} + D_{16}K_{12}\\
M_{22} &= B_{22}E_{22} + B_{21}E_{11} + B_{26}E_{21} + D_{22}K_{22} + D_{21}K_{11} + D_{26}K_{21};\\
H &= B_{16}E_{11} + B_{26}E_{22} + B_{66}E_{12} + D_{16}K_{11} + D_{26}K_{22} + D_{66}K_{12};\\
Q_{13} &= A_{11}E_{13} + A_{12}E_{23};\ Q_{23} = A_{21}E_{13} + A_{22}E_{23}.
\end{aligned} \tag{4.25}$$

Stiffness coefficients, A_{nm}, B_{nm}, C_{nm}, D_{nm}, are determined through layer elastic characteristics and their thickness

$$C_{11} = \sum_{i=1}^{N} \int_{z_{i-1}}^{z_i} B_{11}^{(i)} dz;\quad C_{12} = \sum_{i=1}^{N} \int_{z_{i-1}}^{z_i} B_{12}^{(i)} dz;\quad C_{22} = \sum_{i=1}^{N} \int_{z_{i-1}}^{z_i} B_{22}^{(i)} dz;$$

$$C_{16} = \sum_{i=1}^{N} \int_{z_{i-1}}^{z_i} B_{16}^{(i)} dz; \quad C_{26} = \sum_{i=1}^{N} \int_{z_{i-1}}^{z_i} B_{26}^{(i)} dz; \quad C_{66} = \sum_{i=1}^{N} \int_{z_{i-1}}^{z_i} B_{66}^{(i)} dz;$$

$$B_{11} = \sum_{i=1}^{N} \int_{z_{i-1}}^{z_i} B_{11}^{(i)} z dz; \quad B_{12} = \sum_{i=1}^{N} \int_{z_{i-1}}^{z_i} B_{12}^{(i)} z dz; \quad B_{22} = \sum_{i=1}^{N} \int_{z_{i-1}}^{z_i} B_{22}^{(i)} z dz;$$

$$B_{16} = \sum_{i=1}^{N} \int_{z_{i-1}}^{z_i} B_{16}^{(i)} z dz; \quad B_{26} = \sum_{i=1}^{N} \int_{z_{i-1}}^{z_i} B_{26}^{(i)} z dz; \quad B_{66} = \sum_{i=1}^{N} \int_{z_{i-1}}^{z_i} B_{66}^{(i)} z dz;$$

$$D_{11} = \sum_{i=1}^{N} \int_{z_{i-1}}^{z_i} B_{11}^{(i)} z^2 dz; \quad D_{12} = \sum_{i=1}^{N} \int_{z_{i-1}}^{z_i} B_{12}^{(i)} z^2 dz; \quad D_{22} = \sum_{i=1}^{N} \int_{z_{i-1}}^{z_i} B_{22}^{(i)} z^2 dz;$$

$$D_{16} = \sum_{i=1}^{N} \int_{z_{i-1}}^{z_i} B_{16}^{(i)} z^2 dz; \quad D_{26} = \sum_{i=1}^{N} \int_{z_{i-1}}^{z_i} B_{26}^{(i)} z^2 dz; \quad D_{66} = \sum_{i=1}^{N} \int_{z_{i-1}}^{z_i} B_{66}^{(i)} z^2 dz;$$

$$A_{11} = k^2 \sum_{i=1}^{N} \int_{z_{i-1}}^{z_i} B_{44}^{(i)} dz; \quad A_{22} = k^2 \sum_{i=1}^{N} \int_{z_{i-1}}^{z_i} B_{55}^{(i)} dz; \tag{4.26}$$

$$A_{12} = k^2 \sum_{i=1}^{N} \int_{z_{i-1}}^{z_i} B_{45}^{(i)} dz; \quad A_{21} = k^2 \sum_{i=1}^{N} \int_{z_{i-1}}^{z_i} B_{54}^{(i)} dz.$$

Coefficients $B_{nm}^{(i)}$ are expressed through physicomechanical characteristics of the ith layer material in coordinate system of elasticity principal directions x_i, y_i, z by formulas

$$B_{11}^{(i)} = E_{xi} \cos^4 \phi_i + 2(E_{xi} \nu_y^{(i)} + 2G_{xy}^{(i)}) \sin^2 \phi_i \cos^2 \phi_i + E_{yi} \sin^4 \phi_i;$$

$$B_{12}^{(i)} = B_{21}^{(i)} = E_{xi} \nu_y^{(i)} + [E_{xi} + E_{yi} - 2(E_{xi} \nu_y^{(i)} + 2G_{xy}^{(i)})] \sin^2 \phi_i \cos^2 \phi_i;$$

$$B_{22}^{(i)} = E_{xi} \sin^4 \phi_i + 2(E_{yi} \nu_x^{(i)} + 2G_{xy}^{(i)}) \sin^2 \phi_i \cos^2 \phi_i + E_{yi} \cos^4 \phi_i; \tag{4.27}$$

$$B_{16}^{(i)} = 0,5[E_{xi} \cos^2 \phi_i - E_{yi} \sin^2 \phi_i - (E_{xi} \nu_y^{(i)} + 2G_{xy}^{(i)}) \cos 2\phi_i] \sin 2\phi_i;$$

$$B_{26}^{(i)} = 0,5[E_{xi} \sin^2 \phi_i - E_{yi} \cos^2 \phi_i + (E_{yi} \nu_x^{(i)} + 2G_{xy}^{(i)}) \cos 2\phi_i] \sin 2\phi_i;$$

$$B_{66}^{(i)} = G_{xy}^{(i)} + [E_{xi} + E_{yi} - 2(E_{xi} \nu_y^{(i)} + 2G_{xy}^{(i)})] \sin^2 \phi_i \cos^2 \phi_i;$$

$$B_{44}^{(i)} = G_{xz}^{(i)} \cos^2 \phi_i + G_{yz}^{(i)} \sin^2 \phi_i; \quad B_{55}^{(i)} = G_{xz}^{(i)} \sin^2 \phi_i + G_{yz}^{(i)} \cos^2 \phi_i;$$

$$B_{45}^{(i)} = B_{54}^{(i)} = (G_{xz}^{(i)} - G_{yz}^{(i)}) \sin \phi_i \cos \phi_i,$$

where

$$E_{xi} = \frac{E_x^{(i)}}{1-v_{xy}^{(i)}v_{yx}^{(i)}}; \quad E_{yi} = \frac{E_y^{(i)}}{1-v_{yx}^{(i)}v_{xy}^{(i)}}; \quad E_x^{(i)}v_{yx}^{(i)} = E_y^{(i)}v_{xy}^{(i)}. \tag{4.28}$$

In relations (Eqs. 4.27 and 4.28) $E_x^{(i)}, E_y^{(i)}, v_{xy}^{(i)}, v_{yx}^{(i)}, G_{xy}^{(i)}, G_{xz}^{(i)}, G_{yz}^{(i)}$ are Young's moduli, Poisson's coefficients, and shear moduli of the *i*th layer material in coordinate system *x, y, z*. From the point of view of technical applications symmetric about middle surface scheme of layer picking, when each layer reinforced at angle $+\varphi_i$ corresponds to an angle reinforced at angle, φ_i, is of ultimate interest. The majority of automated technological processes of multilayer package forming provide exactly symmetric picking or mutual adhesion of neighboring symmetric layers with angles $\pm\varphi_i$. In this case for formulas simplification, the middle surface should be taken as a coordinate one, so that in the case of even number of layers, it passes along the border of neighboring layers, and in the case of odd number of layers, it passes through the center of middle layer, which conventionally is divided by the coordinate surface in two layers. Formulas for stiffness characteristics entering in elasticity relations (Eq. 4.25) determination in this case have the form

$$C_{nm} = 2\sum_{i=1}^{N/2} B_{nm}^{(i)}(z_i - z_{i-1}); \quad D_{nm} = \frac{2}{3}\sum_{i=1}^{N/2} B_{nm}^{(i)}(z_i^3 - z_{i-1}^3), \tag{4.29}$$

where *n, m* = 1,2. Other stiffness parameters are equal to zero. In the equalities (Eq. 4.29), the summation is carried out over the layers lying on the one side of coordinate surface.

4.7 Physically Nonlinear Problems

Within the framework of deformation plasticity theory relations (theory of small elastoplastic deformations) considered solid is supposed to be isotropic, whereas the middle stress, σ, and the volume deformation, θ, are connected by linear dependence

$$\sigma = K\theta, \tag{4.30}$$

where

$$K = \frac{E}{3(1-2v)}, \tag{4.31}$$

and where *K* is modulus of volume compression; and *E,* are elasticity modulus and Poisson's coefficient at elastic strains. Change of medium element form takes place on account of shear deformations, then strain (Eq. 4.3) and stress (Eq. 4.16) deviators are proportional. Plastic deformations take place in the shell under Mises yield criterion $\sigma_i \geq \sigma_T$ ($e_i \geq \varepsilon_T$); and σ_T, ε_T are creep stress and creep deformation at one-dimensional tension-compression deformation. The process of active loading is determined by the condition $d\sigma_i > 0$ ($de_i > 0$); at $d\sigma_i < 0$ ($de_i < 0$), material unloading begins, which is supposed to be linear elastic [5].

At plastic stage, the dependencies between stresses and strains for the shell in the domain of active loading can be represented as

$$\sigma_{11} = \left(K + \frac{4E_c}{9}\right)E_{11} + \left(K - \frac{2E_c}{9}\right)(E_{22} + E_{33});$$
$$\sigma_{22} = \left(K + \frac{4E_c}{9}\right)E_{22} + \left(K - \frac{2E_c}{9}\right)(E_{11} + E_{33}); \tag{4.32}$$
$$\tau_{12} = E_c E_{12} / 3; \quad \tau_{13} = E_c E_{13} / 3; \quad \tau_{23} = E_c E_{23} / 3,$$

where $E_c = \sigma_i / e_i$ is secant modulus of the diagram $\sigma_i(e_i)$.

If the dependence $\sigma_i(e_i)$ is approximated by the diagram with linear hardening (Figure 4.7), then the secant modulus is

$$E_c = \frac{\sigma_T}{e_i} + E_1\left(1 - \frac{\varepsilon_T}{e_i}\right), \tag{4.33}$$

where E_1 is tangent modulus at $\sigma_i > \sigma_T$ ($e_i > \varepsilon_T$) (Figure 4.7).

Strain E_{33} is determined from the equation $\sigma_{33} = 0$ as

$$E_{33}^{el} = -\frac{\nu}{1-\nu}(E_{11} + E_{22}); \quad E_{33}^{pl} = -\frac{9K - 2E_c}{9K + 4E_c}(E_{11} + E_{22}), \tag{4.34}$$

where E_{33}^{el} and E_{33}^{pl} are drafting at elastic and plastic deformations correspondingly. At unloading, stresses are expressed through the strains by formulas

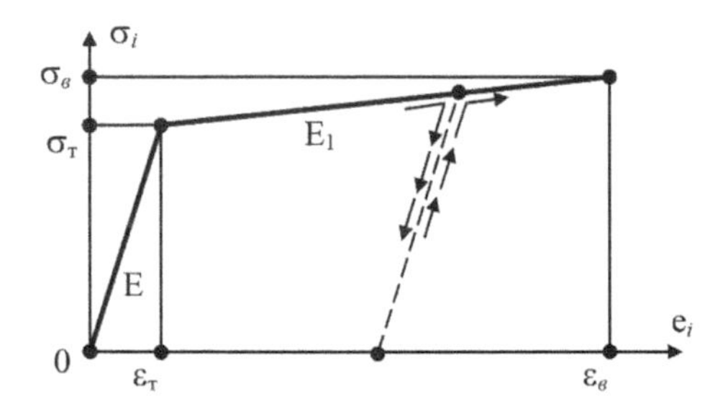

FIGURE 4.7
Diagram with linear hardening for the dependence $\sigma_i(e_i)$.

$$\sigma_{11} = \sigma_{11}^* + \frac{E}{1-\nu^2}[(E_{11} - E_{11}^*) + \nu(E_{22} - E_{22}^*)];$$

$$\sigma_{22} = \sigma_{22}^* + \frac{E}{1-\nu^2}[(E_{22} - E_{22}^*) + \nu(E_{11} - E_{11}^*)]; \quad (4.35)$$

$$\tau_{12} = \tau_{12}^* + G(E_{12} - E_{12}^*); \quad \tau_{13} = \tau_{13}^* + G(E_{13} - E_{13}^*); \quad \tau_{23} = \tau_{23}^* + G(E_{23} - E_{23}^*).$$

Asterisk in (Eq. 4.35) marks stresses and strains at the moment of unloading beginning. The expressions for strains (Eq. 4.4) and stresses (Eq. 4.17) intensities for the shells obtain the form

$$e_i = \frac{2}{3}\sqrt{E_{11}^2 + E_{22}^2 + E_{33}^2 - E_{11}E_{22} - E_{22}E_{33} - E_{33}E_{11} + \frac{3}{4}(E_{12}^2 + E_{13}^2 + E_{23}^2)};$$

$$\sigma_i = \sqrt{\sigma_{11}^2 + \sigma_{22}^2 - \sigma_{11}\sigma_{22} + 3(\tau_{12}^2 + \tau_{13}^2 + \tau_{23}^2)}. \quad (4.36)$$

Formulas for forces and moments are obtained after substitution of Equations 4.32 and 4.35 in Equation 4.18 and integration over shell thickness.

4.8 Equilibrium Equations

In the general case coordinate surface of area, F, of the shell can be constrained by some smooth boundary contour, G (Figure 4.8). Surface load acting on the shell with components $q_1 = q_1(\alpha_1,\alpha_2)$, $q_2 = q_2(\alpha_1,\alpha_2)$, $q_3 = q_3(\alpha_1,\alpha_2)$, $m_1 = m_1(\alpha_1,\alpha_2)$, and $m_2 = m_2(\alpha_1,\alpha_2)$ can be distributed over the whole surface as well as locally (Figure 4.8). At shell contour or at its part external force factors are normal force, T_{ii}^*; shear forces, T_{ij}^* and Q_{ii}^*; bending, M_{ii}^*; and torsion, H^* moments, as well as the values of components of coordinate surface points displacements, u^*,v^*,w^* and normal rotation angles, γ_1^*, γ_2^* $(i, j = 1,2)$. We shall restrict ourselves by the case when the contour, G, consists of the parts, G_1, G_2, G_3, G_4, for which $\alpha_1 = \text{const}$ (contour G_1, G_3) and $\alpha_2 = \text{const}$ (for contour G_2, G_4). Positive directions for external boundary loads are considered coinciding with positive directions for internal force factors (Figure 4.8).

To get equilibrium equations and natural boundary conditions, we use Lagrange variational principle in the form

$$\delta Э = \delta \Pi - \delta A = 0, \quad (4.37)$$

where Π is deformation potential energy, and A is external loads work. We shall represent elementary work of external loads as a sum

$$A = A_F + A_G, \quad (4.38)$$

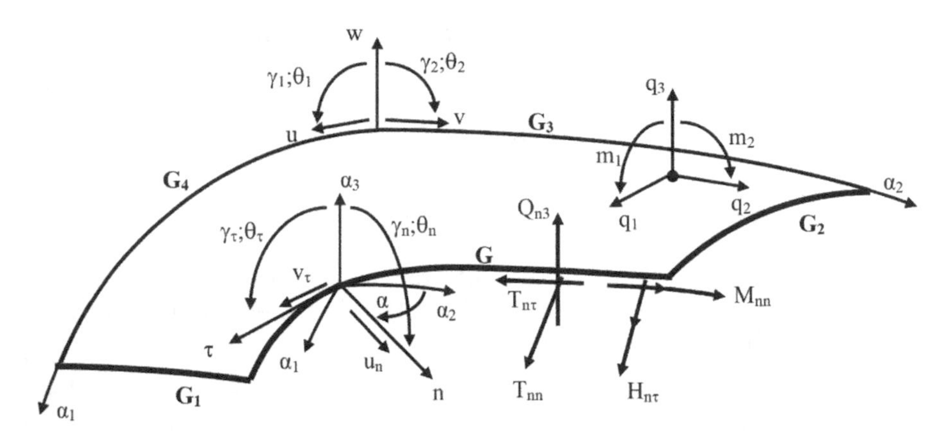

FIGURE 4.8
Shell coordinate surface construction and positive directions for external boundary loads and internal force factors.

where A_F is surface load work, and A_G is boundary load work. The expressions for Π and A_F have the form [1–4]

$$\Pi = \iint_F (T_{11}E_{11} + T_{22}E_{22} + SE_{12} + M_{11}K_{11} + M_{22}K_{22} + HK_{12} + $$
$$+ Q_{13}E_{13} + Q_{23}E_{23})A_1A_2d\alpha_1d\alpha_2; \tag{4.39}$$
$$A_F = \iint_F (q_1u + q_2v + q_3w + m_1\gamma_1 + m_2\gamma_2)A_1A_2d\alpha_1d\alpha_2.$$

Boundary load work A_G is

$$A_G = \int_{G_1} (T_{11}^*u + T_{12}^*v + Q_{11}^*w + M_{11}^*\gamma_1 + H^*\gamma_2)A_2d\alpha_2 -$$
$$-\int_{G_3} (T_{11}^{**}u + T_{12}^{**}v + Q_{11}^{**}w + M_{11}^{**}\gamma_1 + H^{**}\gamma_2)A_2d\alpha_2 +$$
$$+\int_{G_2} (T_{21}^*u + T_{22}^*v + Q_{22}^*w + H^*\gamma_1 + M_{22}^*\gamma_2)A_1d\alpha_1 - \tag{4.40}$$
$$-\int_{G_4} (T_{21}^{**}u + T_{22}^{**}v + Q_{22}^{**}w + H^{**}\gamma_1 + M_{22}^{**}\gamma_2)A_1d\alpha_1.$$

In Equation 4.40, one asterisk marks given boundary loads on the contours G_1 and G_2, and two asterisks mark the ones on the contours G_3 and G_4.

At independent variations of generalized displacements in the area, F, and at the contours G_i equilibrium equations for shells follow from variational equation (Eq. 4.37) [1–4]

$$\begin{aligned}
&\frac{1}{A_1A_2}\frac{\partial}{\partial\alpha_1}(A_2T_{11})-\psi T_{22}+\frac{1}{A_2}\frac{\partial S}{\partial\alpha_2}+k_1\left(Q_{11}+\frac{1}{A_2}\frac{\partial H}{\partial\alpha_2}\right)+q_1=0;\\
&\frac{1}{A_2}\frac{\partial T_{22}}{\partial\alpha_2}+\frac{1}{A_1A_2^2}\frac{\partial}{\partial\alpha_1}(A_2^2S)+2\psi k_1H+k_2\left(Q_{22}+\frac{1}{A_1}\frac{\partial H}{\partial\alpha_1}\right)+q_2=0;\\
&\frac{1}{A_1A_2}\frac{\partial}{\partial\alpha_1}(A_2Q_{11})+\frac{1}{A_2}\frac{\partial Q_{22}}{\partial\alpha_2}-k_1T_{11}-k_2T_{22}+q_3=0;\\
&\frac{1}{A_1A_2}\frac{\partial}{\partial\alpha_1}(A_2M_{11})-\psi M_{22}+\frac{1}{A_2}\frac{\partial H}{\partial\alpha_2}-Q_{13}+m_1=0;\\
&\frac{1}{A_2}\frac{\partial M_{22}}{\partial\alpha_2}+\frac{1}{A_1A_2^2}\frac{\partial}{\partial\alpha_1}(A_2^2H)-Q_{23}+m_2=0,
\end{aligned} \tag{4.41}$$

as well as static boundary conditions:

- contour G_1:

$$T_{11}=T_{11}^*;\quad S+k_2H=T_{12}^*;\quad Q_{11}=Q_{11}^*;$$
$$M_{11}=M_{11}^*;\quad H=H^*;$$

- contour G_2:

$$\begin{aligned}
&S+k_1H=T_{21}^*;\quad T_{22}=T_{22}^*;\quad Q_{22}=Q_{22}^*;\\
&H=H^*;\quad M_{22}=M_{22}^*.
\end{aligned} \tag{4.42}$$

Static boundary conditions at contours G_3, G_4 are formulated similarly to Equation 4.42. Generalized shear forces Q_{11} and Q_{22} are determined as

$$Q_{11}=Q_{13}-T_{11}\theta_1-S\theta_2;\quad Q_{22}=Q_{23}-S\theta_1-T_{22}\theta_2. \tag{4.43}$$

4.9 Motion Equations

For equations of motion describing the transit of the shell from one state to the other at time interval $[t_0, t_1]$ obtaining Hamilton-Ostrogradsky variational equation is used

$$\delta I=\int_{t_o}^{t_1}(\delta K-\delta\Pi+\delta A)dt=0. \tag{4.44}$$

Deformation potential energy, Π, and external forces work, A, are determined by formulas (Eqs. 4.39 and 4.40). Kinetic energy of the shell is

$$K = \frac{1}{2}\iint_F \left[\rho h(\dot{u}^2 + \dot{v}^2 + \dot{w}^2) + \frac{\rho h^3}{12}(\dot{\gamma}_1^2 + \dot{\gamma}_2^2) \right] A_1 A_2 d\alpha_1 d\alpha_2, \tag{4.45}$$

where $\rho = \rho(\alpha_1,\alpha_2)$ is material density; the dot denotes time, t, differentiation. After carrying out corresponding transformations equations of the shell, motion can be represented in the form

$$L_k = m_k \ddot{u}_k, \tag{4.46}$$

where L_k denoted left sides of the equations (4.41); $m_k = \rho h$ ($k = 1,2,3$) and $m_k = \rho h^3/12$ ($k = 4,5$). Surface and boundary loads are functions not only of coordinates α_1, α_2, but also of time, t. For multilayer shells made of composite materials, the parameters of mass characteristics, m_k, are determined as

$$m_k = \sum_{i=1}^{N} \rho_i h_i, \ (k = 1,2,3);$$
$$m_k = \sum_{i=1}^{N} \rho_i (J_{0i} + z_i^2 F_i), \ (k = 4,5), \tag{4.47}$$

where N is layers number, z_i is the distance between the center of the ith layer and coordinate surface of the shell, and J_{0i} and F_i are layer moment of inertia and cross section area taken relative to the length unit of the corresponding coordinate line. From the variational equation (Eq. 4.44) also dynamic force boundary conditions follow, which are written similarly to Equation 4.42. At dynamic high-speed deforming of shells change of physicomechanical material properties takes place, which can be considered for calculating corresponding parameters.

Timoshenko-Reissner shells equations of motion referring to hyperbolic type describe propagation of both coordinate surface deformation waves and bending-shear waves. The equation of motion, Equation 4.46, and the equilibrium equation, Equation 4.41, are obtained in projections on axes connected with undeformed coordinate system, which allows simple formulation of the problem for the case of composite thin-walled structures.

4.10 Boundary and Initial Conditions

In real shell structures, different types of supports can be met, which leads to a variety of their mathematical models boundary conditions. For example, at the boundary $\alpha_1 = \text{const}$ (contour G_1,G_3) the following boundary conditions can be formulated:

- *rigid fixing:*

$$u = v = w = \gamma_1 = \gamma_2 = 0; \tag{4.48}$$

- *pinning:*

$$u = v = w = M_{11} = \gamma_2 = 0; \tag{4.49}$$

- *rigid fixing:*

$$T_{11} = T_{11}^*; \quad v = w = \gamma_1 = \gamma_2 = 0; \tag{4.50}$$

- *movable pivot point:*

$$T_{11} = T_{11}^*; \quad v = w = M_{11} = \gamma_2 = 0. \tag{4.51}$$

Boundary conditions on the boundary α_2 = const are formulated similar to Equations 4.48 through 4.51. For the case of nonhomogeneous kinematic boundary conditions, boundary generalized displacements can be given in the contours of the shell: $u = u^*, v = v^*, w = w^*, \gamma_1 = \gamma_1^*, \gamma_2 = \gamma_2^*$.

If the shell is partially or completely closed, then boundary conditions in Equations 4.48 through 4.51 are replaced by generalized displacements and forces periodicity conditions. When calculating plates and panels, boundary conditions are formulated both on the contour α_1 = const and α_2 = const. In a number of cases, the solution of the problem can be symmetric relative to coordinate lines α_1 = const, α_2 = const. Along such lines, corresponding symmetry conditions are formulated, which allows constructing the solution not for all the area of parameters change α_1, α_2, but only for some characteristic part of it. Equations 4.48 through 4.51 do not cover all the variety of shell boundaries fixing variants, but on their basis, the majority of met in practice variants of boundary conditions can be described by the way of selecting the conditions not contradicting each other.

Initial conditions for equation system (Eq. 4.46) are set for generalized displacements u_k and their velocities $\dot{u}_k$

$$u_k\big|_{t=0} = u_k^0; \quad \frac{\partial u_k}{\partial t}\bigg|_{t=0} = \dot{u}_k^0, \tag{4.52}$$

where $u_k^0 = u_k^0(\alpha_1,\alpha_2)$ and $\dot{u}_k^0 = \dot{u}_k^0(\alpha_1,\alpha_2)$ are given initial values of generalized displacements and their velocities.

4.11 Multiconnected Problems

Thin-walled–bearing elements of airframes often have cutouts of various forms made for exploitation, constructional, or technological reasons. Cutouts (holes) can appear as a result of emergency conditions. As computational region for plates and shells with cutouts is multiconnected, boundary conditions must be formulated both on external and internal contours (cutouts' contours). On the internal contour, as on the external, boundary conditions can be given in kinematic, static, or mixed form. From the point of view of practical applications, nonhomogeneous boundary conditions of a free boundary type are of ultimate interest. Without loss of generality, we shall restrict ourselves to the case when the cutout contour (or its part G_i) coincides with coordinate lines α_1 = const and α_2 = const (Figure 4.3). In this case, natural boundary conditions are formulated in the form (Eq. 4.42).

Form boundary conditions (Eq. 4.42), a number of dependencies between deformation components, E_{11}, E_{22} and K_{11}, K_{22} follows, which must be fulfilled on cutout contours G_i. In particular, on the cutout contour G_1 (α_1 = const), the force T_{22} and bending moment M_{22} can be expressed only through deformation components E_{22} and K_{22} in coordinate direction α_2. For practically important case of one-layer orthotropic shell, for which physical relations (Eqs. 4.23 and 4.24 are fulfilled, on the cutout contour α_1 = const we have

$$\begin{aligned} T_{11}^* &= B_{11}E_{11} + B_{12}E_{22}; \quad M_{11}^* = D_{11}K_{11} + D_{12}K_{22}; \\ T_{22} &= B_{22}E_{22} + B_{21}E_{11}; \quad M_{22} = D_{22}K_{22} + D_{21}K_{11}. \end{aligned} \tag{4.53}$$

Defining E_{11} and K_{11} from the first two relations (Eq. 4.53), we obtain

$$T_{22} = B_{22}E_{22} + \frac{B_{21}(T_{11}^* - B_{12}E_{22})}{B_{11}}; \quad M_{22} = D_{22}K_{22} + \frac{D_{21}(M_{11}^* - D_{12}K_{22})}{D_{11}}. \tag{4.54}$$

Similarly, on the cutout contour G_2 (α_2 = const)

$$T_{11} = B_{11}E_{11} + \frac{B_{12}(T_{22}^* - B_{21}E_{11})}{B_{22}}; \quad M_{11} = D_{11}K_{11} + \frac{D_{12}(M_{22}^* - D_{21}K_{11})}{D_{22}}. \tag{4.55}$$

Relations (Eqs. 4.54 and 4.55) allow circumventing difficulties of mathematical character when numerically realizing boundary conditions (Eq. 4.42) about angular points of the cutout. Basic surface of shells with cutouts is a multiconnected region, so continuity equations, which are necessary and sufficient conditions of continuity for a simply-connected region, must be completed by the requirements of generalized displacements functions increments equality to zero when circulating about the contour G.

4.12 Difference Scheme

For constructing discrete models of original integro-differential initial-boundary problems in nonlinear mechanics of thin-walled structures, finite difference method (FDM) and finite element method (FEM) are most widely used. Each of these methods of discretization has its characteristic advantages and disadvantages, which are revealed when solving problems of the corresponding class [6–8]. The effectiveness of FDM is connected with using the simplest formulas of numerical differentiation, as well as minimum amount of integration points when approximating corresponding functional in variational equations. The advantages of the FEM are connected mainly with smaller in comparison to FDM method sensitivity to the form of external or internal boundary of the shell when calculating nonregular structures of complicated form. This justifies commercial success of such well-known finite-element software as ANSYS and NASTRAN. However when calculating thin-walled structures on the basis of Timoshenko-Reissner models containing singular terms in the original equations, problem discretization using FEM can present difficulties in carrying out calculations. Among other things, this accounts for peculiarities of machine arithmetic when carrying out operations with matrices and can often lead to significant computational errors. Taking into account transverse shear brings about the effect of "false" shear, which reveals itself in extra influence of transverse shear deformations contribution into potential energy of the element; moreover, with shell relative thickness reduction this influence increases.

In the present work when passing from the original continual problem (Eqs. 4.1 through 4.55), formulated in computational domain $L(\alpha_1, \alpha_2)$ in functions of continuous change of coordinates α_1, α_2 to discrete one in functions of discrete analogues of the arguments α_1 and α_2, we use FDM method for discretization in space, α_1, α_2, and time, t, coordinates. Developed on the basis of variational-difference method (VDM) conservative difference schemes (DS) both for Kirchhoff-Love and Timoshenko-Reissner models combined with quasi-dynamical form of relaxation method allow significant widening of thin-walled structures nonlinear deformation peculiarities applied investigations area, including multiconnected problems for arbitrary form of external and internal contour (cutout contour).

When using the FDM method, we introduce two orthogonal uniform meshes on the plane of main coordinates of the shell α_1, α_2: the main mesh, which nodes have integer indices, i, j, and auxiliary mesh, which nodes lie in the middle between the nodes of the main mesh and have fractional indices $(i \pm 1/2, j \pm 1/2)$; $(i \pm 1/2, j)$; and $(i, j \pm 1/2)$ (Figure 4.9).

In the nodes of the main mesh instead of functions of generalized displacements (α_1, α_2), we introduce mesh functions u_k(i,j).

Differential operators are approximated by difference operators of the second order of accuracy $O(\lambda_1^2 + \lambda_2^2)$ with piecewise linear interpolation of functions inside mesh cell based on the values $u_k(i,j)$ in the nodes of the main mesh, where $\lambda_1 = const$, $\lambda_2 = const$ are mesh parameters which are determined as

$$\lambda_1 = \frac{\Lambda_1}{N}; \quad \lambda_2 = \frac{\Lambda_2}{M}, \tag{4.56}$$

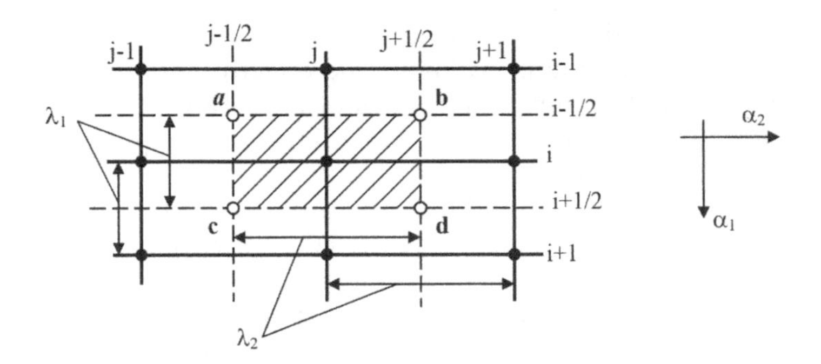

FIGURE 4.9
Main and auxiliary meshes used in finite-difference scheme.

where $0 \leq \alpha_1 \leq \Lambda_1$ and $0 \leq \alpha_2 \leq \Lambda_2$ are arguments α_1 and α_2 domains in computational domain $L(\alpha_1,\alpha_2)$, $0 \leq i \leq N, 0 \leq j \leq M$ is the number of discretization points in α_1,α_2. Both main and auxiliary meshes are considered coinciding with the mesh formed be the lines of principal curvature of the shell. Coordinates of the main mesh nodes are

$$\alpha_1(i) = i \cdot \lambda_1; \quad \alpha_2(j) = j \cdot \lambda_2. \tag{4.57}$$

When using the FDM method, original integral-differential equations and boundary conditions are approximated by difference equations the solution of which depends on the steps λ_1, λ_2 as on the parameters. Approximation accuracy can be risen either by increasing the number of discretization points N, M, or by approximation by difference operators of the higher order of accuracy. However, using higher order approximation leads to increasing the number of nodes when forming difference stencil and loss of simplicity of DS. Besides, if the solution of the original differential problem has derivatives of the order not higher than k, then there is no sense in using approximations of the order higher than $(k + 1, k + 2,...)$. Differential operators used are approximated by difference ones of the second order of approximation $O(\lambda_1^2 + \lambda_2^2)$, which allows getting numerical solutions of static and dynamic problems of shells theory with sufficient accuracy using relatively simple structures of difference equations.

When solving complicated nonlinear multiconnected problems of Timoshenko shells theory numerically, the most effective difference scheme is the one in which all the parameters of tangential E_{11}, E_{22}, E_{12}, bending K_{11}, K_{22}, K_{12}, and transversal deformation E_{13}, E_{23} of the shell coordinate surface are approximated in the nodes of the auxiliary mesh $(i \pm 1/2, j), (i, j \pm 1/2)$ for Lagrange functional discretization. When using finite-difference approximation of deformed state parameters and using the following empiric rule, which we do below: *when constructing DS it is not allowed to open the brackets for nothing and use the formula of product differentiation* [8]. This rule using allows skipping unnecessary complication of finite-difference analogues of equilibrium (Eq. 4.41) and movement (Eq. 4.46) equations structure. The required functions of generalized equations are approximated by piecewise linear ones with function interpolation inside the mesh cell $u_k(i,j)$ in the nodes of the main mesh, for example,

Point $(i + 1/2, j + 1/2)$:

$$\theta_1 = \frac{1}{(A_1)_{i+1/2,j+1/2}} \frac{w_{i,j} + w_{i,j+1} - w_{i+1,j} - w_{i+1,j+1}}{2\lambda_1} +$$

$$+ (k_1)_{i+1/2,j+1/2} \frac{u_{i,j} + u_{i,j+1} + u_{i+1,j} + u_{i+1,j+1}}{4};$$

$$\theta_2 = \frac{1}{(A_2)_{i+1/2,j+1/2}} \frac{w_{i,j} + w_{i+1,j} - w_{i,j+1} - w_{i+1,j+1}}{2\lambda_2} +$$

$$+ (k_2)_{i+1/2,j+1/2} \frac{v_{i,j} + v_{i,j+1} + v_{i+1,j} + v_{i+1,j+1}}{4}; \tag{4.58}$$

$$E_{11} = \frac{1}{(A_1)_{i+1/2,j+1/2}} \frac{u_{i+1,j} + u_{i+1,j+1} - u_{i,j} - u_{i,j+1}}{2\lambda_1} +$$

$$+ (k_1)_{i+1/2,j+1/2} \frac{w_{i,j} + w_{i,j+1} + w_{i+1,j} + w_{i+1,j+1}}{4} + \frac{1}{2}(\theta_1^2)_{i+1/2,j+1/2};$$

$$E_{22} = \frac{1}{(A_2)_{i+1/2,j+1/2}} \frac{v_{i,j+1} + v_{i+1,j+1} - v_{i,j} - v_{i+1,j}}{2\lambda_2} +$$

$$+ \psi_{i+1/2,j+1/2} \frac{u_{i,j} + u_{i,j+1} + u_{i+1,j} + u_{i+1,j+1}}{4} +$$

$$+ (k_2)_{i+1/2,j+1/2} \cdot \frac{w_{i,j} + w_{i,j+1} + w_{i+1,j} + w_{i+1,j+1}}{4} + \frac{1}{2}(\theta_2^2)_{i+1/2,j+1/2};$$

As differential operators in Equations 4.7 through 4.13 where approximated by central finite differences with generalized displacements $u_k(i,j)$ mesh functions interpolation from the nodes of the main mesh to the corresponding nodes of the auxiliary mesh, then difference operators (4.58) approximate differential relations (Eqs. 4.7 through 4.13) with the second order of accuracy $O(\lambda_1^2 + \lambda_2^2)$ [6,8].

4.13 Difference Analogues of Equilibrium and Movement Equations

Shell differential equilibrium equations result from integral conservation equations. For mesh analogues of equilibrium equations obtained when constructing corresponding DS, they must fulfill mesh analogues of conservation equations. Conservative difference schemes fulfilling these requirements allow separating from all the mathematically permitted numerical solutions—the generalized one which is physically correct. For obtaining conservative difference schemes in complicated problems (e.g., for equations with discontinuous coefficients) and multiconnected regions, they use integral-interpolation and variational-difference methods [6,8]. Constructing difference schemes using variational-difference method is based on using extreme variational principles in shells theory. Lagrange functional for the shell is written in finite-difference form; here differential expressions for the components of the shell coordinate surface deformation are replaced by finite-difference expressions, and integration is replaced by summation over mesh cells. Variational-difference scheme is obtained by varying discretized functional.

We shall represent discretized Lagrange functional (Eq. 4.37) for the shell computational domain

$L(\alpha_1,\alpha_2)$ as a sum

$$Э_{\Sigma}=\sum_{i}\sum_{j}Э_{i,j}, \tag{4.59}$$

where

$$Э_{i,j}=\Pi_{i,j}-A_{i,j}, \tag{4.60}$$

and where $\Pi_{i,j}$ and $A_{i,j}$ are deformation potential energy and external forces work for elementary domain of the shell, which is represented in the mesh area as a rectangle with the dimensions $\Delta F_{i,j}=(A_1A_2\lambda_1\cdot\lambda_2)_{i,j}$ (crosshatched area on Figure 4.9).

In Equation 4.59, the summation is carried out over these node points of DS in which corresponding generalized displacements $u_k(i,j)$ are varied. When carrying out numerical integration operations, we use formulas of the second order of accuracy (rectangular formulas). Mesh analogues of subintegral functions in Equations 4.39 and 4.40 are approximated in the corresponding node points of main and auxiliary meshes, which are central points (diagonals intersection points) of mesh cells of area $\Delta F_{i,j}$.

Difference equations follow from the conditions of functional minimization in the form

$$\frac{\partial Э_{\Sigma}}{\partial u_k(i,j)}=0. \tag{4.61}$$

Depending on the notation of the discretized functionals $Э_{i,j}$, VDS obtained have different convergence and accuracy. Analysis of different VDS showed that when calculating complicated nonhomogeneous structures with special features the most effective and economic scheme is VDS, for which all the parameters of stress-strain state are calculated in similar points of auxiliary mesh $(i \pm 1/2, j \pm 1/2)$ [9–12]. When constructing VDS, we suppose that basic surface of the shell is constrained by the cells of external and internal contour (cutout contour) coinciding with coordinate lines $\alpha_1 = \text{const}$ and $\alpha_2 = \text{const}$. We shall represent deformation potential energy $\Pi_{i,j}$ as a sum

$$\Pi_{i,j} = \Pi_{i,j}(E_{11}, E_{22}) + \Pi_{i,j}(E_{12}, K_{12}) + \Pi_{i,j}(E_{13}, E_{23}) + \Pi_{i,j}(K_{11}, K_{22}), \quad (4.62)$$

where

$$\Pi_{i,j}(E_{11}, E_{22}) = 0{,}25\{b_1^*[(A_1A_2)(T_{11}E_{11} + T_{22}E_{22})]^a + b_2^*[(A_1A_2)(T_{11}E_{11} + T_{22}E_{22})]^b + \\ + b_3^*[(A_1A_2)(T_{11}E_{11} + T_{22}E_{22})]^c + b_4^*[(A_1A_2)(T_{11}E_{11} + T_{22}E_{22})]^d\}\lambda_1\lambda_2;$$

$$\Pi_{i,j}(E_{12}, K_{12}) = 0{,}25\{b_1^*[(A_1A_2)(SE_{12} + HK_{12})]^a + b_2^*[(A_1A_2)(SE_{12} + HK_{12})]^b + \\ + b_3^*[(A_1A_2)(SE_{12} + HK_{12})]^c + b_4^*[(A_1A_2)(SE_{12} + HK_{12})]^d\}\lambda_1\lambda_2;$$

$$\Pi_{i,j}(E_{13}, E_{23}) = 0{,}25\{b_1^*[(A_1A_2)(Q_{13}E_{13} + Q_{23}E_{23})]^a + b_2^*[(A_1A_2)(Q_{13}E_{13} + Q_{23}E_{23})]^b + \\ + b_3^*[(A_1A_2)(Q_{13}E_{13} + Q_{23}E_{23})]^c + b_4^*[(A_1A_2)(Q_{13}E_{13} + Q_{23}E_{23})]^d\}\lambda_1\lambda_2;$$

$$\Pi_{i,j}(K_{11}, K_{22}) = 0{,}25\{b_1^*[(A_1A_2)(M_{11}K_{11} + M_{22}K_{22})]^a + \\ + b_2^*[(A_1A_2)(M_{11}K_{11} + M_{22}K_{22})]^b + b_3^*[(A_1A_2)(M_{11}K_{11} + M_{22}K_{22})]^c + \\ + b_4^*[(A_1A_2)(M_{11}K_{11} + M_{22}K_{22})]^d\}\lambda_1\lambda_2. \quad (4.63)$$

Elementary work of external forces in discrete form $A_{i,j}$ is represented as

$$A_{i,j} = [c^* A_1A_2(q_1u + q_2v + q_3w + m_1\gamma_1 + m_2\gamma_2)]_{i,j}\lambda_1\lambda_2 + \\ + [d_1^* A_2(T_{11}^* u + T_{12}^* v + Q_{11}^* w + M_{11}^* \gamma_1 + H^* \gamma_2)]_{i,j}\lambda_2 + \\ + [d_2^* A_1(T_{21}^* u + T_{22}^* v + Q_{22}^* w + H^* \gamma_1 + M_{22}^* \gamma_2)]_{i,j}\lambda_1 - \quad (4.64) \\ - [d_3^* A_2(T_{11}^{**} u + T_{12}^{**} v + Q_{11}^{**} w + M_{11}^{**} \gamma_1 + H^{**} \gamma_2)]_{i,j}\lambda_2 - \\ - [d_4^* A_1(T_{21}^{**} u + T_{22}^{**} v + Q_{22}^{**} w + H^{**} \gamma_1 + M_{22}^{**} \gamma_2)]_{i,j}\lambda_1.$$

In the formulas (Eqs. 4.63 and 4.64) points *a, b, c, d* correspond to the points of auxiliary mesh with indices $(i-1/2, j-1/2)$; $(i-1/2, j+1/2)$; $(i+1/2, j-1/2)$;

$(i+1/2,j+1/2)$ (Figure 4.9), b_k^*, d_k^*, and c^* are weight numbers taking into account integration area when representing corresponding part of basic surface or contour of the shell in the vicinity of node point i, j on mesh area $\Delta F_{i,j}$; and $k = 1,2,3,4$ (Figure 4.9) [6,8,11]. In regular node points of difference, mesh weight numbers are equal to unity: $b_k^* = d_k^* = c^* = 1$, and in irregular points $0 < b_k^*, d_k^*, c^* < 1$.

As it follows from Equations 4.63 and 4.64, for difference analogues of equilibrium equations in regular node point i, j construction nine-point difference stencil must be used. For difference equations relative to node point i, j obtaining one must vary only the part of $\text{Э}(i,j)$ from Equation 4.59, which contains non-zero variations relative to $u_{i,j}$, $v_{i,j}$, $w_{i,j}$, $\gamma_1(i,j)$, $\gamma_2(i,j)$

$$\begin{aligned} \text{Э}(i,j) = 4[\Pi_{i,j}(E_{11},E_{22}) + \Pi_{i,j}(E_{13},E_{23}) + \\ + \Pi_{i,j}(K_{11},K_{22}) + \Pi_{i,j}(E_{12},K_{12})] - A_{i,j}. \end{aligned} \tag{4.65}$$

The variational-difference equation (Eq. 4.61) taking into account Equation 4.65 will be written as follows:

$$\begin{aligned} &\frac{\partial \text{Э}(i,j)}{\partial u_{i,j}} = 0; \quad \frac{\partial \text{Э}(i,j)}{\partial v_{i,j}} = 0; \quad \frac{\partial \text{Э}(i,j)}{\partial w_{i,j}} = 0; \\ &\frac{\partial \text{Э}(i,j)}{\partial \gamma_1(i,j)} = 0; \quad \frac{\partial \text{Э}(i,j)}{\partial \gamma_2(i,j)} = 0. \end{aligned} \tag{4.66}$$

Finite-difference analogues of equilibrium equations relative to the point i, j following from Equation 4.66 after carrying out variation operation are quite lengthy when written in extended form, so they can be represented in operator format as

$$[L_{\lambda_1,\lambda_2}(U_k)]_{i,j} + (Q_k)_{i,j} = 0, \tag{4.67}$$

where $[L_{\lambda_1,\lambda_2}]_{i,j}$ are corresponding finite-difference operators for the vector U_k of mesh functions of generalized displacements $u_k(i, j)$ relative to the node point i, j; $(Q_k)_{i,j}$ are mesh functions of generalized surface and boundary loads components. Depending on the values weight numbers in the expressions in Equations 4.63 through 4.67 take, the VDS developed describe deformation of shells with peculiarities and nonhomogeneities of various character. From Equations 4.57 through 4.67, as a special case of finite-difference approximations of integral-differential relations are obtained for axisymmetric problems and plates and panels of different form. When calculating shells with cutouts difference, mesh is constructed in such a way that the line depicting the line of cutout contour on the plane of principle coordinates α_1, α_2 passed through node points of the main mesh. Mesh functions of generalized displacements and their velocities in node points on cutout contour are defined from the solution of the main equations approximated in these nodes.

Variational-difference formulation of the original initial boundary problem allows constructing conservative difference schemes providing convergence of numerical solutions $u_k(i, j)$ to the exact solution $u_k(\alpha_1,\alpha_2)$ when condensating mesh, which is especially important when solving multiconnected problems when boundary conditions on external and internal boundaries (cutouts' boundaries) are given in natural form [6,8].

Relations (Eqs. 4.63 through 4.67) analysis shows that conservative difference schemes can be obtained using the method of difference approximation only for a number of special cases of shells with simply-connected basic surface, deformation of which is described by the equations with constant coefficients at kinematic boundary conditions of the contour.

When solving nonstationary problems numerically in the domain of continuous time change $t \leq 0$, we introduce two uniform meshes with constant time step, Δt and the main mesh with integer indices, n, and layers

$$t^{(n)} = \Delta t \cdot n, \quad n = 0,1,2,3,..., \tag{4.68}$$

as well as auxiliary mesh with fractional indices ($n \pm 1/2$) and time layers $t^{(n\pm 1/2)}$ (Figure 4.10).

Mesh functions of generalized displacements $u_k(i,j)$ are aligned with the nodes of the main mesh, and mesh functions of the velocities $\dot{u}_k(i,j)$ are aligned with the nodes of auxiliary mesh. Differential operators for the velocities $\dot{u}_k(\alpha_1,\alpha_2)$ are approximated by difference operators of the second-order of approximation $O(\Delta t^2)$

$$[\dot{u}_k]_{i,j}^{(n-1/2)} = \frac{[u_k]_{i,j}^{(n)} - [u_k]_{i,j}^{(n-1)}}{\Delta t}; \quad [\dot{u}_k]_{i,j}^{(n+1/2)} = \frac{[u_k]_{i,j}^{(n+1)} - [u_k]_{i,j}^{(n)}}{\Delta t}. \tag{4.69}$$

In the mesh, domain $\Delta F_{i,j}$ shell kinetic energy (Eq. 4.45) is discretized in the form

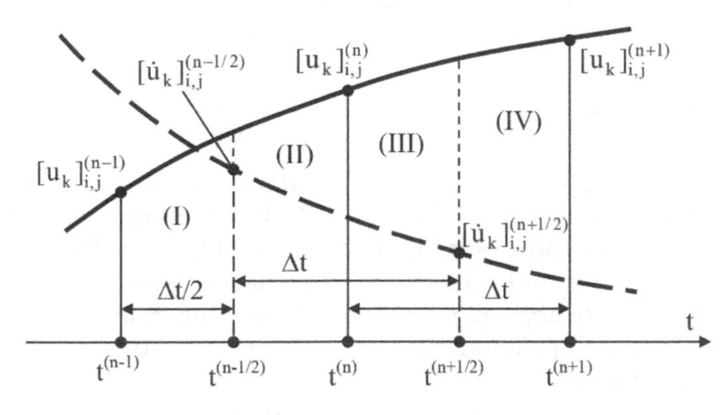

FIGURE 4.10
Two uniform finite-difference meshes used for the numerically solving of nonstationary problems.

$$K_{\Sigma} = \sum_{i}\sum_{j} K_{i,j}, \tag{4.70}$$

where $K_{i,j}$ is kinetic energy of the shell elementary domain corresponding to the mesh cell $\Delta F_{i,j}$

$$K_{i,j} = 0,5\left[\sum_{k} c^{*}_{i,j} m_k(i,j)\dot{u}^2_k(i,j)\right](A_1A_2)_{i,j}\lambda_1\lambda_2. \tag{4.71}$$

Replacing integration in Equation 4.44 by summation over mesh domain $t^{(n)}$, the functional $\boldsymbol{I}$ taking into account Equation 4.59 can be represented in the discrete form as

$$I_{\Sigma} = \sum_{n} I^{(n)}, \tag{4.72}$$

where

$$I^{(n)} = \{0,5[(f_2K_{\Sigma})^{(n-1/2)} + (f_3K_{\Sigma})^{(n+1/2)}] - (f^{*}Э_{\Sigma})^{(n)}\}\Delta t, \tag{4.73}$$

and where $Э_{\Sigma}^{(n)}$ is discretized Lagrange functional (Eq. 4.59) expressed in the values of mesh functions of generalized displacements $[u_k]_{i,j}^{(n)}$, physicomechanical characteristics and loads for the nth time layer, f_1, f_2, f_3, f_4 are weight numbers correspondingly for subdomains (I), (II), (III), (IV) (Figure 4.10), and f^{*} is the resulting weight number for subdomains (II) and (III): $0 \leq f_{1,2,3,4} \leq 1$; $0 \leq f^{*} \leq 1$. Finite-difference analogues of motion equations (Eq. 4.46) follow from variational-difference equations of the form

$$\frac{\partial\, I_{\Sigma}}{\partial [u_k]_{i,j}^{(n)}} = 0. \tag{4.74}$$

Carrying out the operations of varying in Equation 4.74, taking into account Equations 4.69 and 4.71, finite-difference analogues of the shell equations of motion can be represented as

$$\{f^{*}[L_{\lambda_1,\lambda_2}(U_k) + Q_k]\}_{i,j}^{(n)} = \frac{[f^{*}_{22}c^{*}m_k\dot{u}_k]_{i,j}^{(n+1/2)} - [f^{*}_{11}c^{*}m_k\dot{u}_k]_{i,j}^{(n-1/2)}}{\Delta t}, \tag{4.75}$$

where $f^{*}_{11} = (f_1 + f_2)/2$, $f^{*}_{22} = (f_3 + f_4)/2$. For regular in the mesh domain $t^{(n)}$ node point with the index $n > 0$: $f^{*}_{11} = f^{*}_{22} = f^{*} = 1$. Mesh functions of mass characteristics are given in the nodes of auxiliary mesh ($n \pm 1/2$) as $[m_k]_{i,j}^{(n\pm1/2)}$, and mesh functions of boundary and surface loads are aligned with the nth time layer. In the general case for approximating complicated function of nonstationary load cubic spline-interpolation can be used by representing a load plot as a number of characteristic points.

Finite-difference analogues of the shell equations of motion (Eq. 4.75) allow describing transient processes in shell structures for not only loads being time-varying, but also physic-mechanical characteristics of the material, mass characteristics, and geometry parameters of the shells.

4.14 Calculation Technologies of Nonstationary Problems Solution

Establishing (stabilization) method is a widely known method of mathematical physics stationary problems solution. For stationary problem solution some nonstationary process is constructed, steady state of which in equilibrium state determines the solution of the original stationary problem. When solving the system of algebraic equations of the form

$$AU - B = 0, \tag{4.76}$$

using establishing method, equations (Eq. 4.76) are replaced by nonstationary (evolution) equations of the form

$$AU - B = \frac{d^2U}{d\tau^2} + \varepsilon \frac{dU}{d\tau}, \tag{4.77}$$

or

$$AU - B = \varepsilon \frac{dU}{d\tau}, \tag{4.78}$$

where $\varepsilon > 0$ is scalar multiplier, τ is time which can be considered in the general case as time dummies. Representation of establishing method equations in the form (Eq. 4.77) corresponds to optimal linear iteration process, which significantly rises its economic feasibility and effectiveness compared to the form (Eq. 4.78) [8].

Equations (Eq. 4.77) approximation on time mesh with constant time step $\Delta\tau$ with replacing derivatives in time by central finite differences leads to iteration process of finding $U^{(n+1)}$

$$U^{(n+1)} = \frac{4}{2+\varepsilon\Delta\tau} U^{(n)} - \frac{2-\varepsilon\Delta\tau}{2+\varepsilon\Delta\tau} U^{(n-1)} + \frac{2\Delta\tau^2}{2+\varepsilon\Delta\tau} [AU^{(n)} - B], \tag{4.79}$$

where n is iteration number. Equations (Eq. 4.79) are written for the case of stationary iteration process. So if for some function $U = U(x_1,x_2,\tau)$, which is the solution of the system (Eq. 4.76), the following limits exist

$$\tilde{U}(x_1,x_2) = \lim_{\tau\to\infty} U(x_1,x_2,\tau); \quad \lim_{\tau\to\infty} \frac{\partial U(x_1,x_2,\tau)}{\partial \tau} = 0, \tag{4.80}$$

then the function $\tilde{U} = \tilde{U}(x_1,x_2)$ is the solution of the original stationary problem (Eq. 4.76).

The values of iteration parameters—time step $\Delta\tau$ and multiplier ε—must fulfill the requirements of iteration process convergence and minimum of arithmetic operations $n(\delta)$ for obtaining the solution of original stationary problem (Eq. 4.76) with given accuracy δ. Optimal values of parameters ε and $\Delta\tau$ are determined by formulas [8]

$$\varepsilon = 2\sqrt{\frac{\mu_1\mu_2}{\mu_1 + \mu_2}}; \quad \Delta\tau = \frac{2}{\sqrt{\mu_1 + \mu_2}}, \tag{4.81}$$

where μ_1 and μ_2 are minimum and maximum eigenvalues of matrix A respectively. Numerical method efficiency is usually described by the number of iterations $n(\delta)$ required for obtaining solutions with given accuracy, δ, and depending on the convergence rate of iteration process $v_q = \ln(1/q)$. Parameter q for establishing method in the form (Eq. 4.77) is estimated as

$$q = \frac{\mu_2 - \mu_1}{\mu_1 + \mu_2}. \tag{4.82}$$

For ill-conditioned matrices, when the relation μ_2/μ_1 is large, the parameter q is close to unity, which leads to iteration process convergence deterioration and increasing of the number of iterations $n(\delta)$. For establishing method in the form (Eq. 4.77) for optimal values of iteration parameter q is [8]

$$q = \frac{\sqrt{\mu_2} - \sqrt{\mu_1}}{\sqrt{\mu_1} + \sqrt{\mu_2}}. \tag{4.83}$$

When solving static problems of shells theory numerically usually the relation μ_2/μ_1 is large, so establishing method in the form (Eq. 4.77) allows reducing the number of iterations $n(\delta)$ approximately in $\sqrt{\mu_2/\mu_1}$ timed compared to numerical methods for which estimation (Eq. 4.82) is correct.

Let us consider the adaptation of quasi-dynamic form of establishing method to the solution of static problems of shells theory taking into account difference scheme of FDM constructing. Passing to evolutionary problem, we shall replace equilibrium equations (Eq. 4.67) by the equations coinciding in the form with the shell motion in viscous medium equations. Then evolutionary equations of establishing method will be written as follows

$$[L_{\lambda_1,\lambda_2}(U_k)]_{i,j} + (Q_k)_{i,j} = (c^* m_k \ddot{u}_k)_{i,j} + (c^* \varepsilon_k \dot{u}_k)_{i,j}, \tag{4.84}$$

where $\varepsilon_k(i,j)$ are parameters of the medium relative viscosity. For nonstationary equations (Eq. 4.84) solution explicit two-layer difference scheme in time of the second-order of accuracy with time derivatives approximation by difference operators of the shape (Eq. 4.69) is used. Equation 4.84 approximating on time mesh with the step Δt = const and considering mass characteristics parameters to be constant in the mesh domain $t^{(n)}$, equations (Eq. 4.84) can be represented in the form similar to Equation 4.75.

$$[L_{\lambda_1,\lambda_2}(U_k)+Q_k]_{i,j}^{(n)} = c_{i,j}^{*}\left\{(m_k)_{i,j}\frac{[\dot{u}_k]_{i,j}^{(n+1/2)}-[\dot{u}_k]_{i,j}^{(n-1/2)}}{\Delta t}+\frac{[\varepsilon_k\dot{u}_k]_{i,j}^{(n-1/2)}+[\varepsilon_k\dot{u}_k]_{i,j}^{(n+1/2)}}{2}\right\}. \tag{4.85}$$

From the equations (Eq. 4.85), it is simple to obtain expressions for velocities $\left[\dot{u}_k\right]_{i,j}^{(n+1/2)}$ at time layer $t^{(n+1/2)}$ in explicit form

$$\left[\dot{u}_k\right]_{i,j}^{(n+1/2)} = \frac{\left[2m_k-\varepsilon_k^{(n-1/2)}\Delta t\right]_{i,j}}{\left[2m_k+\varepsilon_k^{(n+1/2)}\Delta t\right]_{i,j}}\cdot\left[\dot{u}_k\right]_{i,j}^{(n-1/2)}+\frac{2\Delta t\left[L_{\lambda_1,\lambda_2}(U_k)+Q_k\right]_{i,j}^{(n)}}{c_{i,j}^{*}\left[2m_k+\varepsilon_k^{(n+1/2)}\Delta t\right]_{i,j}}. \tag{4.86}$$

Then carrying out integration from Equation 4.69, we determine mesh functions of generalized displacements $\left[u_k\right]_{i,j}^{(n+1)}$ at time layer $t^{(n+1)}$

$$\left[u_k\right]_{i,j}^{(n+1)} = \left[u_k\right]_{i,j}^{(n)}+\Delta t\cdot\left[\dot{u}_k\right]_{i,j}^{(n+1/2)}. \tag{4.87}$$

So difference approximation (Eq. 4.85) of nonstationary equations (Eq. 4.84) leads to iteration process (Eqs. 4.86 and 4.87) of finding solution of the original stationary problem (Eq. 4.67). The only condition set for quasi-dynamic system of equations is existence of finite limits for the displacements and angles of rotation (when solving the problem in displacements), satisfying the conditions (Eq. 4.80). Establishing method using in the form (Eq. 4.84) allows bringing solution of the original nonlinear static problem (Eq. 4.67) to the solution of quasi-dynamic one (Eq. 4.84), which significantly simplifies the construction and practical realization of calculation algorithm of static problem solution.

Program realization of quasi-dynamic form of establishing method on ECM is relatively simple. In accordance with the used VDS difference stencil of equilibrium equations (Eq. 4.67) left-hand sides calculation is built and then layer-by-layer cycle of velocities and displacements mesh functions calculation according to formulas (Eqs. 4.86 and 4.87) is organized for those node points of the mesh in which corresponding equilibrium equations must be satisfied. Calculation process starts from time point $t^{(0)} = 0$ at some initial (in the special case—zero) approximation for displacements and zero values of the velocities. Iteration process step is finished after satisfying given boundary conditions on the external and internal contours of the shell. Iteration process is carried out until reaching some given measure of establishing, when characteristic values of the shell stress-strain state stop varying within the limits of given accuracy, and maximum errors of equilibrium equations satisfaction, determined as residual of left-hand sides of equations (Eq. 4.84), become less than some given value δ. Sometimes some fixed number of iterations is made, and then the error of the solution obtained is analyzed. As it follows from iteration process equations (Eqs. 4.86 and 4.87), when using establishing method it is not necessary to form and store in ECM memory coefficients $[L_{\lambda_1,\lambda_2}]_{i,j}$ matrix, and consequently to carry out matrix operations usually used for the solution of algebraic equations systems, often leading, as it was mentioned, to errors in results caused by peculiarities of machine arithmetic. For ECM program realization of the method, it is enough to describe only the arrays of displacements and velocities, which are renewed at the following time layers $t^{(n+1/2)}$ and $t^{(n+1)}$ using formulas (Eqs. 4.86 and 4.87) as a result of step-by-step equations (Eq. 4.84) integration. It is reasonable to foresee the printout of some parameters of stress-strain state of the shell for one or several characteristic node points in a certain number of iterations during calculations—forces, moments, displacements, and velocities, as well as maximum and medium (over all the nodes) errors in satisfying equilibrium equations, maximum and medium displacements, and corresponding velocities. This will allow controlling convergence of iteration process and estimating the error of the solution obtained.

The establishing method allows constructing uniform iteration process for solving both linear and nonlinear boundary problems, as well as for solving equations of shells theory obtained on the basis of different hypotheses—both Kirchhoff-Love and Timoshenko-Reissner ones. The number of iterations required for obtaining static problem solution with maximum errors not larger than some percentages, depending on the degree of nonlinearity and physico-mechanical peculiarities of the shell—usually is $n(\delta) = (0{,}5 \div 10)$ K, where K is the general number of the unknown quantities. The experience in establishing method using showed its high efficiency and economic feasibility relative to computing time requirements [9–12].

Nonstationary equations (Eq. 4.84) are written in general quite formally. Time step, Δt, mass characteristics, m_k, and artificial viscosity, ε_k, parameters can be considered only as iteration parameters not having any physical sense, optimum values of which provide the convergence of iteration process (Eqs. 4.86 and 4.87). However, the taken form of the equations (Eq. 4.84) corresponds to some physical model of evolutionary process with energy dissipation and establishing in stationary state. Such a physical model using the method as its basis is an important factor because for nonlinear problems, unlike linear ones, in most cases establishing method is not based on the proofs of convergence and uniqueness of the solution [13]. At the same time, the establishing method in the form (Eqs. 4.84 through 4.87) is free of the main disadvantage of the majority of other nonlinear equations solution methods—high sensitivity to the choice of initial approximation, as introduction of damping terms ε_k leads to the method self-correcting. Evidently, the first work in which nonstationary method was used for solution of theory of shells static problems was V. I. Feodosiev's work published in 1963 [14]. It should be mentioned that when using iteration methods similar to quasi-dynamic form of establishing method considered here, the term "dynamic relaxation" is also used, although the authors that use such a term do not specify that it does not correspond to the essence of the method [15–17].

Quasi-dynamic form of the establishing method can also be realized in mathematical technologies of calculation algorithms construction when using discretization of the original continuous problem using FEM [17]. When constructing difference scheme using FEM nonstationary equations of establishing method in matrix form have the shape

$$M\ddot{U} + C\dot{U} + KU - F = 0, \tag{4.88}$$

where M is mass matrix, K is shell stiffness matrix, C is damping (viscous resistance) matrix, U is matrix of generalized displacements of finite-element model nodes, and F is loads matrix.

Let us denote $U^{(n-1)}, U^{(n)}, U^{(n+1)}$—values of generalized displacements matrix correspondingly at time points $(t-\Delta t), t, (t+\Delta t)$; and Δt is time step. Relating Equation 4.88 to time point, t, using FEM in shells calculations iteration process similar to Equations 4.86 and 4.87 can be represented as

$$\left(M + \frac{\Delta t}{2}C\right)U^{(n+1)} = \Delta t^2[F - KU^{(n)}] + M[2U^{(n)} - U^{(n-1)}] + \frac{\Delta t}{2}CU^{(n-1)}. \tag{4.89}$$

For iteration process (Eq. 4.84) realization, it is necessary to know $U^{(n-1)}$ at the initial time point $t = 0$ ($n = 0$). As initial values for the velocities $\dot{U}$ are considered to be zero, then $U^{(n-1)} = U^{(n+1)}$. Then the first step of iteration process will be written as follows

$$MU^{(n+1)} = \frac{\Delta t^2}{2}[F - KU^{(0)}] + MU^{(0)}, \tag{4.90}$$

Where $U^{(0)}$ is initial (in the special case—zero) approximation of the matrix of generalized displacements. Optimum values of time step, Δt, as well as matrix C elements, are determined similarly to Eq. 4.81.

4.15 Iteration Process Parameters

Iteration process parameters—medium relative viscosity, $\varepsilon_k(i,j)$, and time step, Δt—are determined from the condition of increasing convergence and stability of difference scheme. Formulas (Eq. 4.81) taking into account equations (Eq. 4.84) structure and supposing $[\varepsilon_k]_{i,j}^{(n)} = \varepsilon_k$, can be written in the form

$$\varepsilon_k = 2a_{\varepsilon,(k)}\sqrt{\frac{m_k \mu_{1,(k)}\mu_{2,(k)}}{\mu_{1,(k)} + \mu_{2,(k)}}}; \quad \Delta t_k = 2a_{t,(k)}\sqrt{\frac{m_k}{\mu_{1,(k)} + \mu_{2,(k)}}}, \tag{4.91}$$

where

$$\mu_{1,(k)} = \min_{i,j}[\mu_{1,(k)}]_{i,j}; \quad \mu_{2,(k)} = \max_{i,j}[\mu_{2,(k)}]_{i,j}, \tag{4.92}$$

and where $\mu_{1,(k)}$ and $\mu_{2,(k)}$ are minimum and maximum eigenvalues for corresponding difference operators in the equations (Eq. 4.84); $a_{\varepsilon,(k)}$ and $a_{t,(k)}$ are correction factors close to unity. Time step, Δt, for all the DS as a whole is determined from the condition

$$\Delta t = \min_k \Delta t_k. \tag{4.93}$$

For nonlinear problems, as well as for the problems with singularities and nonhomogeneities exact determination of difference operators spectra limits is connected with significant mathematical difficulties, so $\mu_{1,(k)}$ and $\mu_{2,(k)}$ are estimated within the framework of linear relations at corresponding simplifications if the original equations. Linearized equilibrium equations (Eq. 4.41) after neglecting unimportant terms can be represented in the form

$$\begin{gathered}
\frac{1}{A_1}\frac{\partial T_{11}}{\partial \alpha_1} + \frac{1}{A_2}\frac{\partial S}{\partial \alpha_2} + q_1 = 0; \\
\frac{1}{A_2}\frac{\partial T_{22}}{\partial \alpha_2} + \frac{1}{A_1}\frac{\partial S}{\partial \alpha_1} + q_2 = 0; \\
\frac{1}{A_1^2}\frac{\partial^2 M_{11}}{\partial \alpha_1^2} + \frac{1}{A_2^2}\frac{\partial^2 M_{22}}{\partial \alpha_2^2} + \frac{2}{A_1 A_2}\frac{\partial^2 H}{\partial \alpha_1 \partial \alpha_2} - k_1 T_{11} - k_2 T_{22} + q_3 = 0; \\
\frac{1}{A_1}\frac{\partial M_{11}}{\partial \alpha_1} + \frac{1}{A_2}\frac{\partial H}{\partial \alpha_2} - Q_{13} + m_1 = 0; \\
\frac{1}{A_2}\frac{\partial M_{22}}{\partial \alpha_2} + \frac{1}{A_1}\frac{\partial H}{\partial \alpha_1} - Q_{23} + m_2 = 0.
\end{gathered} \tag{4.94}$$

Without loss of generality, we set $q_1 = \text{const}$; $q_2 = \text{const}$; $q_3 = \text{const}$; $m_1 = \text{const}$; and $m_2 = \text{const}$. Taking into account geometry and physical relations (Eqs. 4.7 through 4.24), we shall represent the equations (Eq. 4.94) in functions of generalized displacements, leaving only corresponding components of solutions u_k in each of them

$$\begin{gathered}
-\frac{B_{11}}{A_1^2}\frac{\partial^2 u}{\partial \alpha_1^2} - \frac{B_{33}}{A_2^2}\frac{\partial^2 u}{\partial \alpha_2^2} = q_1; \\
-\frac{B_{22}}{A_2^2}\frac{\partial^2 v}{\partial \alpha_2^2} - \frac{B_{33}}{A_1^2}\frac{\partial^2 v}{\partial \alpha_1^2} = q_2; \\
\frac{D_{11}}{A_1^4}\frac{\partial^4 w}{\partial \alpha_1^4} + \frac{D_{22}}{A_2^4}\frac{\partial^4 w}{\partial \alpha_2^4} + \frac{2D_{12} + 4D_{33}}{A_1^2 A_2^2} \cdot \frac{\partial^4 w}{\partial \alpha_1^2 \partial \alpha_2^2} +
\end{gathered}$$

$$+(k_1^2B_{11}+k_1k_2B_{12}+k_1k_2B_{21}+k_2^2B_{22})w=q_3; \tag{4.95}$$

$$-\frac{D_{11}}{A_1^2}\frac{\partial^2\gamma_1}{\partial\alpha_1^2}-\frac{D_{33}}{A_2^2}\frac{\partial^2\gamma_1}{\partial\alpha_2^2}=m_1;$$

$$-\frac{D_{22}}{A_2^2}\frac{\partial^2\gamma_2}{\partial\alpha_2^2}-\frac{D_{33}}{A_1^2}\frac{\partial\gamma_2}{\partial\alpha_1^2}=m_2.$$

Approximating equations (Eq. 4.95) on uniform mesh with steps $h_x = A_1\lambda_1$, $h_y = A_2\lambda_2$ in coordinate directions $x = \alpha_1$, $y = \alpha_2$, mesh analogues of the equations (Eq. 4.95) can be represented in canonic form [6,8]

$$-B_{11}\frac{u_{i-1,j}-2u_{i,j}+u_{i+1,j}}{h_x^2}-B_{33}\frac{u_{i,j+1}-2u_{i,j}+u_{i,j-1}}{h_y^2}=q_1;$$

$$-B_{22}\frac{v_{i,j+1}-2v_{i,j}+v_{i,j-1}}{h_y^2}-B_{33}\frac{v_{i-1,j}-2v_{i,j}+v_{i+1,j}}{h_x^2}=q_2;$$

$$D_{11}\frac{w_{i-2,j}-4w_{i-1,j}+6w_{i,j}-4w_{i+1,j}+w_{i+2,j}}{h_x^4}+$$

$$+D_{22}\frac{w_{i,j-2}-4w_{i,j-1}+6w_{i,j}-4w_{i,j+1}+w_{i,j+2}}{h_y^4}+$$

$$+D_{13}\frac{w_{i+1,j+1}-2w_{i,j+1}+w_{i-1,j+1}-2w_{i+1,j}+4w_{i,j}-2w_{i-1,j}+w_{i+1,j-1}-2w_{i,j-1}+w_{i-1,j-1}}{h_x^2h_y^2}+$$

$$+D_{23}\cdot w_{i,j}=q_3; \tag{4.96}$$

$$-D_{11}\frac{(\gamma_1)_{i-1,j}-2(\gamma_1)_{i,j}+(\gamma_1)_{i+1,j}}{h_x^2}-D_{33}\frac{(\gamma_1)_{i,j-1}-2(\gamma_1)_{i,j}+(\gamma_1)_{i,j+1}}{h_y^2}=m_1;$$

$$-D_{22}\frac{(\gamma_2)_{i,j-1}-2(\gamma_2)_{i,j}+(\gamma_2)_{i,j+1}}{h_y^2}-D_{33}\frac{(\gamma_2)_{i-1,j}-2(\gamma_2)_{i,j}+(\gamma_2)_{i+1,j}}{h_x^2}=m_2,$$

where $i = 0,1,2,\ldots,N$; $j = 0,1,2,\ldots,M$; $D_{13} = 2D_{12}+4D_{33}$; $D_{23} = k_1^2B_{11} + 2k_1k_2B_{12} + k_2^2B_{22}$.

Mesh parameters λ_1,λ_2 are determined by the formulas (Eq. 4.56). We suppose that basic surface of the shell when depicting it on the plane of principal coordinates α_1,α_2 is restricted by the area in the shape of rectangle with the sides Λ_1,Λ_2 at homogeneous boundary conditions on the boundaries $x = 0;X$ and $y = 0;Y$ of the form

$$u\big|_{x=0}=u\big|_{x=X}=0;\quad v\big|_{x=0}=v\big|_{x=X}=0;$$

$$w\big|_{x=0} = w\big|_{x=X} = \left.\frac{\partial^2 w}{\partial x^2}\right|_{x=0} = \left.\frac{\partial^2 w}{\partial x^2}\right|_{x=X} = 0; \tag{4.97}$$

$$\gamma_1\big|_{x=0} = \gamma_1\big|_{x=X} = 0; \quad \gamma_2\big|_{x=0} = \gamma_2\big|_{x=X} = 0,$$

and

$$u\big|_{y=0} = u\big|_{y=Y} = 0; \quad v\big|_{y=0} = v\big|_{y=Y} = 0;$$

$$w\big|_{y=0} = w\big|_{y=Y} = \left.\frac{\partial^2 w}{\partial y^2}\right|_{y=0} = \left.\frac{\partial^2 w}{\partial y^2}\right|_{y=Y} = 0; \tag{4.98}$$

$$\gamma_1\big|_{y=0} = \gamma_1\big|_{y=Y} = 0; \gamma_2\big|_{y=0} = \gamma_2\big|_{y=Y} = 0.$$

Let us determine eigenvalues λ_{nm} and corresponding eigenvectors $(u_k)_{nm}$ of difference operators by the example of the first end third mesh equations of finite-difference scheme (Eq. 4.96). The problem of finding difference operators spectra is connected with determination of eigenvalues and eigenvectors of the corresponding matrix A (i.e., such values of λ), for which different from zero solutions of the following equation system exist:

$$Au = \lambda u. \tag{4.99}$$

Transforming the first equations of the system (Eq. 4.96) taking into account Equation 4.99, we have

$$B_{11}\frac{u_{i-1,j} - 2u_{i,j} + u_{i+1,j}}{h_x^2} + B_{33}\frac{u_{i,j+1} - 2u_{i,j} + u_{i,j-1}}{h_y^2} + \lambda u_{i,j} = 0, \tag{4.100}$$

or after corresponding transformations

$$\begin{aligned} &B_{11}h_y^2(u_{i-1,j} + u_{i+1,j}) + B_{33}h_x^2(u_{i,j+1} + u_{i,j-1}) - \\ &-2(B_{11}h_y^2 + B_{33}h_x^2 - 0,5h_x^2h_y^2 \cdot \lambda)u_{i,j} = 0. \end{aligned} \tag{4.101}$$

The solution of the problem (Eq. 4.101) is sought in the form

$$u_{i,j} = \sin\alpha x_i \cdot \sin\beta y_j, \tag{4.102}$$

where α and β are some constants determined below. Substituting Equation 4.102 in Equation 4.101, we have

$$\begin{aligned} &B_{11}h_y^2\left[\sin\alpha(x_i - h_x) + \sin\alpha(x_i + h_x)\right]\sin\beta y_j + \\ &+B_{33}h_x^2\left[\sin\beta(y_j - h_y) + \sin\beta(y_j + h_y)\right]\sin\alpha x_i - \\ &-2\left(B_{11}h_y^2 + B_{33}h_x^2 - 0,5h_x^2h_y^2 \cdot \lambda\right)\sin\alpha x_i \cdot \sin\beta y_j = 0. \end{aligned} \tag{4.103}$$

Using known trigonometric relations

$$\begin{aligned} &\sin\alpha\left(x_i - h_x\right) + \sin\alpha\left(x_i + h_x\right) = 2\sin\alpha x_i \cdot \cos\alpha h_x; \\ &\sin\beta\left(y_j - h_y\right) + \sin\beta\left(y_j + h_y\right) = 2\sin\beta y_j \cdot \cos\beta h_y, \end{aligned} \tag{4.104}$$

Equation (4.103) can be represented in the form

$$\begin{aligned} &2B_{11}h_y^2\cos\alpha h_x \cdot \sin\alpha x_i \cdot \sin\beta y_j + 2B_{33}h_x^2\cos\beta h_y \cdot \sin\alpha x_i \cdot \sin\beta y_j - \\ &-2\left(B_{11}h_y^2 + B_{33}h_x^2 - 0,5h_x^2h_y^2 \cdot \lambda\right)\sin\alpha x_i \cdot \sin\beta y_j = 0. \end{aligned} \tag{4.105}$$

Taking into account representation (Eq. 4.102), finally we have

$$2\left[B_{11}h_y^2\cos\alpha h_x + B_{33}h_x^2\cos\beta h_y - \left(B_{11}h_y^2 + B_{33}h_x^2 - 0,5h_x^2h_y^2 \cdot \lambda\right)\right]u_{i,j} = 0. \tag{4.106}$$

As only non-zero solutions of the problem (Eq. 4.100) with boundary conditions (Eqs. 4.97 and 4.98) are of interest, from Equation 4.106, it follows

$$B_{11}h_y^2\cos\alpha h_x + B_{33}h_x^2\cos\beta h_y - B_{11}h_y^2 - B_{33}h_x^2 + 0,5h_x^2h_y^2 \cdot \lambda = 0. \tag{4.107}$$

From the last equation, it is easy to obtain

$$\lambda = \frac{2B_{11}}{h_x^2}\left(1 - \cos\alpha h_x\right) + \frac{2B_{33}}{h_y^2}\left(1 - \cos\beta h_y\right) \tag{4.108}$$

or

$$\lambda = \frac{4B_{11}}{h_x^2}\sin^2\frac{\alpha h_x}{2} + \frac{4B_{33}}{h_y^2}\sin^2\frac{\beta h_y}{2}. \tag{4.109}$$

The values of constants α and β are chosen so that the function $u_{i,j}$ in Equation 4.102 satisfied corresponding boundary conditions in Equations 4.97 and 4.98, from which it follows

$$\alpha = \frac{n\pi}{X}; \quad \beta = \frac{m\pi}{Y}, \tag{4.110}$$

where $n = 1,2,\ldots,N-1$; $m = 1,2,\ldots,M-1$. So we obtain $(N-1) \times (M-1)$ eigenvalues λ_{nm} and corresponding eigenvectors u_{nm}

$$\begin{aligned} u_{nm} &= \sin\frac{n\pi x}{X}\sin\frac{m\pi y}{Y}; \\ \lambda_{nm} &= \frac{4B_{11}}{h_x^2}\sin^2\frac{n\pi h_x}{2X} + \frac{4B_{33}}{h_y^2}\sin^2\frac{m\pi h_y}{2Y}. \end{aligned} \tag{4.111}$$

From Equation 4.111, it is easy to obtain the estimation of eigenvalues spectrum boundaries

$$0 < \lambda_1 \le \lambda_{nm} \le \lambda_2, \tag{4.112}$$

where $\lambda_1 = \lambda_{\min}$, $\lambda_2 = \lambda_{\max}$ are minimum and maximum eigenvalues of mesh operator

$$\lambda_1 = \frac{4B_{11}}{h_x^2}\sin^2\frac{\pi h_x}{2X} + \frac{4B_{33}}{h_y^2}\sin^2\frac{\pi h_y}{2Y};$$

$$\begin{aligned} \lambda_2 &= \frac{4B_{11}}{h_x^2}\sin^2\frac{(N-1)\pi h_x}{2X} + \frac{4B_{33}}{h_y^2}\sin^2\frac{(M-1)\pi h_y}{2Y} \\ &= \frac{4B_{11}}{h_x^2}\cos^2\frac{\pi h_x}{2X} + \frac{4B_{33}}{h_y^2}\cos^2\frac{\pi h_y}{2Y}. \end{aligned} \tag{4.113}$$

In particular, from Equation 4.113, it follows that all eigenvalues of the problem (Eq. 4.100) are positive. We shall bring the third equation of finite-difference system (Eq. 4.96) to canonical form

$$\begin{aligned} &\frac{D_{11}}{A_1^4}\frac{\partial^4 w}{\partial \alpha_1^4} + \frac{D_{22}}{A_2^4}\frac{\partial^4 w}{\partial \alpha_2^4} + \frac{2D_{12}+4D_{33}}{A_1^2A_2^2}\cdot\frac{\partial^4 w}{\partial \alpha_1^2 \partial \alpha_2^2} + \\ &+(k_1^2B_{11} + k_1k_2B_{12} + k_1k_2B_{21} + k_2^2B_{22})w = q_3. \end{aligned} \tag{4.114}$$

The problem of the corresponding finite-difference problem eigenvalues and eigenvectors determination is brought to the solution of the equation

$$\begin{aligned} &D_{11}\frac{w_{i-2,j} - 4w_{i-1,j} + 6w_{i,j} - 4w_{i+1,j} + w_{i+2,j}}{h_x^4} + \\ &+D_{22}\frac{w_{i,j-2} - 4w_{i,j-1} + 6w_{i,j} - 4w_{i,j+1} + w_{i,j+2}}{h_y^4} + \\ &+D_{13}\frac{w_{i+1,j+1} - 2w_{i,j+1} + w_{i-1,j+1} - 2w_{i+1,j} + 4w_{i,j} - 2w_{i-1,j} + w_{i+1,j-1} - 2w_{i,j-1} + w_{i-1,j-1}}{h_x^2h_y^2} + \\ &+D_{23}\cdot w_{i,j} - \lambda\cdot w_{i,j} = 0, \end{aligned} \tag{4.115}$$

at boundary conditions (Eqs. 4.97 and 4.98). Equation 4.115 solution is sought in the form

$$w_{i,j} = \sin\alpha x_i \cdot \sin\beta y_j, \tag{4.116}$$

from which it follows

$$D_{11}h_y^4\left[\sin\alpha(x_i - 2h_x) + \sin\alpha(x_i + 2h_x)\right]\sin\beta y_j -$$

$$-4D_{11}h_y^4\left[\sin\alpha(x_i - h_x) + \sin\alpha(x_i + h_x)\right]\sin\beta y_j + 6D_{11}h_y^4\cdot\sin\alpha x_i\sin\beta y_j +$$

$$+D_{22}h_x^4\left[\sin\beta(y_j - 2h_y) + \sin\beta(y_j + 2h_y)\right]\sin\alpha x_i -$$

$$-4D_{22}h_x^4\left[\sin\beta\left(y_j-h_y\right)+\sin\beta\left(y_j+h_y\right)\right]\sin\alpha x_i+6D_{22}h_x^4\sin\alpha x_i\sin\beta y_j+$$

$$+D_{13}h_x^2h_y^2[\sin\alpha(x_i+h_x)\sin\beta(y_j+h_y)+\sin\alpha(x_i-h_x)\sin\beta(y_j+h_y)+$$

$$+\sin\alpha(x_i+h_x)\sin\beta(y_j-h_y)+\sin\alpha(x_i-h_x)\sin\beta(y_j-h_y)]-$$

$$-2D_{13}h_x^2h_y^2[\sin\alpha(x_i+h_x)+\sin\alpha(x_i-h_x)]\sin\beta y_j-$$

$$-2D_{13}h_x^2h_y^2[\sin\beta(y_j-h_y)+\sin\beta(y_j+h_y)]\sin\alpha x_i+4D_{13}h_x^2h_y^2\sin\alpha x_i\sin\beta y_j+$$

$$+D_{23}h_x^4h_y^4\sin\alpha x_i\sin\beta y_j-\lambda\cdot h_x^4h_y^4\sin\alpha x_i\sin\beta y_j=0. \tag{4.117}$$

After corresponding transformations using trigonometric relations (Eq. 4.104) and expression (Eq. 4.116) we have

$$2D_{11}h_y^4\cos 2\alpha h_x\cdot w_{i,j}-8D_{11}h_y^4\cos\alpha h_x\cdot w_{i,j}+6D_{11}h_y^4\cdot w_{i,j}+$$

$$+2D_{22}h_x^4\cos 2\beta h_y\cdot w_{i,j}-8D_{22}h_x^4\cos\beta h_y\cdot w_{i,j}+6D_{22}h_x^4\cdot w_{i,j}+$$

$$+4D_{13}h_x^2h_y^2\cos\alpha h_x\cos\beta h_y\cdot w_{i,j}-$$

$$-4D_{13}h_x^2h_y^2\cos\alpha h_x\cdot w_{i,j}-4D_{13}h_x^2h_y^2\cos\beta h_y\cdot w_{i,j}+$$

$$+4D_{13}h_x^2h_y^2\cdot w_{i,j}+D_{23}h_x^4h_y^4\cdot w_{i,j}-\lambda\cdot h_x^4h_y^4\cdot w_{i,j}=0. \tag{4.118}$$

As we are seeking non-zero solution of the problem (Eq. 4.115), we have

$$2D_{11}h_y^4\cos 2\alpha h_x-8D_{11}h_y^4\cos\alpha h_x+6D_{11}h_y^4+$$

$$+2D_{22}h_x^4\cos 2\beta h_y-8D_{22}h_x^4\cos\beta h_y+6D_{22}h_x^4+$$

$$+4D_{13}h_x^2h_y^2\cos\alpha h_x\cos\beta h_y-4D_{13}h_x^2h_y^2\cos\alpha h_x-4D_{13}h_x^2h_y^2\cos\beta h_y+$$

$$+4D_{13}h_x^2h_y^2+D_{23}h_x^4h_y^4-\lambda\cdot h_x^4h_y^4=0. \tag{4.119}$$

Using relations $\cos 2\xi = 1-2\sin^2\xi$; $\sin 2\xi = 2\sin\xi\cos\xi$, Equation 4.119 can be transformed to the form

$$16D_{11}h_y^4\sin^2\frac{\alpha h_x}{2}-16D_{11}h_y^4\sin^2\frac{\alpha h_x}{2}\cdot\cos^2\frac{\alpha h_x}{2}+$$

$$+16D_{22}h_x^4\sin^2\frac{\beta h_y}{2}-16D_{22}h_x^4\sin^2\frac{\beta h_y}{2}\cdot\cos^2\frac{\beta h_y}{2}+$$

$$+16D_{13}h_x^2h_y^2\sin^2\frac{\alpha h_x}{2}\sin^2\frac{\beta h_y}{2}+D_{23}h_x^4h_y^4-\lambda\cdot h_x^4h_y^4=0. \tag{4.120}$$

From the Equation 4.120, we obtain

$$\lambda = 16\frac{D_{11}}{h_x^4}\sin^4\frac{\alpha h_x}{2} + 16\frac{D_{22}}{h_y^4}\sin^4\frac{\beta h_y}{2} + 16\left(\frac{2D_{12}+4D_{33}}{h_x^2 h_y^2}\right)\sin^2\frac{\alpha h_x}{2}\sin^2\frac{\beta h_y}{2} +$$

$$+k_1^2 B_{11} + 2k_1 k_2 B_{12} + k_2^2 B_{22}. \quad (4.121)$$

As boundary conditions for $w_{i,j}$ in Equations 4.97 and 4.98 are also satisfied at the values (Eq. 4.110) of parameters α and β (Eq. 4.110), finally we have the following estimation for eigenvalues λ_{nm} and eigenvectors w_{nm}

$$w_{nm} = \sin\frac{n\pi x}{X}\sin\frac{m\pi y}{Y};$$

$$\lambda_{nm} = \frac{16D_{11}}{h_x^4}\sin^4\frac{n\pi h_x}{2X} + \frac{16D_{22}}{h_y^4}\sin^4\frac{m\pi h_y}{2Y} + \frac{32D_{12}+64D_{33}}{h_x^2 h_y^2}\sin^2\frac{n\pi h_x}{2X}\sin^2\frac{m\pi h_y}{2Y} +$$

$$+k_1^2 B_{11} + 2k_1 k_2 B_{12} + k_2^2 B_{22}. \quad (4.122)$$

So after corresponding transformations for other equations of finite-difference system (Eq. 4.96) at boundary conditions (Eqs. 4.97 and 4.98) evaluation formulas for $\mu_{1,(k)}$ and $\mu_{2,(k)}$ can be represented in the following way:

- the smallest eigenvalues

$$\mu_{1,(1)} = 4\left[\frac{B_{11}}{(A_1\lambda_1)^2}\cdot\sin^2\frac{\pi}{2}\frac{\lambda_1}{\Lambda_1} + \frac{B_{33}}{(A_2\lambda_2)^2}\cdot\sin^2\frac{\pi}{2}\frac{\lambda_2}{\Lambda_2}\right];$$

$$\mu_{1,(2)} = 4\left[\frac{B_{22}}{(A_2\lambda_2)^2}\cdot\sin^2\frac{\pi}{2}\frac{\lambda_2}{\Lambda_2} + \frac{B_{33}}{(A_1\lambda_1)^2}\cdot\sin^2\frac{\pi}{2}\frac{\lambda_1}{\Lambda_1}\right];$$

$$\mu_{1,(3)} = 16\left[\frac{D_{11}}{(A_1\lambda_1)^4}\sin^4\frac{\pi}{2}\frac{\lambda_1}{\Lambda_1} + \frac{D_{22}}{(A_2\lambda_2)^4}\sin^4\frac{\pi}{2}\frac{\lambda_2}{\Lambda_2} +\right.$$

$$\left.+\frac{2D_{12}+4D_{33}}{(A_1\lambda_1)^2(A_2\lambda_2)^2}\sin^2\frac{\pi}{2}\frac{\lambda_1}{\Lambda_1}\sin^2\frac{\pi}{2}\frac{\lambda_2}{\Lambda_2}\right] + k_1^2 B_{11} + 2k_1 k_2 B_{12} + k_2^2 B_{22};$$

$$\mu_{1,(4)} = 4\left[\frac{D_{11}}{(A_1\lambda_1)^2}\cdot\sin^2\frac{\pi}{2}\frac{\lambda_1}{\Lambda_1} + \frac{D_{33}}{(A_2\lambda_2)^2}\cdot\sin^2\frac{\pi}{2}\frac{\lambda_2}{\Lambda_2}\right];$$

$$\mu_{1,(5)} = 4\left[\frac{D_{22}}{(A_2\lambda_2)^2}\cdot\sin^2\frac{\pi}{2}\frac{\lambda_2}{\Lambda_2} + \frac{D_{33}}{(A_1\lambda_1)^2}\cdot\sin^2\frac{\pi}{2}\frac{\lambda_1}{\Lambda_1}\right]; \quad (4.123)$$

- the largest eigenvalues

$$\mu_{2,(1)} = 4\left[\frac{B_{11}}{(A_1\lambda_1)^2}\cdot\cos^2\frac{\pi}{2}\frac{\lambda_1}{\Lambda_1}+\frac{B_{33}}{(A_2\lambda_2)^2}\cdot\cos^2\frac{\pi}{2}\frac{\lambda_2}{\Lambda_2}\right];$$

$$\mu_{2,(2)} = 4\left[\frac{B_{22}}{(A_2\lambda_2)^2}\cdot\cos^2\frac{\pi}{2}\frac{\lambda_2}{\Lambda_2}+\frac{B_{33}}{(A_1\lambda_1)^2}\cdot\cos^2\frac{\pi}{2}\frac{\lambda_1}{\Lambda_1}\right];$$

$$\mu_{2,(3)} = 16\left[\frac{D_{11}}{(A_1\lambda_1)^4}\cdot\cos^4\frac{\pi}{2}\frac{\lambda_1}{\Lambda_1}+\frac{D_{22}}{(A_2\lambda_2)^4}\cdot\cos^4\frac{\pi}{2}\frac{\lambda_2}{\Lambda_2}+\right.$$

$$\left.+\frac{2D_{12}+4D_{33}}{(A_1\lambda_1)^2(A_2\lambda_2)^2}\cos^2\frac{\pi}{2}\frac{\lambda_1}{\Lambda_1}\cos^2\frac{\pi}{2}\frac{\lambda_2}{\Lambda_2}\right]+k_1^2B_{11}+2k_1k_2B_{12}+k_2^2B_{22};$$

$$\mu_{2,(4)} = 4\left[\frac{D_{11}}{(A_1\lambda_1)^2}\cdot\cos^2\frac{\pi}{2}\frac{\lambda_1}{\Lambda_1}+\frac{D_{33}}{(A_2\lambda_2)^2}\cdot\cos^2\frac{\pi}{2}\frac{\lambda_2}{\Lambda_2}\right];$$

$$\mu_{2,(5)} = 4\left[\frac{D_{22}}{(A_2\lambda_2)^2}\cdot\cos^2\frac{\pi}{2}\frac{\lambda_2}{\Lambda_2}+\frac{D_{33}}{(A_1\lambda_1)^2}\cdot\cos^2\frac{\pi}{2}\frac{\lambda_1}{\Lambda_1}\right], \tag{4.124}$$

where k_1,k_2 and A_1,A_2 are characteristic values of principal curvatures and Lame parameters of the shell. As it follows from the estimation (Eq. 4.112), the matrix of finite-difference system (Eq. 4.96) is positively definite and has non-zero determinant. As the matrix is non-singular, it has inverse matrix and consequently equations system (Eq. 4.96) at boundary conditions (Eqs. 4.97 and 4.98) has unique solution [6,8].

Artificial viscosity parameters ε_k significantly depend on boundary conditions on external and internal boundaries of the shell and weakly depend on mesh parameters. Critical value of time step, Δt, satisfying DS stability condition, significantly depends on mesh parameters and weakly depends on boundary conditions on shell boundaries. For decreasing computing time, the accepted magnitude of time step, Δt, must be fairly close to critical magnitude Δt. For nonlinear equations, Δt magnitude providing stability of iteration process is smaller than that for linear problem, moreover, Δt decreasing is the more significant the larger the contribution of nonlinear terms in the solution is. Because of the exponential character of errors decreasing, evolutionary process establishing takes place quite quickly. For obtaining stationary problem solution with equal errors $\delta_k^{(n)} = \delta$ for different solution components u_k iteration process must be carried out for a number of steps $n_k(\delta)$ up to time point $t_k^{(n)}(\delta)$, which can be estimated lower taking into account (91) at corresponding for each mesh equation (Eq. 4.67) optimum values of iteration parameters as

$$n_k(\delta) \cong \frac{1}{2}\sqrt{\frac{\mu_{2,(k)}}{\mu_{1,(k)}}}\ln\frac{1}{\delta}; \quad t_k^{(n)}(\delta) \cong \sqrt{\frac{m_k}{\mu_{1,(k)}}}\ln\frac{1}{\delta}. \tag{4.125}$$

Computing time is determined by the number of iterations for all the DS as a whole $n(\delta) \geq n_{max}$, necessary for all the components establishing with errors $\delta_k^{(n)} \leq \delta$, as well as corresponding establishing time $t^{(n)}(\delta) \geq t_{max}$, where

$$n_{\max} = \max_k n_k(\delta); \quad t_{\max} = \max_k t_k^{(n)}(\delta); \quad n_{\min} = \min_k n_k(\delta);$$
$$t_{\min} = \min_k t_k^{(n)}(\delta); \quad \Delta t_{\min} = \min_k \Delta t_k; \quad \Delta t_{\max} = \max_k \Delta t_k, \tag{4.126}$$

and where

$$t_k^{(n)}(\delta) = \Delta t_k \cdot n_k(\delta); \quad t^{(n)}(\delta) = \Delta t \cdot n(\delta). \tag{4.127}$$

The number of iterations $n_k(\delta)$ is determined by the relation $\mu_{2,(k)}/\mu_{1,(k)}$, and so it cannot be decreased by means of some optimization of iteration process parameters. As it follows from the formulas (Eqs. 4.91, and 4.123 through 4.125), establishing time $t^{(n)}(\delta)$ is determined by the time of low-frequency component of error damping, for which $t^{(n)}(\delta) > t_{min}$ and $n_k(\delta) < n_{max}$. High-frequency components of error, for which $t^{(n)}(\delta) < t_{max}$, are damped quickly, but corresponding values $n_k(\delta)$ are maximum because of large values of relations $\mu_{2,(k)}/\mu_{1,(k)}$: $n_k(\delta) > n_{min}$. Usually when calculating plates and shells $\min(\Delta x;\Delta y) > h$, so low-frequency component in Equation 4.67, for which $t_k > \Delta t_{min}$ corresponds to equilibrium equation in normal projection, whereas for high-frequency components $\Delta t_k < \Delta t_{max}$ ($k \neq 3$).

When carrying out iteration process up to time point $t^{(n)}(\delta) = t_{max}$ with time step $\Delta t_k = \Delta t_{min}$ the number of iterations for low-frequency component in Equation 4.127 rises from $n_k(\delta)$ to $\tilde{n}_{(k)}(\delta)$, which can be estimated taking into account (Eq. 4.126) as

$$\tilde{n}_k(\delta) \cong \frac{1}{2}\sqrt{\frac{\mu_{\max}}{\mu_{\min}}} \ln \frac{1}{\delta}, \tag{4.128}$$

where

$$\mu_{\min} = \min_k [m_k / \mu_{2,(k)}]; \quad \mu_{\max} = \max_k [m_k / \mu_{1,(k)}]. \tag{4.129}$$

Comparing Equations 4.125 and 4.128, it is easy to see that $\tilde{n}_{(k)}(\delta) > n_{max}$; so for all the DS, $n(\delta) \geq \tilde{n}_{(k)}(\delta)$. So using in Equation 4.127 for all the DS as a whole time step, $\Delta t < \Delta t_{max}$, leads to the number of iterations increasing and, consequently, to increasing of computing time by a factor of $\sim\tilde{n}_{(k)}(\delta)/n_{max}$.

Unlike dynamic problems, when calculating establishing only finite result (steady state) is of interest, and intermediate solutions have no sense. So iteration parameter Δt can have relatively large values for providing minimum number of iterations and at the same time satisfying DS stability condition. The value of density, ρ, used in static problem (Eq. 4.67) solution iteration process does not have such a strict physical sense as for dynamic problems (Eq. 4.75), and parameters m_k can be taken on accounts of convenience and for increasing the critical value of Δt. Let us represent non-stationary equations (Eq. 4.84) in the form

$$[L_{\lambda_1,\lambda_2}(U_k)+Q_k]_{i,j}=(c^* m_k^* \ddot{u}_k)_{i,j}+(c^* \varepsilon_k \dot{u}_k)_{i,j}, \tag{4.130}$$

where $m_k^* = a_k m_k$. Coefficients $a_k \geq 1$ using allows "filtering" high-frequency error components, significantly increasing critical value of Δt for all the PC and decreasing computing time. Formulas (Eqs. 4.83, 4.86, and 4.91) are transformed taking into account (Eq. 4.130) replacing m_k by m_k^*. Coefficients a_k values can be estimated from the condition $\Delta t_k = \Delta t_{\max}$ as

$$a_k = \left(\frac{\Delta t_{\max}}{\Delta t_k}\right)^2. \tag{4.131}$$

Introducing coefficients a_k can be interpreted as introducing dummy densities $\rho_k = a_k\rho$. As it follows from Equation 4.127, the number of iterations $n_k(\delta)$ does not depend on m_k.

When solving physically nonlinear problems described by relations (Eqs. 4.30 through 4.36), difficulties connected with compression deformation E_{33} calculation in plastic zone emerge, as formula (Eq. 4.34) for E_{33}^{pl} contains secant modulus E_c, which, as it follows from Equations 4.34 and 4.36, depends on E_{33}^{pl} itself. Usually for E_{33}^{pl} determination some internal iteration process is constructed, which is not connected with the method used for the general equations solution [2]. In this case, as we carry out calculations of establishing,

$$[u_k]_{i,j}^{(n)} \to [u_k]_{i,j}^{(n+1)}, \tag{4.132}$$

and consequently: $[e_i]_{i,j,l}^{(n)} \to [e_i]_{i,j,l}^{(n+1)}$, where $[e_i]_{i,j,l}^{(n)}$ is mesh function of strain intensity. So when using establishing method for the general equations solution and calculating E_{33}^{pl} at $(n+1)$th step the value $[e_i]_{i,j,l}^{(n)}$ taken from the previous nth step of iteration process can be used, and it is not necessary to construct special iteration process. When carrying out program realization of two-dimensional elasto-plastic problems numerical solutions, we have to introduce not only two-dimensional arrays of generalized displacements and velocities mesh functions but also take into account unloading (Eq. 4.35)—three-dimensional arrays of strains, stresses and strain intensities mesh functions (i.e., to store information concerning stress-strain state parameters for all the volume occupied by the shell). This leads to great memory requirements and computing time increasing. However the main part of computing time is taken by exchange processes connected with seeking and handling elements of multidimensional arrays and not by mathematical operations themselves. Replacing three-dimensional arrays by two-dimensional ones basing on original technologies of algorithm construction allows, as the results of investigations specially carried out showed, decreasing computing time more than twice and significantly reducing required storage space.

4.16 Nonstationary Problems Solution Calculation Technologies

For solution of shell structure equations of motion (Eq. 4.75), an explicit two-layer difference scheme in time is used, similar to Equations 4.85 through 4.87, from which

$$\left[\dot{u}_k\right]_{i,j}^{(n+1/2)} = \frac{\left[f_{11}^{*}c^{*}m_k\right]_{i,j}^{(n-1/2)}}{\left[f_{22}^{*}c^{*}m_k\right]_{i,j}^{(n+1/2)}}\left[\dot{u}_k\right]_{i,j}^{(n-1/2)} + \frac{\Delta t\left\{f^{*}\left[L_{\lambda_1,\lambda_2}(U_k)+Q_k\right]\right\}_{i,j}^{(n)}}{\left[f_{22}^{*}c^{*}m_k\right]_{i,j}^{(n+1/2)}};$$

$$\left[u_k\right]_{i,j}^{(n+1)} = \left[u_k\right]_{i,j}^{(n)} + \Delta t\left[\dot{u}_k\right]_{i,j}^{(n+1/2)}. \qquad (4.133)$$

Dynamic processes in shell structures occur with energy dissipation, which is caused by the work of external and internal friction forces. Degree of oscillations damping is usually estimated by the values of oscillations logarithmic decrement, δ [8]. At corresponding values parameters of artificial viscosity, ε_k, in nonstationary equations (Eqs. 4.84 through 4.86) can be used for taking into account (integrally) energy dissipation in dynamic problems. For determining the correlation between parameters of artificial viscosity, ε_k, and oscillations logarithmic decrement, δ, let us consider the problem concerning one-mass system oscillations, which is described by well-known equation in the form

$$\ddot{x} + 2n\dot{x} + \omega^2 x = 0, \qquad (4.134)$$

where x is generalized displacement, $2n = \varepsilon/m$, $\omega^2 = c/m$, m is mass parameter, ε is viscous drag coefficient, and parameter $c > 0$. For the cases of "small" drag ($n < \omega$), when oscillations are of evanescent character, oscillations logarithmic decrement, δ, is determined as: $\delta = n\tau_1$, where τ_1 is period of damped periodic vibration: $\tau_1 = \tau[1 + (\delta/2\pi)^2]^{1/2}$, τ is period of free vibration: $\tau = 2\pi/\omega$. Approximately setting $\tau_1 \cong \tau$, we can obtain the estimation for dynamic viscosity ε_{dyn}

$$\varepsilon_{\text{dyn}} = 2\delta\frac{m}{\tau}. \qquad (4.135)$$

For the case of limit aperiodic mode ($n = \omega$), which is optimal from the point of view of static problem solution using establishing method, the parameter of artificial viscosity, $\varepsilon_{\text{stat}}$, is determined as

$$\varepsilon_{\text{stat}} = 4\pi \frac{m}{\tau}. \tag{4.136}$$

Then

$$\varepsilon_{\text{dyn}} = \varepsilon_{cm} \frac{\delta}{2\pi}. \tag{4.137}$$

So taking into account formulas (Eqs. 4.91 and 4.137), for nonstationary problems parameters of artificial viscosity, ε_k, corresponding to the given value of δ_k, can be estimated as

$$\varepsilon_k = a_{\varepsilon,(k)} \frac{\delta_k}{\pi} \sqrt{\frac{m_k \mu_{1,(k)} \mu_{2,(k)}}{\mu_{1,(k)} + \mu_{2,(k)}}}, \tag{4.138}$$

where $a_{\varepsilon,(k)}$ are close to unity correction factors.

From comparison of formulas (Eq. 4.133 and Eqs. 4.86 and 4.87), it follows that the taken form of nonstationary equations (Eq. 4.84) in the establishing method leads to a uniform difference scheme for solution of both static and dynamic problems. This effectively allows studying geometrically and physically nonlinear stress-strain state of shell structures at combined loading without reconstructing calculation algorithm and only changing corresponding DS parameters (e.g., studying transient processes in statically preloaded structures, determining residual deflections, and strains).

Explicit difference scheme in time used for shell structure equations of motion solution is conditionally stable [6,8]. The value of time step, Δt, is determined from the condition of difference scheme stability, which for dynamic problems has a form of Courant condition and taking into account equations (Eq. 4.75) structure is written in the form

$$\Delta t \leq \frac{\Delta s}{c}, \tag{4.139}$$

where $\Delta s = \min[\Delta x; \Delta y]$; c is strain waves propagation in the shell velocity; and Δx, Δy are mesh steps. It can be shown that critical value of Δt satisfies the condition

$$\Delta t \leq \frac{T_{\min}}{\pi}, \tag{4.140}$$

where $T_{\min}$ is minimal period of free vibrations described by finite-difference model of the shell.

The developed applied mathematical technologies of nonlinear initial-boundary problems of thin structures mechanics solution can be used not only for investigation of aerospace and machine-building structures strength reliability parameters, but also for studying reinforced concrete structures at different types of external loads, including seismic ones [18–20].

References

1. Novozhilov V.V. *[Osnovy nelinejnoj teorii uprugosti] Fundamentals of the Nonlinear Theory of Elasticity*. Editorial URSS, Moscow, 2003. (In Russ.).
2. Karmishin A.V., Lyaskovec V.A., Myachenkov V.I., Frolov A.N. *[Statika i dinamika tonko-stennyh obolochechnyh konstrukcij] Static and Dynamics of Thin-Walled Shell Structures*. Mashinostroenie, Moscow, 1975. (In Russ.).
3. Vol'mir A.S. *[Nelinejnaya dinamika plastinok i obolochek] Nonlinear Dynamics of Plates and Shells*. Nauka, Moscow, 1972. (In Russ.).
4. Vasil'ev V.V. *[Mekhanika konstrukcij iz kompozicionnyh materialov] Mechanics of Structures Made of Composite Materials*. Mashinostroenie, Moscow, 1988. (In Russ.).
5. Il'yushin A.A. *[Mekhanika sploshnoj sredy] Continuum Mechanics*. MGU, Moscow, 1990. (In Russ.).
6. Samarskij A.A. *[Teoriya raznostnyh skhem] Theory of Difference Schemes*. Nauka, Moscow, 1989. (In Russ.).
7. Bauld N.R., Goree J.G., Tzeng L.S. A comparison of finite-difference and finite-element methods for calculating free edge stresses in composites. *Computers & Structures*, 1985, 20 (5): 897–914.
8. Bahvalov N.S., ZHidkov N.P., Kobel'kov G.M. *[Chislennye metody] Numerical Methods*. Fizmatlit. Laboratoriya Bazovyh Znanij. Moscow, 2001. (In Russ.).
9. Dmitriev V.G., Preobrazhenskij I.N. *[Deformirovanie gibkih obolochek s vyrezami]* Deformation of flexible shells with cuts. Izvestiya AN SSSR. *Solid Mechanics*, 1988, 1: 177–184. (In Russ.).
10. Preobrazhenskij I.N., Golda YU.L., Dmitriev V.G. *[Chislennyj metod issledovaniya napryazhenno-deformirovannogo sostoyaniya gibkih kompozitnyh obolochek vrashcheniya, oslablen-nyh vyrezami razlichnoj formy]* Numerical method for studying the stress-strain state of flexible composite shells of revolution, weakened by cuts of various shapes. *Mechanics of Composite Materials*, 1985, 6: 1030–1035. (In Russ.).
11. Dmitriev V.G., Egorova O.V., Rabinsky L.N. Solution of nonlinear initial boundary-value problems of the mechanics of multiply connected composite material shells on the basis of conservative difference schemes. *Composites: Mechanics, Computations, Applications*, 2015, 4: 265–277.
12. Dmitriev V.G. et al. Nonlinear strain of multilayer composite shells of revolution in large displacements and angles of rotation of the normal. *Izvestiya Vysshikh Uchebnykh Zavedeniy. Aviatsionnaya Tekhnika*, 2017, 2: 8–15.
13. Tarakanov S.I. *[O skhodimosti metoda "dinamicheskaya relaksaciya" v zadachah nagruzheniya uprugih obolochek vrashcheniya]* On the convergence of the "dynamic relaxation" method in problems of loading elastic shells of revolution. *Vestnik MGU: Mat. mekh*, 1984, 5: 90–93. (In Russ.).
14. Feodos'ev V.I. *[Ob odnom sposobe resheniya nelinejnyh zadach ustojchivosti deformiruemyh system]* On one method of solving nonlinear problems of stability of deformable systems. *Applied Mathematics and Mechanics*, 1963, 27 (2): 265–274. (In Russ.).
15. Frieze P.A., Hobbs R.E., Dowling P.J. Application of dynamic relaxation to the large deflection elasto-plastic analysis of plates. *Computers & Structures*, 1978, 8 (2): 301–310.

16. Turvey G.J., Der Avanessian N.G.V. Elastic large deflection of circular plates using graded finite-differences. *Computers & Structures*, 1986, 23 (6): 763–774.
17. Tong P. An adaptive dynamic relaxation method for static problems. *Computational Mechanics 86: Theory and Applications Proceedings of the International Conference, Tokyo*, 1986, 1: II/89-II/101.
18. Dmitriev V.G. Mathematical modelling of non-linear deformation process for frame-type building structures under seismic loads. *International Journal for Computational Civil and Structural Engineering*, 2012, 8 (2): 13–29.
19. Dmitriev V.G., Sudyin A.A. Deformation of reinforced concrete spherical dome with cutouts on the damped foundation beds. *International Journal for Computational Civil and Structural Engineering*, 2009. 1&2 (5): 13–22.
20. Dmitriev V.G., Roffe A.I. Study how viscoelastic damper parameters impact deformation and load-bearing capacity of frame-type reinforced concrete structures under seismic loads. *International Journal for Computational Civil and Structural Engineering*, 2015, 11 (1): 104–114.

5

Some Solutions of Differential Equations in Engineering Calculations

Lubov Mironova and Leonid Kondratenko

CONTENTS

This chapter presents differential equations with respect to a rectangular coordinate system, which are widely used in solving various engineering problems. The classical approaches and methods for solving partial differential equations are also presented. The mathematical formulation gives practical orientation to the problems associated with studies of the fastness (strength) of structures and the dynamic effects of physical processes caused by technological influences In compiling the theoretical material, we give references to authors who conducted research in this area. This chapter reflects the experience of lecturing on mathematical methods of modeling, as well as the personal participation of the authors in the work in this technical field. For a more extensive study of the problems described, literary sources are given at the end of the chapter.

5.1 Classical Problems in Solving Engineering Problems

When solving engineering issues, it is often necessary to consider the problems of the motion of a certain system of material bodies in the chosen observer's reference system. The system of material bodies can be considered a material system or a set of material points, if the movement of each of them individually depends on the movement and position of individual points.

The objectives of studying the motion of any point of such a system may be different. For example, we can carry out calculations for the strength of structural elements, predict or create the mechanical effects we need, or control many movements of deformable bodies to the required degree. As a rule, such problems are based on classical models of continuum mechanics, among which are a model of an elastic body, a model of an ideal fluid and gas, a model of a viscous fluid, a model of conducting fluid in magnetic hydrodynamics, and others.

However, in engineering science, there are many questions that can be answered only on the basis of special knowledge. For example, in the development of innovative technologies for manufacturing parts by metalworking, the actual problems remain the study and analysis of wave movements in the closed system "machine—tool—part." Effects associated with the propagation of waves in solids, lead to a decrease in positioning accuracy, the emergence of additional stress state and deformation, both of the cutting tool, and details.

In the theoretical study of specific issues of engineering, it is advisable to use the classical problems of mechanics. Such an approach leads the researcher to the necessity of choosing reference systems in which the motion of the material system and the state of the studied medium are described. The choice of options may be different.

Note that the motion of the system of material bodies is determined with respect to some coordinate system x^1, x^2, x^3, which can be chosen by the researcher arbitrarily. However, when choosing a coordinate system, specialists will use the methods of various branches of mechanics, for example, vector mechanics [1] (Newton mechanics [2]) and analytical mechanics, of which variational mechanics [1] is an integral part. We give the following explanation.

It should be noted that when solving many problems of dynamics, the chosen coordinate system for describing the initial state of body motion is not suitable for observing the entire body motion. Therefore, sometimes it is convenient to use the coordinate axes, which themselves move in space in such a way that they always retain such positions that are more in tune with the instantaneous position of the body [3].

Or, if the axes with respect to which body movement is considered to be movable, then it is necessary to have a way to determine the position and movement of these axes in space. This approach is implemented with the help of another system of axes, which themselves are fixed in space. With the help of a new reference system, it is possible to consider the movement of moving axes [3].

In vector mechanics, the motion of a system of material bodies relative to inertial coordinate systems moving relative to each other in translational motion with a constant velocity over time is considered [2]. When solving physical problems in practice, the Cartesian coordinate system with the origin at the center of gravity of the solar system, in which distant stars are considered fixed, is often chosen as the initial coordinate system. Also, Newton's theory is based on two main vectors: the "impulse" and the "force" [1].

In analytical mechanics, the motion of a system of material bodies is complemented by an accompanying coordinate system. Along with the coordinates x^1, x^2, x^3 the concept of a system is introduced—"Lagrangian coordinates" of individual points ξ^1, ξ^2, ξ^3, which are considered other coordinates of the same points of space in region *D*.

In the variational mechanics developed by Euler and Lagrange, the theory is based on two scalar quantities: the "kinetic" and "force function." The advantages of variational methods are associated with complete freedom in choosing the coordinate system of the problem under consideration. The reader will find a detailed presentation of the variational principles of mechanics in the literature [1].

Many engineering calculations are associated with the consideration of the movement of a structural element of a complex technical system (the system of material bodies). The motion of a rather complex mechanical system is described by a large number of individual differential equations [1]. Differential formulations are valid in the field of continuous phenomena.

A theoretical study of the functioning of any technical mechanism can be described using universal equations of mechanics, thermodynamics, and electrodynamics. Moreover, for continuous smooth distributions, universal equations can be written in the form of partial differential equations.

In the presence of discontinuous distributions of characteristics, the more general nature of integral formulations manifests itself. At the same time, differential formulations retain their significance in the field of continuous phenomena and are supplemented by conditions on the boundaries of the discontinuity [2].

In many cases, engineering systems consider technical systems consisting of a certain (or large) number of bodies. Such systems can be closed or open. When studying the motion of a system or its elements, it is necessary to rely on the real properties of bodies. It is important to know their structure, the nature of the substances, and the physic mechanical, chemical, and technological properties of the materials. The internal stresses caused by both direct forces of interaction between molecules, and the transfer of a macroscopic amount of movement through an elementary section, caused by the result of thermal movement of molecules.

One of the approaches to the study of the movement of material bodies is the application of a phenomenological macroscopic theory based on general laws and hypotheses consistent with experimental data. This theory includes continuum mechanics. The applied hypothesis of continuity allows the researcher to consider approximately all bodies consisting of a large number of individual particles in a substantial volume, as a medium filling the space in a continuous way [2]. Not only ordinary material bodies, but also various fields (e.g., electromagnetic and temperature) can be considered as a continuous continuum. This idealization makes it possible to use the mathematical apparatus of continuous functions, differential and integral calculus.

Starting to consider the solution of differential equations in technical calculations of engineering, we give the definitions of the classical provisions adopted in mechanics.

Space. By space, we mean a collection of points defined using numbers, which are called "coordinates."

Metric space. Continuous metric manifolds are spaces in which distances between points are defined.

For example, if the coordinates of two points M_1 (x_1, y_1, z_1) and M_2 (x_2, y_2, z_2) are known, then the distance between the points is determined by the formula

$$r = \sqrt{(x_1 - x_2)^2 + (y_1 - y_2)^2 + (z_1 - z_2)^2}. \tag{5.1}$$

Euclidean space. Only such spaces are considered, in each of which you can enter a Cartesian coordinate system that is the same for all points. Physical real space on a small scale can be considered Euclidean with great precision. Vector mechanics are based on Euclidean space.

Space of configurations. It is a set of generalized coordinates of a mechanical system, as which any set of parameters characterizing the position of a mechanical system can be chosen. A complex mechanical system is replaced by a single point, the motion of which is studied in the space of configurations. This is an abstract space, the number of dimensions of which is determined by the conditions of the problem [1]. The space of configuration is widely used in analytical mechanics.

Absolute time. Idealized concept, which is suitable for the correct description of reality in those cases when the effects of the theory of relativity are not taken into account. It is believed that time flows the same for all observers [2].

The phenomenological macroscopic theory of continuum mechanics considers the motion of a material system (continuum) in Euclidean space using absolute time.

In engineering calculations, it remains relevant to study the process of movement of deformable bodies that make up a single material system. The application of the theory based on the fundamental hypotheses of metric space and absolute time is not always justified. The findings are in some cases not consistent with experience. Therefore, the adopted model of space and time can be refined and generalized [2].

5.2 Linear Elastic Body

Calculations in mechanics related to the motion of a material system, as a rule, are based on the application of universal models of either the system itself in the form of a continuum or its structural elements. From the point of view of formalism, the universal model is a conceptual apparatus, with the help of which it is possible to give a mathematical description of the object, process, or phenomenon under investigation. Any model is partially idealized and has a number of assumptions that significantly simplify the mathematical relationships of dependent variables.

The problem of applying models that are more accurately adapted to the conditions of the studied processes in calculations remains relevant today. On the one hand, by applying more accurate mathematical models, we are forced to solve systems with a large number of equations and variables and to look for conditions for the convergence and stability of solutions. At the same time, the obtained approximate solutions may go beyond the specified accuracy and not correspond to the actual results. Therefore, in many cases it is advisable to carry out theoretical studies using classical models of continuum mechanics. We give the definitions of some such models.

Solid body. The material system for which the distance between two points does not change is called a "solid body." The possibilities of motion of particles of a solid body are limited by its rigidity [1].

Elastic body. An elastic body is a medium in which the components of the stress tensor in each particle are functions of the components of the strain tensor, the components of the metric space, temperature, and other parameters of a physicochemical nature [2].

Linear elastic body. For a linear elastic body, the dependence of the total stress on the total strain is represented by the linear relation [4] within the limits of elasticity. Such dependence is formulated by the generalized Hooke law. So, for example, the presence of all stress components along the edges of an infinitely small parallelepiped, cut out near a given point (Figure 5.1), determines the following deformation components [5]

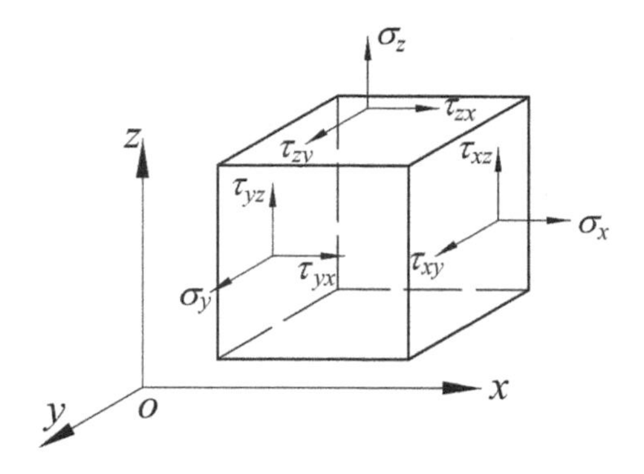

FIGURE 5.1
Components of stress along the edges of an infinitely small parallelepiped, cut out near a given point.

$$\varepsilon_x = \frac{1}{E}\left[\sigma_x - \mu\left(\sigma_y + \sigma_z\right)\right];\ \gamma_{yz} = \frac{\tau_{xy}}{G};$$

$$\varepsilon_y = \frac{1}{E}\left[\sigma_y - \mu\left(\sigma_x + \sigma_z\right)\right];\ \gamma_{xz} = \frac{\tau_{yz}}{G}; \tag{5.2}$$

$$\varepsilon_z = \frac{1}{E}\left[\sigma_z - \mu\left(\sigma_y + \sigma_x\right)\right];\ \gamma_{xy} = \frac{\tau_{zx}}{G}.$$

Here ε_x, ε_y, ε_z are relative linear deformations; σ_x, σ_y, σ_z is the normal stresses in the direction of the axes *x*, *y*, *z*; τ_{xy}, τ_{yz}, τ_{zx} is the tangential stresses; γ_{xy}, γ_{yz}, γ_{zx} is the relative shear deformations; *E* is the modulus of elasticity of the material; μ is the Poisson's ratio; and *G* is the shear modulus of the material.

Isotropic body. Isotropic is an elastic body whose elastic properties are the same in all directions (the number of elastic constants is reduced to two).

A body is called not only isotropic but also homogeneous if the properties of volume elements cut in different parts of the body are the same.

For an isotropic body, expression (Eq. 5.2) is formulated as follows: the components of the strain tensor at the point under consideration are linearly dependent on the components of the stress tensor at the same point.

5.3 Motion of a Material Point and Heat Transfer Processes

The movement of the material point. A moving point at different times is identified with different points in space. The motion of a point is known if the law of motion is known. For any point of the continuum, defined by the coordinates a, b, c, we can write the law of motion, which includes the functions of four variables—the initial coordinates, a, b, c, and time, t

$$x^i = x^i(a, b, c, t), i = 1, 2, 3. \tag{5.3}$$

The coordinates a, b, c or ξ^1, ξ^2, ξ^3, the individualizing points of the continuum, and the time, t, are called Lagrange variables. Functions (Eq. 5.3) constitute the law of motion of the continuum. The Cartesian coordinate system is shown in Figure 5.2.

In applications, the Euler point of view is quite often defined when the motion is considered known if the speed ($\boldsymbol{v} = d\boldsymbol{r}/dt$), acceleration ($\boldsymbol{a} = d\boldsymbol{v}/dt$), or other quantities are given as functions of the coordinates of the space x^1, x^2, x^3 and time, t

$$\boldsymbol{v} = \boldsymbol{v}\ (x^1, x^2, x^3, t);\ \boldsymbol{a} = \boldsymbol{a}\ (x^1, x^2, x^3, t); \tag{5.4}$$

Here the variables $\boldsymbol{r}, \boldsymbol{v}, \boldsymbol{a}$ are vectors.

The geometric coordinates of the space x^1, x^2, x^3 and time, t, are called Euler variables.

The task of the motion of a continuous medium from the points of view of Lagrange and Euler are mechanically equivalent to each other [2]. You can always make the transition from Lagrange variables to Euler variables and vice versa. In the first case, it is necessary to find a solution to the system of differential equations in which the functions are given implicitly. In the second case, the transition from Euler variables to Lagrange variables is associated with the integration of ordinary differential equations.

In engineering calculations, problems are mainly considered, the formulation of which is formulated on the basis of Euler variables. For example, in the Cartesian coordinate system, the velocity components of the moving point u, v, w are known as functions of the Euler variables

$$u = u\ (x, y, z, t);\ v = v\ (x, y, z, t);\ w = w\ (x, y, z, t). \tag{5.5}$$

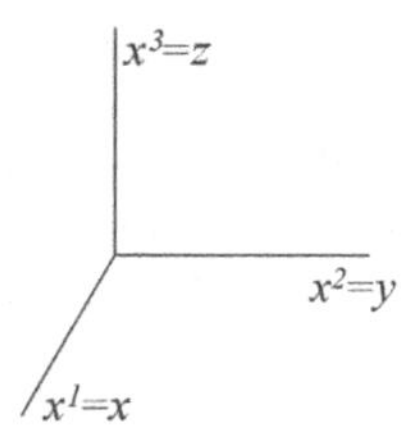

FIGURE 5.2
Cartesian coordinate system.

Then the following differential relations are true

$$\frac{dx}{dt}=u(x,y,z,t);\ \frac{dy}{dt}=v(x,y,z,t);\ \frac{dz}{dt}=w(x,y,z,t). \tag{5.6}$$

Having solved the system of three differential equations, we find the coordinates of the point x, y, z as functions of time, t, and three arbitrary constants, C_1, C_2, C_3, which are determined at some given moment, t_0. The parameters obtained individualize a point of continuous medium and are Lagrange variables. The solution of system (Eq. 5.6) is the law of motion of a material point (Eq. 5.3) in Lagrange variables.

Let us give a general differential equation of the translational motion of a material point under the action of a force that coincides with the direction of motion and changes in time. The equation has the following form

$$ma = F(t). \tag{5.7}$$

Here m is the mass of a point; a is the acceleration with which the point moves; F is the force, under the action of which the point moves.

For the material system consisting of the i-th number of bodies, Equation 5.7 takes the following form

$$m_i a_i = F_i(t) + R_i(t). \tag{5.8}$$

Here is entered the index i is the number of the body; $F_i(t)$, $R_i(t)$ are the resultant of all active forces and reactions, respectively, of the bonds applied to the i-th body.

Because acceleration is a derivative with respect to speed, the researcher needs to consider a system of second-order differential equations. One solution to these problems will be given below.

Heat transfer processes. The temperature distribution (T) in a continuous medium can be defined as from the Lagrange point of view

$$T = T(\xi^1, \xi^2, \xi^3, t), \tag{5.9}$$

and from the Euler point of view

$$T = T(x^1, x^2, x^3, t). \tag{5.10}$$

However, the transition from Euler variables to Lagrange variables is more complicated than the example above. Here it is necessary to use the rule of differentiation of a quite complicated function. Therefore, in practical problems the study of thermal processes should be carried out on the basis of the theory of thermal conductivity.

The transfer of heat from one part of the body to another or from one body to another that is in contact with the first is called "heat conduction."

Analytical thermal conductivity does not consider matter as a collection of discrete particles but as a continuous medium—a continuum. Such a model of heat propagation is accepted if the sizes of the differential volumes are sufficiently large compared with the size of the molecules and the distances between them [6]. It is assumed that the body is uniform and isotropic.

The main physical quantity in the analytical study of thermal conductivity is temperature. This value is a spatio-temporal characteristic of the phenomenon being studied, the definition of which is reduced to finding the dependence

$$T = f(x, y, z, \tau), \tag{5.11}$$

where x, y, z are spatial coordinates in the Cartesian system, and τ is time.

The combination of instantaneous temperature values at all points of the studied space is called a "temperature field." Because temperature is a scalar value, the temperature field is also a scalar field.

The reader can find a detailed account of the theory of heat conduction and mathematical methods for solving problems in the literature [6,7].

Representing a complex technical system as a continuous medium (continuum) and applying classical models of continuum mechanics, the researcher, reaching the goal, has to solve systems of differential equations. We give some of them as the most common in the consideration of applied problems.

5.4 Static and Dynamic Equilibrium State of the Body

In the strength calculations, the researcher has to consider the structural element (material body) in equilibrium. Let us give the differential equations of static and dynamic equilibrium of an elastic solid with respect to the Cartesian coordinate system.

If the body is at rest, then the equilibrium conditions are determined by static equations

$$\begin{aligned}
&\frac{\partial \sigma_x}{\partial x} + \frac{\partial \tau_{xy}}{\partial y} + \frac{\partial \tau_{xz}}{\partial z} + \rho X = 0;\\
&\frac{\partial \tau_{yx}}{\partial x} + \frac{\partial \sigma_y}{\partial y} + \frac{\partial \tau_{yz}}{\partial z} + \rho Y = 0;\\
&\frac{\partial \tau_{zx}}{\partial x} + \frac{\partial \tau_{zy}}{\partial y} + \frac{\partial \sigma_z}{\partial z} + \rho Z = 0.
\end{aligned} \tag{5.12}$$

Here X, Y, Z is the projections of the bulk forces on the x, y, z axes; and ρ is the density of the material.

In the case of a body moving, the right sides of equations (Eq. 5.12) are not zero and, according to Newton's second law (the law of conservation of momentum), are equal to the product of the body mass by the corresponding projection of its acceleration (i.e. associated with the projections of accelerations, which for small displacements u, v, w are expressed through the components of displacements and time, t)

$$\frac{\partial \sigma_x}{\partial x}+\frac{\partial \tau_{xy}}{\partial y}+\frac{\partial \tau_{xz}}{\partial z}+\rho X=\rho\frac{\partial^2 u}{\partial t^2};$$

$$\frac{\partial \tau_{yx}}{\partial x}+\frac{\partial \sigma_y}{\partial y}+\frac{\partial \tau_{yz}}{\partial z}+\rho Y=\rho\frac{\partial^2 v}{\partial t^2}; \tag{5.13}$$

$$\frac{\partial \tau_{zx}}{\partial x}+\frac{\partial \tau_{zy}}{\partial y}+\frac{\partial \sigma_z}{\partial z}+\rho Z=\rho\frac{\partial^2 w}{\partial t^2}.$$

In Equation 5.13, the notation for the components of displacement u, v, w adopted in the theory of elasticity is given.

In assessing the stress-strain state of the structural element, geometrical and physical equations are added to Equations 5.12 and 5.13.

Geometric equations

$$\varepsilon_x=\frac{\partial u}{\partial x},\ \gamma_{xy}=\frac{\partial u}{\partial y}+\frac{\partial v}{\partial x};$$

$$\varepsilon_y=\frac{\partial v}{\partial y},\ \gamma_{yz}=\frac{\partial v}{\partial z}+\frac{\partial w}{\partial y}; \tag{5.14}$$

$$\varepsilon_z=\frac{\partial w}{\partial z},\ \gamma_{zx}=\frac{\partial w}{\partial x}+\frac{\partial u}{\partial z}.$$

Physical equations

$$\sigma_x=2G\varepsilon_x+\lambda\theta,\ \tau_{xy}=G\gamma_{xy};$$

$$\sigma_y=2G\varepsilon_y+\lambda\theta,\ \tau_{yz}=G\gamma_{yz}; \tag{5.15}$$

$$\sigma_z=2G\varepsilon_z+\lambda\theta,\ \tau_{zx}=G\gamma_{zx}.$$

Here λ is the Lame's constant; θ is the relative volume deformation, determined by the ratios

$$\theta = \varepsilon_x + \varepsilon_y + \varepsilon_z; \lambda = \frac{2\mu G}{1-2\mu}. \tag{5.16}$$

The equations are the basic equations of the linear theory of elasticity. Mathematical ratios of conditions on the body surface are added to these equations. The reader will find a complete presentation of the theory in the literature [5,8,9].

5.5 Longitudinal Oscillations of a Direct Elastic Rod

In this section, we consider the problem of longitudinal oscillations of a direct elastic rod in the absence of mass forces and associate this process with the theory of elasticity. In this case, we obtain the relationship between the velocities of the particles of the elementary volume of the elastic rod and the stresses.

Considering the process of loading the rod as a uniaxial stress state, in our case we rewrite the first differential equation of system (Eq. 5.13) in the form

$$\frac{\partial \sigma_x}{\partial x} = \rho \frac{\partial^2 u}{\partial t^2}. \tag{5.17}$$

Or

$$\rho \frac{\partial \upsilon}{\partial t} = \frac{\partial \sigma}{\partial x}, \tag{5.18}$$

where υ is the longitudinal velocity of movement of particles of the elementary volume of an elastic rod along the axis; σ is the longitudinal stress.

Taking into account formulas (Eqs. 5.2 and 5.14), we write for our case the dependence of relative linear deformation and displacement in the form

$$\varepsilon_x = \frac{1}{E}\sigma_x; \ \varepsilon_x = \frac{\partial u}{\partial x}. \tag{5.19}$$

Because the connection between the two material points M_1 and M_2 is made in the form of an elastic rod, and each point under consideration has a different mass; the point M_2 will carry out longitudinal oscillations along the Ox axis. Moreover, the acceleration of the material point of the system will be directly proportional to the second derivative of the point moving along the x coordinate.

Let us write the equation of longitudinal oscillations, which is applicable with sufficient accuracy in practical calculations of engineering problems [10]. Equation has the form

$$\frac{\partial^2 u}{\partial t^2} = \frac{E}{\rho}\frac{\partial^2 u}{\partial x^2}. \tag{5.20}$$

In view of (5.17) we will finally have

$$\frac{\partial \sigma_x}{\partial x} = E\frac{\partial^2 u}{\partial t^2}. \tag{5.21}$$

Or

$$\frac{1}{E}\frac{\partial \sigma_x}{\partial x} = \frac{\partial^2 u}{\partial t^2}. \tag{5.22}$$

We integrate expression (Eq. 5.22) over x, taking the following initial conditions $x = 0, \sigma = \sigma_0$. Taking into account (Eq. 5.19) we get

$$\begin{gathered}\frac{1}{E}\int_0^x \frac{\partial \sigma_x}{\partial x}dx = \int_0^x \frac{\partial^2 u}{\partial x^2}dx;\\ \frac{1}{E}(\sigma_x - \sigma_0) = \left(\frac{\partial u}{\partial x}\right)_x - \left(\frac{\partial u}{\partial x}\right)_{x=0}.\end{gathered} \tag{5.23}$$

The resulting expression is differentiated by t. We have

$$\frac{1}{E}\frac{\partial \sigma_x}{\partial t} = \frac{\partial \upsilon}{\partial t}. \tag{5.24}$$

Taking into account the fact that the surface forces acting on each point of the cross section of an elementary volume are directed in the opposite direction and taking into account Equation 5.15, we will finally have [10]

$$\rho\frac{\partial \upsilon}{\partial t} = -\frac{\partial \sigma}{\partial x}. \tag{5.25}$$

$$\frac{1}{E}\frac{\partial \sigma_x}{\partial t} = -\frac{\partial \upsilon}{\partial t}. \tag{5.26}$$

Equations 5.25 and 5.26 describe the relationship between the velocities of the longitudinal movement of the cross sections of the elementary volume of a direct elastic rod and the changes in the longitudinal stresses with the gradients of changes of the same variables along the length of the rod.

For short bars, the equations can be written in ordinary derivatives [10]

$$\rho \frac{d\upsilon}{dt} = -\frac{d\sigma}{dx}. \tag{5.27}$$

$$\frac{1}{E}\frac{d\sigma_x}{dt} = -\frac{d\upsilon}{dt}. \tag{5.28}$$

Let two bodies of mass, m_1 and m_2, be rigidly interconnected by a direct elastic rod. Taking this system in the form of a continuum, we will present a scheme (Figure 5.3) in which the elastic rod at the initial and final points (M_1, M_2) has m_1 and m_2. Moreover, the mass of the point M_1 is much greater than the mass of the point M_2. And let the rod at the point M_1 be communicated in the direction of the segment M_1 M_2. Place the origin of the coordinate system at point M_1, or we will assume that the rod moves parallel to the chosen reference system.

Such a model can be applied to a two-link mechanism for the transfer of translational motion from the engine to the executive body, Figure 5.4.

Integrating Equation 5.27 over the x coordinate and further differentiating it over t, we obtain the relation for the longitudinal vibrations of the rod

$$\vartheta_n \frac{d\sigma}{dt} = \upsilon_1 - \upsilon_2, \tag{5.29}$$

where ϑ_n is the coefficient of longitudinal elasticity, $\vartheta = l/E$. Considering the elastic modulus for the driving and driven members to be the same and constant values, integrating Equation 5.29, we obtain the dependence of the normal stresses on the displacements of the rod points

$$\vartheta_n \sigma = (\upsilon_1 - \upsilon_2)t. \tag{5.30}$$

Or

$$\frac{l}{E}\sigma = x_1 - x_2, \frac{l}{E}\sigma = \Delta u. \tag{5.31}$$

Ratios (Eq. 5.31)—is Hooke law.

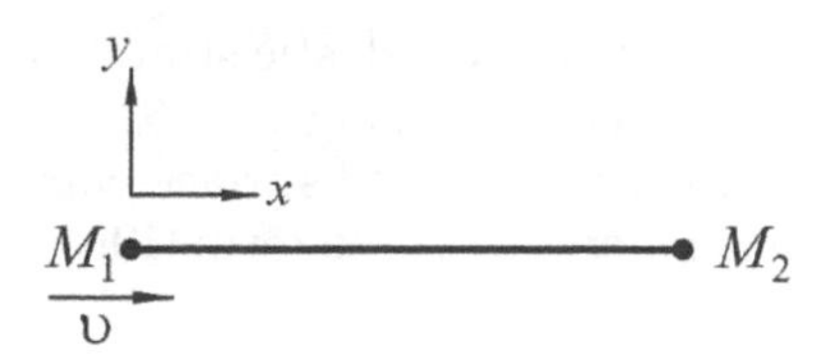

FIGURE 5.3
Chosen reference system.

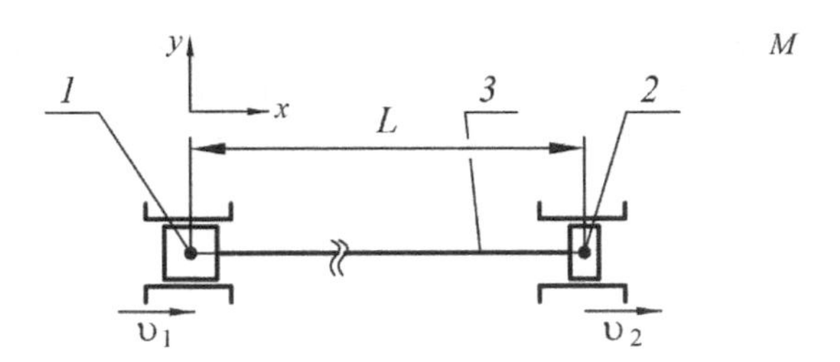

FIGURE 5.4
Scheme the mechanism for the transmission of translational motion: 1 is the leading link (engine); 2 is the slave link (executive body); 3 is the connecting link (trunk); υ_1, and υ_2 is the longitudinal speed of movement of links 1 and 2.

The normal stresses developed in the rod overcome the resistance force arising on the driven link, which is the resultant force of its two components: friction force (F_f) and inertia force (F_i), which are determined by the ratios

$$F_f = k \sin \alpha;\ F_i = -m_2 g/dt. \tag{5.32}$$

Here k is the friction coefficient. F_i is the force is the inertial component.

Without taking into account the direction of the velocity of motion and, taking $F = \sigma f$, where f is the cross-sectional area of the body, we write the following relation [10]

$$\beta\sigma f = F_0(t) + k_1\upsilon_1(t) + k_2\upsilon_2 + m_2 g\frac{d\upsilon_2}{dt}. \tag{5.33}$$

Here β is the proportionality coefficient, which depends on the friction coefficient taking into account the direction of movement of the slave link $\beta = (1-c)$ is the sgn υ_2; and k_1, k_2 is the loss coefficients proportional to the speeds of the driving and driven units.

The solution to this problem is described in detail in the literature [11].

5.6 The Rotation of the Elastic Cylindrical Rod about its Axis

Consider the process of rotation of an elastic straight cylindrical rod, about its axis, in the coordinate system ζ, φ, x, Figure 5.5.

In the absence of mass forces, mass, and surface pairs, we use the differential equation of angular momentum in the form [10]

$$\rho\frac{\partial\left(r^2\Omega\right)}{\partial t} = \frac{\partial(r\tau)}{\partial x}. \tag{5.34}$$

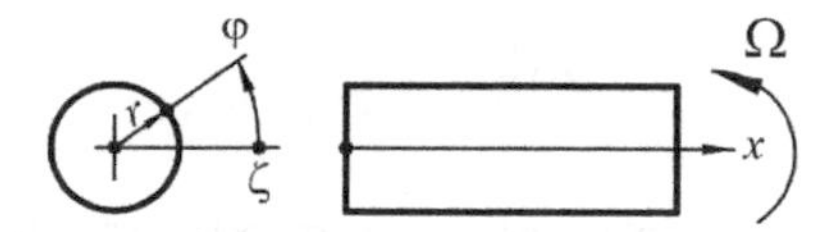

FIGURE 5.5
Rotation of the elastic straight cylindrical rod about the x-axis.

The angular velocity of rotation and angular acceleration are determined by the formulas

$$\Omega = \frac{\partial \varphi}{\partial t}; \quad \omega = \frac{\partial \Omega}{\partial t}. \tag{5.35}$$

In formulas (Eqs. 5.34 and 5.35), r is the radius of the considered point of the cross section of the rod in the coordinate ζ; Ω is the angular velocity of rotation of the rod; τ is the tangential stresses over the cross section of the rod; and ω is the angular acceleration.

For the case when the tangential stresses are maximal on the outer surface with radius, r_0, we write the following boundary conditions

$$r = r_0; \tau(r_0) = \tau_{max}. \tag{5.36}$$

Taking into account the fact that the stresses are directed in the opposite direction from the direction of the speed of rotation of the rod, under boundary conditions (Eq. 5.36), Equation 5.34 takes the form

$$\rho r_0 \frac{\partial \Omega}{\partial t} = -\frac{\partial \tau_{max}}{\partial x}. \tag{5.37}$$

If any external factors affect the rod, leading to the appearance of distributed surface pairs, then it is necessary to consider the differential equation in the form

$$\rho \frac{\partial \left(r^2 \Omega\right)}{\partial t} = \frac{\partial (r\tau)}{\partial x} + \frac{\partial Q}{\partial x}. \tag{5.38}$$

Here Q is the resultant of reactive surface forces.

5.7 Torsional Vibrations of a Straight Cylindrical Rod

For torsional vibrations of a direct cylindrical rod in the absence of mass forces, mass and surface pairs, we rewrite the differential equation of the moment of momentum (Eq. 5.34) in the following form [10]

$$r\rho \frac{\partial \Omega}{\partial t} = -\frac{\partial \tau}{\partial x}, \tag{5.39}$$

Here in after, τ will be under stood as the maximum shear stress.

In the absence of longitudinal oscillations and under the action of only a twisting impulse, twisting motions will arise in the rod, when each point from the adjacent one moves to a distance, dx. Moreover, as shown by the theory of elasticity [2], these sections are rotated as hard disks. Therefore, the points of two sections that are at a distance, r_i, from the axis of the rod when the section is rotated by an angle are displaced relative to each other by a distance

$$dx = r_i \, d\varphi. \tag{5.40}$$

Then the following expression is true [10]

$$\frac{\partial^2 \varphi}{\partial t^2} = \frac{G}{\rho} \frac{\partial^2 \varphi}{\partial x^2}. \tag{5.41}$$

Here G is the shear modulus of the material.

Taking into account Equations 5.35, 5.39, and 5.41, we finally get

$$\frac{1}{Gr} \frac{\partial \tau}{\partial t} = -\frac{\partial \Omega}{\partial x}. \tag{5.42}$$

Equations 5.39 through 5.42 describe the relationship of the shear rate relative to the axis of the rod of flat sections of the elementary volume of a straight cylindrical rod and the rate of change of maximum tangential stresses with gradients of changes of the same variables along the length [10].

For a short rod, Equation 5.39 can be rewritten in ordinary derivatives

$$r_0 \rho \frac{d\Omega}{dt} = -\frac{d\tau}{dx}. \tag{5.43}$$

5.8 Inertial Disc Rotating at the End of the Rod

Equations 5.34 and 5.35 are valid for the model of a cylindrical rod, at the end of which a disk of mass, m, is fixed with a flywheel inertia moment J, Figure 5.6. The system is referred to the coordinates ζ, φ, x.

The rod is driven to rotate. The rotation of the disc is hampered by the moment of resistance, M_r. Such a model can be taken as a simulation model of the work of the machine spindle when machining holes.

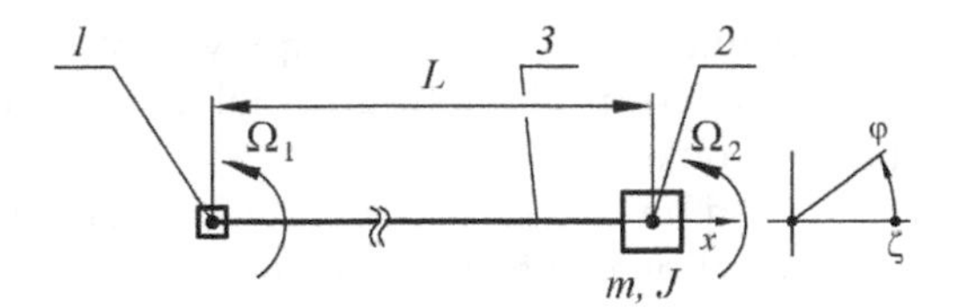

FIGURE 5.6
A rotating cylindrical rod with a disc fixed at the end.

The point of communication of the rotational motion to the rod will be considered the element of the leading link (position 1), and the driven link will be the disk fixed at the end of the rod (position 2). Denote the angular velocities of these elements as Ω_1 and Ω_2, respectively. Obviously, their values will be different in magnitude.

Integrating equation 5.42 over the x coordinate and further differentiating it over t, we obtain the relation for torsional vibrations of a rotating disk

$$\vartheta_k \frac{d\tau}{dt} = \Omega_1 - \Omega_2, \tag{5.44}$$

where ϑ_k is the coefficient of torsional elasticity, $\vartheta_k = \frac{l}{\rho GW}$; Ω_1, Ω_2 are the angular velocities of rotation of the cross sections of the rod and the disk, respectively; W is the geometrical moment of the resistance of the cross section of the disk, $W = \pi d^3/16$.

Integrating Equation 5.44, we obtain the dependence of tangential stresses on the change in the twisting angle

$$\upsilon_{1k}\tau = \varphi_1 - \varphi_2, \; \vartheta_{1k} = \frac{l}{\rho G}. \tag{5.45}$$

Tangential stresses that develop in a rod overcome the resistance of forces $M_r = M_{r0} + h_k\Omega_2$ arising on the disk (the driven link), as well as the forces of inertia

$M_d = J\, d\Omega_2/dt$. Here h_k is the proportional loss factor of the angular velocity of rotation of the disk. Taking into account that $\tau = M/W$, we finally get

$$\tau(t)\; W = M_{r0}(t) + h_x\Omega_2(t) + J\frac{d\Omega_2}{dt}. \tag{5.46}$$

Sections 5.5 through 5.7 present differential equations that relate the velocities of the particles of an elementary volume of a cylindrical rod to stresses resulting from factors of influence. Such an approach made it possible to develop a new method for studying the dynamics of rotating structural elements proposed by L. Kondratenko to assess the resource on strength of an instrumental-technological complex in engineering technologies.

The equations obtained make it possible to apply a motion transmission scheme and, on its basis, to study the longitudinal and torsional vibrations of a moving technological object.

The Kondratenko method has been successfully implemented in the study of wave processes and nonlinear effects in complex mechanisms for transferring motion from drive to tool in metalworking and well-drilling technologies. The reader can find a detailed exposition of this question in the literature [10–20].

5.9 Differential Heat Equation

To solve problems associated with finding the temperature field, it is necessary to solve the differential heat equation, which in the Cartesian coordinate system has the following form [6]

$$\frac{\partial T}{\partial \tau} = a\left(\frac{\partial^2 T}{\partial x^2} + \frac{\partial^2 T}{\partial y^2} + \frac{\partial^2 T}{\partial z^2}\right) = a\nabla^2 T. \tag{5.47}$$

Here T is the temperature; τ is the time; a is the coefficient of thermal diffusivity; and ∇^2 is the Laplace operator.

$$\nabla^2 = \frac{\partial^2}{\partial x^2} + \frac{\partial^2}{\partial y^2} + \frac{\partial^2}{\partial z^2}. \tag{5.48}$$

If there are sources of heat inside the body, the differential equation of heat conduction has the form

$$\frac{\partial T}{\partial \tau} = a\nabla^2 T + \frac{w}{c\rho}. \tag{5.49}$$

Here w is the specific power of the absorbed (emitted) heat per unit time; c is the specific heat capacity of the body; and ρ is the density of the material.

Many calculations in engineering problems are carried out using the model of a finite cylinder because a solution to the differential heat equation is greatly simplified. In the polar coordinate system, the equation has the form

$$\frac{\partial T}{\partial \tau} = a\left(\frac{\partial^2 T}{\partial r^2} + \frac{1}{r}\frac{\partial T}{\partial r} + \frac{1}{r^2}\frac{\partial^2 T}{\partial \varphi^2} + \frac{\partial^2 T}{\partial z^2}\right). \tag{5.50}$$

Here is the polar angle (the angle between the radius-vector, r, and the x-axis).

The relationship between the polar and rectangular coordinate systems is always defined. The reader will find examples of solving such problems in the literature [21]. Equation 5.50 has been called the differential equation of heat conduction in cylindrical coordinates. The differential heat conduction equation in spherical coordinates is not given in this section.

The differential heat conduction equation establishes a relationship between temporal and spatial temperature changes for an infinitely small volume of a body. To solve a differential equation, one must know the temperature distribution inside the body at the initial moment of time (initial condition), the geometric shape of the body and the law of interaction between the environment and the surface of the body (boundary condition).

The combination of the initial and boundary conditions is called "boundary conditions." The initial condition is called the "temporal boundary condition"; the boundary condition is the spatial boundary condition.

The initial condition is determined by setting the law of temperature distribution inside the body at the initial moment of time, that is,

$$T\,(x, y, z, 0) = f\,(x, y, z), \tag{5.51}$$

Where $f\,(x, y, z)$ is a known function.

For a uniform temperature distribution at the initial time, the initial condition takes the form

$$T\,(x, y, z, 0) = T_0 = \text{const.} \tag{5.52}$$

In the theory of heat conduction, boundary conditions are formulated depending on the physical basis of heat transfer between bodies and flows of liquids and gases.

5.10 Differential Equation with Constant Coefficients for Some Function of Two Variables

We first consider a linear homogeneous second-order equation with constant coefficients for a function of one variable $f(x)$, which is called an "ordinary differential equation" and has the form

$$y'' + p\,y' + q\,y = 0, \tag{5.53}$$

where p and q is the constant real numbers.

Taking private solutions in the form $y = e^{kx}$, $k = \text{const}$, we get $y' = ke^{kx}$, $y'' = k^2e^{kx}$.

Substituting the obtained expressions into Equation 5.53, we get

$$e^{kx}\left(k^2+pk+q\right)=0. \tag{5.54}$$

It is obvious that

$$k^2+pk+q=0. \tag{5.55}$$

If k satisfies Equation 5.55, then e^{kx} will be a solution to Equation 5.53.

Equation 5.55 is called the "characteristic equation" of the differential equation (Eq. 5.53). The characteristic equation is a quadratic equation with two roots, k_1 and k_2. Depending on the value of the discriminate, the roots can be real or complex conjugate.

For the case when the roots of the characteristic equation are real and not equal to each other, the general solution (Eq. 5.53) will be

$$y_1=e^{k_1x};\ y_2=e^{k_2x}. \tag{5.56}$$

Since these solutions are linearly independent, then the general solution of Equation 5.53 will be

$$y_1=C_1\,e^{k_1x}+C_2\,e^{k_2x}, \tag{5.57}$$

where C_1, C_2 are arbitrary constants.

It should be noted that solutions (Eq. 5.57) of the differential equation (Eq. 5.53) have the property of superposition, which follows from the theorem: *if y_1 and y_2 have two linearly independent solutions of equation* (Eq. 5.53), *then*

$$y=C_1\,y_1+C_2\,y_2, \tag{5.58}$$

where C_1, C_2 is the arbitrary constants, *there is its general solution* [22].

Two solutions of the equation y_1 and y_2 are called linearly independent if the relation

$$\frac{y_1}{y_2}\neq \text{const}. \tag{5.59}$$

We will give without conclusions the general solution of Equation 5.53 for the case when the roots of the characteristic equation (Eq. 5.55) are complex conjugate

$$k_1=\alpha+\beta j,\ k_1=\alpha-\beta j. \tag{5.60}$$

Here j is the imaginary unit.

The general integral of the differential equation (Eq. 5.53) will be

$$y = e^{\alpha x}\left(A\cos\beta x + B\sin\beta x\right). \tag{5.61}$$

Here A and B are the arbitrary constants.

Using the method of calculation in simpler tasks, it is easier for the reader to master the method of calculation and establish the advantages and disadvantages of one or another method of obtaining the desired solution.

For example, the differential heat equation without heat sources (Eq. 5.47) for some function, T, of two variables, ξ and η, can be represented as a linear homogeneous differential equation of the second order with constant coefficients (a, b, c, d, f, g)

$$a\frac{\partial^2 T}{\partial \xi^2} + b\frac{\partial^2 T}{\partial \xi \partial \eta} + c\frac{\partial^2 T}{\partial \eta^2} + d\frac{\partial T}{\partial \xi} + f\frac{\partial T}{\partial \eta} + gT = 0. \tag{5.62}$$

Solutions of this equation have the property of superposition similar to solutions of an ordinary homogeneous differential equation. If T_1 and T_2 are two partial solutions of Equation 5.62, then the expression $T = C_1T_1 + C_2T_2$ is a general solution of Equation 5.62 for arbitrary values of C_1, and C_2, since $T_1/T_2 \neq \text{const}$.

Taking private decisions in the form

$$T = Ce^{k\xi + l\eta}, \tag{5.63}$$

We have

$$\frac{\partial T}{\partial \xi} = kCe^{k\xi+l\eta}; \quad \frac{\partial T}{\partial \eta} = lCe^{k\xi+l\eta}; \quad \frac{\partial^2 T}{\partial \xi \partial \eta} = klCe^{k\xi+l\eta};$$

$$\frac{\partial^2 T}{\partial \xi^2} = k^2Ce^{k\xi+l\eta}; \quad \frac{\partial^2 T}{\partial \eta^2} = l^2Ce^{k\xi+l\eta}. \tag{5.64}$$

Substituting the expressions (Eqs. 5.63 and 5.64) into Equation (5.62) and reducing all terms by $Ce^{k\xi + l\eta}$, we obtain the equation of coefficients, which is the characteristic equation of the differential equation (Eq. 5.62)

$$ak^2 + bk\,l + cl^2 + dk + f\,l + g = 0. \tag{5.65}$$

Relation (Eq. 5.63) is a particular solution for those values of k and l that satisfy the characteristic equation (Eq. 5.65). If we take an arbitrary value of one of these two coefficients, then the second coefficient is determined from Equation 5.65 (i.e., we get countless private solutions).

Equation 5.65 is a quadratic equation. Assuming k to be variable, and l to be a constant value, then, similarly to the reasoning used before, when solving Equation 5.65 for k, we can get two roots, k_1 and k_2, either real or complex conjugate.

The result obtained for the roots of k depends on the physical essence of the process under study, described by the differential equation (Eq. 5.62) [6].

Relation (Eq. 5.63) can be written as a product of two functions, one of which depends on the variable, ξ, and the other on the variable, η. For example,

$$T = Ce^{k\xi+l\eta} = Ce^{k\xi}e^{l\eta} = C\vartheta(\xi)\theta(\eta). \tag{5.66}$$

In this case, we come to the method of separation of variables. However, there are such solutions of Equation 5.62 for which this separation is impossible [6].

As an example, we will give a solution to the simple problem of finding a symmetric temperature field in an unbounded plate with the given cooling law and initial conditions [22].

Let the unlimited plate be cooled according to a given law.

$$y'' + 2y' + 5y = 0, T = f(y).$$

Find the distribution of the temperature field satisfying the initial conditions:

$$y|_{x=0} = 0; y'|_{x=0} = 1.$$

This task relates to a one-dimensional heat conduction problem (i.e., to finding the temperature along the x coordinate). Here, material properties and cooling medium are not indicated. According to the condition of the problem, the temperature decrease begins at the point $x = 0$ with the speed $dT/dx = 1$ conventional units.

The solution of the problem is reduced to finding a particular solution of an ordinary differential equation. According to Equation 5.55, the characteristic equation can be written as

$$k^2 + 2k + 5 = 0,$$

which has complex-conjugate roots

$$k_1 = -1 + 2j; k_1 = -1 - 2j.$$

In view of Equation 5.58, the general solution of a differential equation is

$$y = e^{-x}\left(A\cos 2x + B\sin 2x\right).$$

The first derivative is calculated by the rule of differentiation of the product of two functions:

$u = e^{-x}$; $v = A \cos 2x + B \sin 2x$. We will finally have

$$y' = e^{-x} \cdot 2B\cos 2x - e^{-x} \cdot 2B\sin 2x.$$

We will find a particular solution satisfying the given initial conditions, and we will determine the corresponding values of A and B. On the basis of the first condition we find

$$0 = e^{-0}\left(A\cos 2\cdot 0 + B\sin 2\cdot 0\right), \text{ whence } A = 0.$$

From the second condition we get $1 = 2\ B$, whence $B = 1/2$.

The sought private solution will be

$$y = \frac{1}{2}e^{-x}\sin 2x.$$

The solution graphic is presented in Figure 5.7.

Similar conditions can be formulated for the problem of determining the longitudinal velocity of an elementary section of a rod under the action of a variable longitudinal force, $F(x, t)$, described by a differential equation

$$\frac{\partial F}{\partial t} = \frac{1}{\rho}\frac{\partial^2 F}{\partial x^2} = \frac{1}{\rho}\frac{\partial \sigma}{\partial x}, \tag{5.67}$$

thus obtaining a relationship between the rates of change of force and the changes in stresses in the rod.

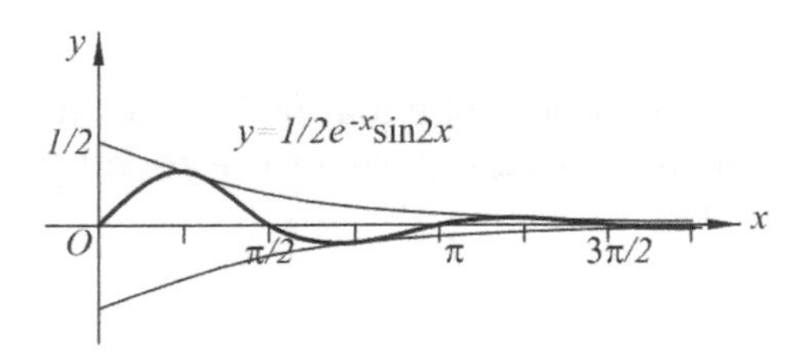

FIGURE 5.7
Graphic representation of the desired private decision.

5.11 Finding a Solution to a Differential Equation by Classical Methods: Fourier Method

Numerous problems of mechanics, thermophysics, and electrodynamics are described by differential equations in partial derivatives of the second order. There are various methods for solving equations, among which are the Fourier method, the Green function method, and the method of characteristics. In this section, we present the classical method of solving partial differential equations, the method of separation of variables (Fourier method).

Consider the differential heat equation without heat sources (Eq. 5.47). The essence of the method lies in the fact that a set of partial solutions of T_n that satisfy the equation and the boundary condition can be found, and then a number of these solutions are compiled using the principle of superposition.

$$T = C_1T_1 + C_2T_2 + ... = \sum_{n=1}^{\infty} C_nT_n. \tag{5.68}$$

A particular solution is sought as the product of two functions, one of which $\theta(\tau)$ depends on time, τ, and the other $\vartheta(x, y, z)$ depends on the coordinates. Ratio is fair

$$T = C\ \theta(\tau)\ \vartheta(x,y,z). \tag{5.69}$$

Here C is an arbitrary constant.

We substitute the solution (Eq. 5.69) into Equation 5.47, we get

$$\theta'(\tau)\vartheta(x,y,z) = a\ \theta(\tau)\nabla^2\vartheta(x,y,z). \tag{5.70}$$

Separating the variables, we rewrite Equation 5.70 in the form

$$\frac{\theta'(\tau)}{\theta(\tau)} = a\frac{\nabla^2\vartheta(x,y,z)}{\vartheta(x,y,z)}. \tag{5.71}$$

Equality of the left and right sides of Equation 5.68 for any values of time and coordinates can occur only if both parts of the equality are equal to some constant value, D, that is,

$$\frac{\theta'(\tau)}{\theta(\tau)} = D = \text{const}; \tag{5.72}$$

$$a\frac{\nabla^2\vartheta(x,y,z)}{\vartheta(x,y,z)} = D = \text{const}. \tag{5.73}$$

We make the transformation left side of the Equation 5.71 or 5.72 and integrate the resulting expression

$$\frac{\partial \theta}{\partial \tau} = D\,\theta(\tau); \int \partial \theta = \int D\,\theta(\tau)\,\partial \tau. \tag{5.74}$$

We obtain a solution that will satisfy Equation 5.69

$$\theta(\tau) = e^{D\tau}. \tag{5.75}$$

The constant D is chosen from physical considerations. For thermal processes seeking to equilibrium, when after a long period of time ($\tau \to \infty$) a certain equilibrium should be established, the value of D will always be negative. Set

$$D = -ak^2, \tag{5.76}$$

where a is the coefficient of thermal diffusivity; and k is a constant determined from boundary conditions.

Substituting Equation 5.76 into Equations 5.72 and 5.75 we get

$$\theta(\tau) = e^{-a\,k^2\tau}. \tag{5.77}$$

$$\nabla^2 \vartheta(x,y,z) + k^2 \vartheta(x,y,z) = 0. \tag{5.78}$$

The differential equation (Eq. 5.78) is often called the Pokel equation, which is well studied in mathematical physics [6]. The solution of the differential heat equation (Eq. 5.47) by the Fourier method is determined by the geometric shape of the body, the initial temperature distribution, and the conditions for heat exchange between the body and the environment.

Let us give an example of solving the heat conduction equation for an unbounded plate (one-dimensional temperature distribution) using the method described. For a function of two variables $T\,(x, \tau)$, this equation has the form

$$\frac{\partial T}{\partial \tau} = a\frac{\partial^2 T(x,\tau)}{\partial x^2}. \tag{5.79}$$

A private decision we believe

$$T = C\,\theta(\tau)\,\vartheta(x). \tag{5.80}$$

When we substitute Equation 5.80 into Equation 5.79, we have

$$\frac{\theta'(\tau)}{\theta(\tau)} = a\frac{\vartheta''(x)}{\vartheta(x)} = -ak^2 = \text{const}. \tag{5.81}$$

Or

$$\frac{d\theta}{d\tau} = -ak^2\theta(\tau); \tag{5.82}$$

$$\frac{d^2\vartheta}{dx^2} = -k^2\vartheta(x). \tag{5.83}$$

The solution of Equation 5.82 will be the expression (Eq. 5.77). Equation 5.83 has two solutions

$$\vartheta_1(x) = \sin kx, \text{ or } \vartheta_2(x) = \cos kx, \tag{5.84}$$

because the second derivative of the function $\vartheta(x)$ must be equal to the function itself, multiplied by some quantity (k^2)

$$\begin{aligned} \vartheta_1'(x) &= k\cos kx;\ \vartheta_1''(x) = -k^2 \sin kx = -k^2\vartheta_1(x), \\ \vartheta_2'(x) &= -k\sin kx;\ \vartheta_2''(x) = -k^2 \cos kx = -k^2\vartheta_2(x). \end{aligned} \tag{5.85}$$

These solutions are linearly independent. The general solution of Equation 5.83 will be

$$\vartheta(x) = C\vartheta_1(x) + D\vartheta_2(x) = C\sin kx + D\cos kx. \tag{5.86}$$

Taking into account Equations 5.77, 5.80, and 5.86, we write the particular solution of the heat equation for our case

$$T(x,\tau) = C\sin kx e^{-ak^2\tau} + D\cos kx e^{-ak^2\tau}. \tag{5.87}$$

The constant k is determined from the boundary conditions. The constants C and D are determined from the initial conditions. They have certain meanings depending on the conditions of the task.

The solution of the problem given in Section 5.10 and the solution (Eq. 5.87) are identical. However, these solutions were obtained by different methods for solving homogeneous differential equations.

In the case of the propagation of heat waves in the body, the quantity D will be an imaginary quantity.

5.12 Application of Methods of Integral Transformation

To solve practical engineering problems, classical methods for solving differential equations cannot be applied because of the arising mathematical difficulties. For these purposes, operational methods solutions are widely used.

Such methods include the Laplace integral transform method. The theoretical aspect of this method the reader will find in the work of G. Deutsch [23].

The method of the Laplace transform is that we are taught to study not the function (the original) but rather its modification (the image). This transformation is carried out by multiplying the integrand by the exponential function and integrating the new integrand within certain limits. The Laplace transform symbolically is written as

$$L\left[f(t)\right] = F(s) = \int_0^{\infty} f(t)e^{-st}dt. \tag{5.88}$$

Writing $L[f(t)]$ means L-transform.

Here, the function $F(s)$ is a function of the complex variable $s = x + jy$.

The Laplace transform associates the one-valued function $F(s)$ of the complex variable s (*image*) with the corresponding function $f(t)$ of the real variable t (*original*).

For the image to exist, the integral (Eq. 5.88) must converge. The convergence condition imposes restrictions on the function $F(s)$. Readers may familiarize themselves with the conditions of limitation in [11].

If the problem is solved in the images, then the finding of the original in the image (inverse transformation) is performed according to the inversion formula. Such a transition is called the "inverse Laplace transform," symbolically denoted by L^{-1}, and corresponds to

$$f(t) = L^{-1}[F(s)] = \frac{1}{2\pi}\int_{x-j\infty}^{x+j\infty} F(s)e^{st}ds, \ t > 0. \tag{5.89}$$

The right side of Equation 5.89 is called the "inversion of the Laplace integral" and is a complex integral. If the ratio (Eq. 5.89) is rewritten as

$$\begin{aligned} &\int_{-\infty}^{+\infty} e^{jyt}(x + jy)dy = 2\pi e^{-xt}f(t) \quad \text{at } t < 0; \\ &\int_{-\infty}^{+\infty} e^{jyt}(x + jy)dy = 0 \ \text{ at } \ t < 0, \end{aligned} \tag{5.90}$$

then formulas (Eqs. 5.88 and 5.90) have physical meaning

If the image is a fractional function, s, and has the form

$$F\left(s\right) = \frac{\varphi(s)}{\psi(s)} = \frac{A_0 + A_1 s + A_2 s^2 + \ldots}{B_1 s + B_2 s^2 + \ldots}. \tag{5.91}$$

Then applying the decomposition theorem, we get

$$f(t) = L^{-1}\left[\frac{\varphi'(s)}{\psi(s)}\right] = \sum_{n=1}^{\infty} \frac{\varphi(s_n)}{\psi'(s_n)} e^{snt}. \tag{5.92}$$

Here s_n is the simple roots of the function $\psi(s)$. The denominator has a countable set of simple roots and does not contain a free term, provided that $A_0 \neq 0$. In this case, the integral (Eq. 5.89) exists.

We will outline the benefits of the operational method of solving problems compared to classical methods.

First, the process of applying the integral transformation is typical for problems of various physical nature (e.g., heat transfer and mechanical motion, vibrations, electrical conductivity).

Secondly, the operational method is convenient for solving boundary value problems. The method allows solving problems with different boundary conditions in the same way.

Thirdly, the application of simple operational calculus theorems allows one to obtain solutions in a form convenient for calculation in small and large time intervals.

Fourthly, the efficiency of solving various tasks using the Laplace transform method is determined by the presence of image tables.

The disadvantage of using this method is the limitations to solving problems with spatial coordinates.

We apply the Laplace transform to the solution of the differential heat equation for a one-dimensional heat flux in the plate, which has the form

$$\frac{\partial T(x,\tau)}{\partial \tau} = a\frac{\partial^2 T(x,\tau)}{\partial x^2}. \tag{5.93}$$

We write the Laplace transform for the variable s

$$L\left[\frac{\partial T(x,\tau)}{\partial \tau}\right] = aL\left[\frac{\partial^2 T(x,\tau)}{\partial x^2}\right]. \tag{5.94}$$

We have a second-order differential equation with constant coefficients relative to the image

$$a\frac{d^2 T(x,s)}{dx^2} - sT(x,s) + u(x) = 0. \tag{5.95}$$

We write the initial conditions for the case when at the initial moment of time the temperature at all points of the body is the same and equal to zero

$$T(x, 0) = u(x) = 0. \tag{5.96}$$

In view of Equation 5.94, Equation 5.95 in the images will take the form

$$T''(x,s) - \frac{s}{a}T(x,s) = 0. \tag{5.97}$$

The solution to Equation 5.97 will be

$$T(x,s) = A \cdot ch\sqrt{\frac{s}{a}}x + B \cdot sh\sqrt{\frac{s}{a}}x = A_1 e^{\sqrt{\frac{s}{a}}x} + B_1 e^{-\sqrt{\frac{s}{a}}x}. \tag{5.98}$$

Here A, B, A_1, B_1 are constant coefficients with respect to x. The coefficients are related by

$$A_1 = \frac{A+B}{2}, \ A_1 = \frac{A-B}{2}. \tag{5.99}$$

The coefficients are from boundary conditions. Then, using the image tables we find the original $T\ (x, s)$.

Consider this problem when the initial temperature distribution is given by some function $u(x)$, and

$$T\ (x, 0) = u(x). \tag{5.100}$$

Then Equation 5.97 takes the form

$$T''(x,s) - \frac{s}{a}T(x,s) + u(x) = 0. \tag{5.101}$$

If at the initial moment of time the temperature in all currents is the same and is equal to

$$u(x) = T_0 = \text{const}, \tag{5.102}$$

then the solution of Equation 5.101 will be

$$T(x,s) - \frac{T_0}{s} = A \cdot ch\sqrt{\frac{s}{a}}x + B \cdot sh\sqrt{\frac{s}{a}}x = A_1 e^{\sqrt{\frac{s}{a}}x} + B_1 e^{-\sqrt{\frac{s}{a}}x}. \tag{5.103}$$

$$A_1 = \frac{A+B}{2}, \ A_1 = \frac{A-B}{2}. \tag{5.104}$$

Constant coefficients A and B each time are determined from the boundary conditions.

It is known from mathematical physics that the heat equation is a parabolic equation. Let us show that the process of applying an integral transform is also universal for solving second-order hyperbolic differential equations.

The equation of a hyperbolic type is the wave equation, which in the case of a one-dimensional distribution is

$$\frac{\partial^2 u}{\partial t^2} = a^2 \frac{\partial^2 u}{\partial x^2}. \tag{5.105}$$

This second-order differential equation describes the free vibrations of a string attached to the ends. We write:

border conditions

$$u\,|_{x=0} = 0,\ u\,|_{x=l} = 0; \tag{5.106}$$

and initial conditions

$$u\,|_{t=0} = \varphi_0(x),\ \frac{\partial u}{\partial t}\,|_{t=0} = \varphi_1(x); \tag{5.107}$$

The solution of the differential equation (Eq. 5.105) with the boundary and initial conditions (Eqs. 5.106 and 5.107) is well known. Using the Fourier method (the method of separation of variables, Section 5.10), we obtain formulas for the expansion of given functions $\varphi_0(x)$ and $\varphi_1(x)$ into a Fourier series in sines in the interval $(0, l)$. These formulas are

$$\varphi_0(x) = \sum_{k=1}^{\infty} a_k \sin\frac{k\pi x}{l},\quad \varphi_1(x) = \sum_{k=1}^{\infty} \frac{k\pi a}{l} b_k \sin\frac{k\pi x}{l}. \tag{5.108}$$

The coefficients of the expansion a_k and b_k are calculated by the known formulas

$$a_k = \frac{2}{l}\int_0^l \varphi_0(x)\sin\frac{k\pi x}{l}dx,\quad b_k = \frac{2}{k\pi a}\int_0^l \varphi_1(x)\sin\frac{k\pi x}{l}dx. \tag{5.109}$$

For comparison, we present the solution of a practical engineering problem obtained by the method of integral transform. We investigate the oscillations of the mechanism that drives the body with the help of a spring.

5.13 Oscillations in a Nonlinear Spring-Loaded Mechanism with One Degree of Freedom

Devices with spring-loaded parts are used in many mechanisms. When analyzing work of these mechanisms, usually investigated is movement. However, during oscillations, especially forced ones, which are often the main mode of operation, it is important to know not only the peculiarities of movement, but also the laws of voltage variation, which have a significant impact on the working capacity of the product. For these purposes, we apply the Kondratenko method [10,11]. We formulate the following statement of the problem.

Consider a device (Figure 5.8), where the leading rod (pusher) moves at a speed, υ_1. The rod moves the body, having mass, m. This body moves along the x-axis from a spring. The body moves along the guides of the cylinder with a speed, υ_2. The body is affected by the resistance force, F, and the friction force. We believe that the spring works within the limits of elasticity and fluctuations at the end of the driven body do not affect the movement of the pusher.

The friction force in the mechanism will be proportional to the velocity of the body and is calculated by the formula

$$F_m = h\,\upsilon_2, \tag{5.110}$$

where h is the coefficient of friction loss.

Let in the static the spring have a characteristic

$$\Delta x\, C = F_0, \tag{5.111}$$

where Δx is the displacement; C is the stiffness coefficient; and F_0 is the force.

Equation 5.111 is rewritten as

$$\Delta x = \frac{lF_0}{fE_y},\; f = \frac{\pi\left(D_{ex}^2 - D_{in}^2\right)}{4},\; E_y = \frac{lC}{f}. \tag{5.112}$$

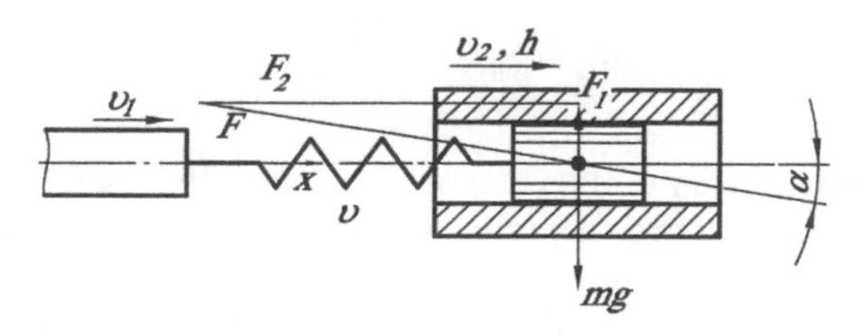

FIGURE 5.8
Scheme of the spring-loaded mechanism.

Here l is the length of the spring; f is the sectional area of the spring; E_y is the conditional modulus of elasticity of the spring; and D_{ex}, D_{in} is the external and internal diameters of a spring.

It is known that under the action of longitudinal loads the wire (diameter, d) is subject to torsion. In the spring, shear stresses arise, which can be determined using the relations

$$\tau_{\max} = \frac{4F}{\pi d^2}\left(1 + \frac{2D_0}{d}\right). \tag{5.113}$$

Here D_0 is the average diameter of the spring.

Spring parameters can be described by the formula [10,24]

$$K = \frac{4c-1}{4c-4} + \frac{0{,}615}{c}, \quad c = \frac{D_0}{d}. \tag{5.114}$$

If we consider a spring as a rod with a cross-sectional area, f, then under the action of a longitudinal force, it will develop conditional compressive-tensile stresses (σ_y)

$$\sigma_y = \frac{F}{f} = \frac{\Delta x E_y}{l}. \tag{5.115}$$

Taking into account Equations 5.112 through 5.114, we have

$$\sigma_y = \frac{d^3}{2KD_0\left(D_{\mathcal{H}}^2 - D_{\mathcal{B}}^2\right)}\tau_{\max} = K_1\tau_{\max}. \tag{5.116}$$

Let us assume that the spring works within the limits of elasticity, and after compressing with some static force, F_0, the body is affected by the variable component, F_a, of the force with small amplitude. Under such conditions, it is possible to consider the conditional density of the material of the spring constant and equal to

$$\rho_y = k_n \rho_{\mathcal{M}} \frac{l_1 d^2}{D_{\mathcal{H}}^2 - D_{\mathcal{B}}^2}. \tag{5.117}$$

Here ρ_y is the real density of the wire material; k_n is the coefficient taking into account the change in wire diameter at the ends of the spring; and l_1 is the length of the wire.

Under these assumptions, within the framework of the one-dimensional model, the known equation of longitudinal elastic oscillations (Eq. 5.105) for displacements can be written in the form

$$\frac{\partial^2 u}{\partial t^2} = \frac{E_y}{\rho_y} \frac{\partial^2 u}{\partial x^2}. \tag{5.118}$$

Equation 5.118, we supplement by the equation of momentum in differential form (Eq. 5.25) [10,25]

$$\rho_y \frac{\partial \upsilon}{\partial t} = -\frac{\partial \sigma_y}{\partial x}. \tag{5.119}$$

From do we have

$$E_y \frac{\partial^2 u}{\partial x^2} = -\frac{\partial \sigma_y}{\partial x}. \tag{5.120}$$

After differentiation along the x coordinate and integration with respect to t of Equation 5.120, we obtain Equation 5.26 given in Section 5.5. We have

$$E_y \frac{\partial \upsilon}{\partial x} = -\frac{\partial \sigma_y}{\partial t}. \tag{5.121}$$

Equations 5.119 and 5.120 describe the elastic oscillations in a spring relatively completely [10]. These equations are similar to the equations obtained for metal rods (Section 5.5). As a result of the transition under zero initial conditions, we have a one-dimensional distribution of the function and two first-order differential equations. We apply the Laplace transform to the solution of Equations 5.119 and 5.120:

$$L\left[\rho_y \frac{\partial \upsilon}{\partial t}\right] = \rho_y s \upsilon(s); \; L\left[-\frac{\partial \sigma_y}{\partial t}\right] = -s\sigma_y(s). \tag{5.122}$$

As a result of the transition under zero initial conditions, we have a one-dimensional distribution of the function and two ordinary differential equations of the first order

$$\rho_y s \upsilon(s) = -\frac{d\sigma_y(s)}{dx}. \tag{5.123}$$

$$E_y \frac{d\upsilon(s)}{dx} = -s\sigma_y(s). \tag{5.124}$$

The solution of Equations 5.123 and 5.124 are form

$$\sigma_y(s,x) = \sigma_{1y}(s,0) ch[\theta_n(s)x] - \frac{\theta_n(s)}{s} E_y \upsilon_1 sh[\theta_n(s)x]. \tag{5.125}$$

$$\upsilon(s,x)=\upsilon_1(s,0)ch\left[\theta_n(s)x\right]-\frac{s}{\theta_n(s)E_n}\sigma_{1y}(s,0)sh[\theta_n(s)x]. \quad (5.126)$$

Here θ_n (s) is the operator coefficient of waves propagation, which is determined by the formula

$$\theta_n(s)=\pm\sqrt{\frac{s}{E_y}\left[\rho_y s+\psi(s)\right]}. \quad (5.127)$$

Here, ψ is a function that takes into account friction losses during the movement of coils of a spring.

The functions σ_y (s, x) and υ (s, x) describe the oscillations of the speeds of movement and stresses in a power elastic line.

Assuming that the amount of movement from the pusher to the body is transmitted completely (i.e., there are no reflected waves, and the friction of the turns is insignificant), we obtain the equation of motion in Laplace images in the form [10]

$$\upsilon_2(s)\left[1-c+h_n\vartheta_n(s)s+m\vartheta_n(s)s^2\right]=\frac{1-c}{ch[\theta(s)l]}\upsilon_1(s)-F(s)\vartheta_n(s)s. \quad (5.128)$$

Equation of oscillations of conditional stress

$$\sigma_y(s)=\frac{1}{f(1-c)\left[1-c+h_n\vartheta_n(s)s+m\vartheta_n(s)s^2\right]}\left[F(s)+\frac{\upsilon_1(s)(h_n+ms)}{ch[\theta_n(s)l]}\right]. \quad (5.129)$$

Here

$$\vartheta_n(s)=\frac{l}{fE_y}Z(s);\ Z(s)=\frac{th\left[l\theta_n(s)\right]}{l\theta_n(s)};\ \theta_n(s)=\pm s\sqrt{\frac{\rho_y}{E_y}}. \quad (5.130)$$

Here Z(s) is the function elastic of the spring.

We substitute in the resulting Equations 5.128 through 5.130

$$s=j\omega. \quad (5.131)$$

We get real functions

$$Z(j\omega)=\frac{tg\alpha}{\alpha}=Z(\alpha);\ ch\alpha=\cos\alpha;\ \alpha=l\omega\sqrt{\frac{\rho_y}{E_y}}. \quad (5.132)$$

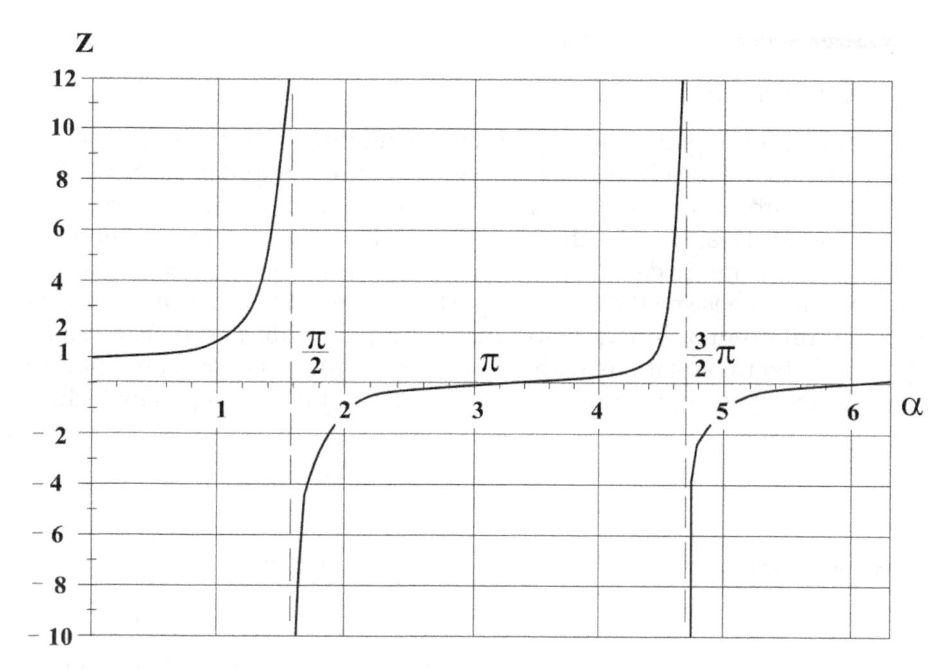

FIGURE 5.9
Change of function Z depending on the dimensionless parameter α.

Here ω is the circular frequency of harmonic oscillations.

The graph of changes in the function $Z(\alpha)$ is shown in Figure 5.9.

This graph clearly defines the zones of stable and unstable operation of the mechanism.

It can be seen from the graph that when this parameter tends to zero, the function Z tends to unity, that is,

$$\alpha \to 0, Z \to 1. \tag{5.133}$$

In cases of variation of α in the intervals $\pi/2 + k\pi > \alpha > \pi + k\pi$, the function Z takes negative values.

Thus, the proposed method for studying the mechanism, and the solution obtained using the Laplace transform allow us to determine the law of the oscillations of the bar (slave link) stresses caused by the longitudinal movement of the spring (the leading link).

This approach allows us to obtain a mathematical model for the study of tangential stresses in the case of the transmission of rotational motion from the leading link to the driven link. The reader will find a solution to this problem in the literature [15].

5.14 Conclusion

Note the following. Nowadays, numerical methods for solving differential equations are well developed, which, using computer-mathematical programs, are successfully used in solving practical engineering problems [26–28]. Despite this, the classical methods for solving differential equations and methods of integral transformation remain relevant at the present time. These methods are simple to perform, allow finding solutions of linear differential equations (ordinary and partial derivatives) and some types of integral equations with the help of algebraic actions, are vivid in describing physical processes, and are in demand in solving many technical problems.

References

1. Lanczos C. *The Variational Principles of Mechanics*. University of Toronto Press, Toronto, Canada, 1949.
2. Sedov L.I. *Continuum Mechanics*, Vol. 1, 2. Nedra, Moscow, 1970. (In Russ.).
3. Routh E.J. *Dynamics of a System of Rigid Bodies*. Macmillan and Company, London, UK, 1905.
4. Reiner M. Rheology. In *Handbuch Der Physik*, Vol. 4. S. Flugge (Ed.), Springer, Berlin, Germany, 1958.
5. Bezuhov N.I. [Osnovy teorii uprugosti, plastichnosti i polzuchesti] *Fundamentals of the Theory of Elasticity, Plasticity and Creep*. Moscow, Vysshaya shkola, 1968. (In Russ.).
6. Lykov A.V. [Teoriya teploprovodnosti] *Theory of Heat Conductivity*. Moscow, Vysshaya shkola, 1967. (In Russ.).
7. Carslaw H.S. and Jaeger J.C. *Conduction of Heat in Solids*. Clarendon Press, Oxford, 1959.
8. Muskhelishvili N.I. [Nekotorye osnovnye zadachi matematicheskoj teorii uprugosti] *Some Basic Problems of the Mathematical Theory of Elasticity*. Nauka, Moscow, 1966. (In Russ.).
9. Love A.E.H. *A Treatise on the Mathematical Theory of Elasticity*, 4th ed. Cambridge University Press, 1927. First American printing 1944.
10. Kondratenko L.A. [Raschet kolebanij v detalyah i uzlah mashin] *Calculation of Velocity Variations and Stresses in Machine Assemblies and Components*. Moscow, Sputnik, 2008. (In Russ.).
11. Mironova L., Kondratenko L. Application of the Laplace transform in problems of studying the dynamic properties of a material system and in engineering technologies. Chapter 1 Application of the Laplace transform in problems of studying the dynamic properties of a material system and in engineering technologies. *Advanced Mathematical Techniques in Engineering Sciences*. M. Ram and J. P. Davim (Eds.), CRC Press, Taylor & Francis Group, Boca Raton, FL, 2018.

12. Kondratenko L.A. [Kolebaniya i metody upravleniya skorost'yu dvizheniya tekhnologicheskih ob"ektov] *Vibrations and Speed Regulation Methods of Movement of Technological Objects*. Moscow, MRSU, 2005. (In Russ.).
13. Kondratenko L.A. [Mekhanika rolikovogo val'cevaniya teploobmennyh trub] Mechanics roller rolling heat exchange tubes. Moskow, Sputnik, 2015. (In Russ.).
14. Kondratenko L., Terekhov V., Mironova L. The aspects of roll-forming process dynamics. AT the 22end International Conference on Vibroegineering, Moscow. *Vibroengineering PROCEDIA*, 2016: 460–465.
15. Kondratenko L.A., Terekhov V.M., Mironova L.I. [Ob odnom metode issledovaniya krutil'nyh kolebanij sterzhnya i ego primenenii v tekhnologiyah mashinostroeniya] About one method of research torsional vibrations of the core and his application in technologies of mechanical engineering. *Journal of Engineering & Automation Problems*, 2017, 1: 133–137. (In Russ.).
16. Kondratenko L., Mironova L., Terekhov V. Investigation of vibrations during deepholes machining. *Vibroengineering Procedia*, 2017, 11: 7–11. doi:10.21595/vp.2017.18285.
17. Kondratenko L., Mironova L., Terekhov V. On the question of the relationship between longitudinal and torsional vibrations in the manufacture of holes in the details. *Vibroengineering Procedia*, 2017, 12: 6–11. doi:10.21595/vp.2017.18461.
18. Kondratenko L., Dmitriev V., Mironova L. Simulation of a drive with a long connecting link. *Vibroengineering Procedia*, 2017, 12: 231–236. doi:10.21595/vp.2017.18461.
19. Kondratenko L.A., Mironova L.I. [Imitacionnoe modelirovanie nelinejnyh privodov s raspredelennymi parametrami silovyh magistralej] Imitation of nonlinear drives with distributed parameters of power lines. *Journal of Engineering & Automation Problems*, 2018, 1: 92–97. (In Russ.).
20. Kondratenko L.A., Terechov V.M., Mironova L.I. [Poterya ustojchivosti raboty privodov vrashcheniya s dlinnymi stal'nymi silovymi liniyam] Loss of stability of the drives of rotation with long steel force lines. *Journal of Engineering & Automation Problems*, 2018, 2: 40–46. (In Russ.).
21. Mironova L.I., Dmitriev V.G., Kondratenko L.A. [Zadachi dlya prakticheskih zanyatij po matematike. Analiticheskaya geometriya na ploskosti] Tasks for practical classes in mathematics. *Analytical Geometry on the Plane*. Moscow, Sputnik, 2018. (In Russ.).
22. Piskunov N.S. [Differencial'noe i integral'noe ischisleniya dlya vtuzov]. *Differential and Integral Calculus for Technical Colleges*. Moscow, Fizmatgiz. (In Russ.).
23. Doetsch G. Anleitung zum praktiscen gebrauch der Laplace-transformation. R. Oldenbourg, München, 1961.
24. Mironova L., Kondratenko L. Oscillations in a nonlinear spring-loaded mechanism with one degree of freedom [Kolebaniya v nelinejnom podpruzhinennom mekhanizme s odnoj stepen'yu svobody]. In the collection: Nonlinear dynamics of machines School-NDM, Moscow, 2017, 356–363. doi:10.18411/a-2017-048. (In Russ.).
25. Kondratenko L., Mironova L. Features of loss of stability of the work of two-link mechanisms that have an infinite number of degrees of freedom. *International Journal of Mathematical, Engineering and Management Sciences*, 2018, 3(4): 315–334.

26. Nakamura T., Yamamoto H., Xiao X. Fast calculation methods for reliability of connected-(r, s)-out-of-(m, n):F lattice system in special cases. *International Journal of Mathematical, Engineering and Management Sciences*, 2018, 3(2): 113–122.
27. Singh U.P., Medhavi A., Gupta R.S., Shankar Bhatt S. Theoretical study of heat transfer on peristaltic transport of Non-Newtonian fluid flowing in a channel: Rabinowitsch fluid model. *International Journal of Mathematical, Engineering and Management Sciences*, 2018, 3(4): 450–471.
28. Ramadevi B., Ramana Reddy J.V., Sugunamma V. Influence of thermodiffusion on time dependent Casson fluid flow past a wavy surface. *International Journal of Mathematical, Engineering and Management Sciences*, 2018, 3(4): 472–490.

6

Application of Transmuted Gumbel Copula for Energy Modeling

Alok Dhaundiyal, S. B. Singh, and Muammel M. Hanon

CONTENTS

6.1 Introduction

Bioenergy plays a crucial role in the economic activities, which is incessantly burgeoning in a competitive atmosphere of industrialization and development. As the energy consumption is strongly correlated with economy of country, hence the wide range of alternative sources is to be examined for meeting the requirement of urbanization. Fossil fuels has been considered as solution of solving dearth of energy sources for many centuries. The irrevocable effect of remnants released alongside with the main products on the finite atmosphere has made necessary analyzing the alternative renewable forms of energy. In this manner, exploration of the new renewable forms of energy has begun to subdue the ecological disturbance. Thus, application of biological material as the new renewable energy source is becoming center of attention for the power-generation industries. Plants encompass the huge area of Earth and use of solar energy at the colossal level for conversion of water and carbon into vital organic compounds. Therefore, it is necessary to use fecundity of organic matter, which can be later used as a source of clean energy without disturbing the carbon-cycle of ecosystem.

For the same reason, in Dhaundiyal and Tewari (2015) conducted a comparative evaluation of fossil fuel with forest waste. It was concluded that pine needles are cost effective for off-grid areas of country. Moreover, the feedstock production cost of pine needles is relatively good to coal. However, there is a fluctuation in overall cost of power generation as a result of carbon taxation, yet an optimal yield of 2.335 Mg hm^{-2} yr^{-1} would make a forest waste far more competitive than fossil fuel. To know holistic view of pine waste, the experiments are conducted on 10 kWh gasifier, and it has been observed that specific biomass consumption of pine waste is 1.59 kg per kWh at 60% load (Dhaundiyal and Tewari 2016). Groundnut shell, wood, and agro-residue briquettes have a feed rate of 50 kg/h at the rated capacity of 625 MJh^{-1} (Bhoi et al. 2006). In the joint venture of Indian and the US Agency for International Development to reduce dependency on imported oil, Talib et al. (1986) reported that the average replacement of diesel plant from Babul, eucalyptus, and teak was 81.8%, 68%, and 67.6%, respectively. So, the overall statistical data favors implementation of advance technology to curtail consumption of the conventional fuels.

But conversion technology demands an analytical insight of chemical as well as thermal interaction of biomass within programmed thermal history. Pyrolysis is one such process that causes thermal decomposition of biomass in the absence of air. However, pyrolysis is a broad class of thermal conversion, but its primal tenet is similar for all the cases except the imposed thermal conditions. This process usually occurs in the temperature range of 200°C–400°C (Dhaundiyal and Tewari 2017), but it depends on the substrate used for analysis purpose. The maximum decomposition rate of cellulose occurs within range of 300°C–327°C for pine needles, whereas at the same heating rate, decomposition perk shows the slightest negative drift for another class of coniferous species, *Cedrus deodar*, from pine needles (Dhaundiyal and Hanon 2018). However, hemicellose decomposition peak shows a drastic variation of decomposition rates among pine needles, *C. deodar*, and the weed (Dhaundiyal et al. 2018b). Difference in thermal decomposition peaks is mainly correlated to chemical composition and kinetics of concurrent reactions to provide desirous products for energy generation. The yield of the secondary fuel derived through pyrolysis is governed by independent parallel reactions; therefore, the inclusion of kinetic models becomes a necessary part to analyze the pyrolysis problem. Hence, a concise and robust mathematical scheme acts as an interface between kinetic reactions and the user, and a complete analysis could be materialized.

Hence, a new theory is proposed to analyze thermal conversion process critically. The objective of this study is to demarcate pyrolysis as a physical phenomenon that is divided into two subindependent events. Here, these events are classified as primary and secondary pyrolysis, which can influence each other but are mutually exclusive. Multireaction model is subjected to revise through bringing a concept of bivariate distribution function of activation energies. Approximation is based on the relatively narrow distribution of bivariate activation energies to temperature integral. Marginal probability density functions of the Rayleigh and the Weibull distributions are integrated together in the Gumbel-Hougaard Copula to solve thermal conversion problem.

6.2 Material and Methods

6.1.1 Thermochemical Process

Demarcation of thermal conversion is based on the thermal constraints imposed on a process. Categories of thermal conversion are: gasification, pyrolysis, and direct liquefaction. In general, pyrolysis does not require any catalyst for obtaining the liquid products. The light decomposed molecules are converted into oily compounds via homogenous reactions in the gas phase. However, liquefaction is thermal decomposition performed in the presence of suitable catalyst. The liquid by products of low molecular weight is obtained at the end of liquefaction, which are highly unstable and reactive. Repolymerization of lighter molecular compounds into oily compounds makes it suitable for use (Molten et al. 1983). It has been reported that variation of heat of reaction changed the pathway of pyrolysis process. It is observed that several endothermic and exothermic peaks are obtained for biomass pyrolysis (Dhaundiyal and Tewari 2017; Dhaundiyal and Hanon 2018; Dhaundiyal et al. 2018a, 2018b). According to various studies, cellulose pyrolysis is reported to follow endothermic course (Brown et al. 1952; Cho et al. 2010; Zhang et al. 2018), whereas lignin pyrolysis has exothermic heat of reaction (Roberts 1971; Ibbett et al. 2011). Mostly, pyrolysis is seen as an integral part of gasification process, which has relatively low stoichiometric ratio to combustion process. Thus, pyrolysis is a nascent stage of carrying out gasification of any biomass (Dhaundiyal and Gupta 2014). The carbonaceous compounds break into small molecules in the gaseous form (i.e., condensable vapor (tars and oils) and solid charcoal under pyrolysis conditions). Nonsynchronous form of decomposition of biomass is reported by Lanzetta and Blasi (1998). They stated that the major fraction of volatile released during the initial stage of pyrolysis (250°C–300°C) (Lanzetta and Di Blasi 1998). It indicates that the series of transformation occurring during thermal conversion are incoherent and assumed to be independent to each other.

The overall reaction scheme of pyrolysis is yet to comprehend extensively, but simplifying schemes have been proposed to serve the purpose of modeling pyrolysis of biomass. The pyrolysis of biomass is bifurcated into two independent events, namely primary and secondary pyrolysis. Primary pyrolysis refers to the degradation of major components of biomass (i.e., cellulose, hemicellulose, and lignin). Concurrent and simultaneous decomposition of cellulose, hemicellulose, and lignin at different zones of fuel is function of local temperature and makes the primary reactions regime of biomass pyrolysis. Enthalpy variation within this primary regime is reported to be less (Zaror and Pyle 1984). The final products of the primary reactions (i.e., char and volatiles) are catalyzed in the secondary stage of decomposition, which is referred as the secondary pyrolysis of biomass. Such autocatalytic reactions are triggered when the hot volatile comes in physical contact of untreated biomass. Many researchers assume it as a single event and attempt to unify the conversion process through a single distribution pattern. Although the local temperature is a key factor to determine the extent of this secondary pyrolysis (secondary event), yet it is not proved that these two events strongly correlated to each other. Auto-catalytic secondary reactions are difficult to model since the experimental information related to mechanism of these reactions and reaction rates is anonymous to observer (Sinha et al. 2000). For modeling primary reactions, a one-step first order global reaction has been recommended in various kinetic models (Bamford et al. 1946; Matsumoto et al. 1969; Maa and Bailie 1973; Fan et al. 1977; Kansa et al. 1977). But the most accurate and up-to-date approach for modeling biomass pyrolysis follows the multistep reaction scheme, the distributed activation energy model (DAEM) (Ferdous et al. 2002; Galgano and Blasi 2003; Dhaundiyal and Singh 2017a, 2017b, 2017c). Kansa et al. (1977) and Roberts and Clough (1963) separately brought round the common notion that the modeling of secondary pyrolysis was necessary to bridge the gap between experimental and model predictions. A model of two-step mechanism, where an "Intermediate product" formed during primary pyrolysis involves in the secondary stage of decomposition, proposed by Koufopanos et al. (1991). As it was complicated to establish the relationship of components of transition state experimentally, the proposed model of Koufopanos et al. (1991) was modified later. The concept of multistep reactions has been developed on the basis of Koufopanos's model (Jalan and Srivastava 1999). The two-step reaction scheme imbibes a major role of product of primary pyrolysis in secondary reactions, which modified distribution pattern of the final end product. However, it is failed to address the autocatalytic secondary reactions. A comprehensive outlook of different classes of mechanism proposed for pyrolysis of wood and other cellulosic materials is reported in the review article of Di Blasi (1993). The kinetic mechanism of cellulosic material is demarcated into three groups: one-step global models; single-stage, multireaction models; and two-stage, semi-global models. The first group of models assume the pyrolysis as a single step first-order reaction, and it is believed the reaction is proceed in the same manner as illustrated:

$$\text{Solid} \xrightarrow{k} a \ \text{Gases} + b \ \text{Tars} + c \ \text{Char} \tag{6.1}$$

where k is a rate constant of the reaction, and a, b and c are yield coefficients of the different products of pyrolysis. Di Blasi (1993) quoted values of kinetic triplet for different species of wood, and he reported that the kinetic constants was strongly correlated to the experimental conditions under which the values were obtained. The second class of modeling based on simultaneous and competing first-order reactions in which biomass is decomposed into tar, char, and gases. The third category of models postulates that the pyrolysis is a two-stage process, where the products of the first stage disintegrate further in the presence of each other to initiate the second stage of pyrolysis of biomass. These models are initially propounded for cellulose by Bradbury et al. (1979), for lignin by Antal (1983), and for wood by Koufopanos et al. (1991). However, Koufopanos works was reformed later by Jalan and Srivastava (1999). To explain the secondary pyrolysis, in the experimental work of Lee et al. (1977) who concluded that a single-step reaction model was implausible to provide an accurate model. Thus, they suggested the following multistep scheme that is based on concept of parallel competitive reactions

$$\begin{aligned} &\text{wood} \xrightarrow{\text{heat}} \text{char} + \text{tar} + \text{gas} \\ &\text{tar} \xrightarrow{\text{char}} \text{char} + \text{gas} + \text{heat} \end{aligned} \tag{6.2}$$

The decomposition reaction accelerated by the hot char is responsible for the apparent residence time dependency of exothermic pyrolysis reactions and the secondary char formation. The effectiveness of proposed model of Lee et al. (1977) and inability of single-step scheme is also suggested by other researchers in their work (Akita 1959; Panton and Rittmann 1971; Murty and Blackshear 1967; Broido and Nelson 1975).

The concept of multireaction scheme is, however, very old. It is related to work of Constable (1925), who noticed a compensation effect between A (frequency factor) and E (activation energy) for surface catalyzed reactions. It was proposed that the reactant surface had a variety of sites that arose a notion of distribution of activation energies governing the reaction (Constable 1925). In Vand (1943) coined a reactivity distribution model for irreversible electrical resistance changes upon annealing of evaporated metal films. The concurrent reaction scheme received breakthrough from the commendable work of Pitt (1962) and Hanbaba (1967) in the coal pyrolysis community. Later, the concept of stochastic modeling was introduced to describe activation energies of various parallel reaction as a continuous density function of normal distribution for coal pyrolysis (Anthony and Howard 1976). Another approach to solve complex reaction network is pivoted on the discrete energy distribution model. It had been developed by the workers of the Institute Francais du Petrole for pertroleum generation, and it was documented and published by Ungerer and Pelet (1987). The concept of introducing reactivity distribution originates from the isoconversional methods of Friedman (1964) and Ozawa (1965), which were initially propounded for evaluating a single activation energy and reaction order, but later it has been used for knowing the characteristic of activation energy with respect to conversion. Miura et al. (1995) and Miura and Maki (1998) gave a model that is related to overlapping of parallel reactions. He emphasized that the effective fraction must rather be used than the actual fraction in the isoconversional methods. He consolidated that the reaction profiles were not represented by "Dirac-delta" function; thus, at any given level conversion exhibits the overlapping of reactions with different activation energies. Thereafter, he deduced an expression to demarcate activation energies that have or haven't significant contribution to conversion point. However, it is difficult to obtain a constant frequency factor through Miura's method while predicting solution from the normal distribution model (Cai et al. 2011). But the computed values of parameters compensate for each other in such a manner as mandated by the mathematical solution and should be remembered. The classical form of DAEM or multireaction model (MRM) comprises a univariate form of density function that defines the activation energy of the whole process to be governed by same marginal distribution function. The nonisothermal nth-order DAEM equation is defined as:

$$1-\alpha=\int_{0}^{\infty} exp\left[\int_{T_0}^{T} \frac{A}{\beta} exp\left(\frac{-E}{RT}\right) dT\right] f(E)dE \tag{6.3}$$

where, E is the activation energy, A is the frequency factor, R is the ideal gas constant and δ is the heating rate, n is the reaction order, T is the absolute temperature, T_0 is the initial reaction temperature, α is the conversion, and $f(E)$ is the density function of activation energies.

De Caprariis et al. (2012) observed that the experimental TG data of coal provided two different slopes, which indicated that two different mechanism occur at same conversion point, therefore, a single Gaussian model (1-DAEM) is not enough to simulate biomass pyrolysis. Thus, they proposed an implementation of double Gaussian DAEM model to describe thermal decomposition of coal. The remedial measure of overcoming shortcomings of the classical DAEM, Zhang et al. (2014) have also adopted the multi-Gaussian approach to describe thermal conversion process.

The classical form of DAEM can be modified by considering pyrolysis process as series of two independent events occurring independently; hence, the proposed form of DAEM can be given as:

$$1-\alpha=\int_{0}^{\infty} exp\left[\int_{T_0}^{T} \frac{A}{\beta} exp\left(\frac{-(f_1E_1+f_2E_2)}{RT}\right) dT\right] \left(H^t(E_1,E_2)\right)\left(f_1dE_1+f_2dE_2\right) \tag{6.4}$$

In this equation, the function $H^t(E_1,E_2)$ represents the transmuted bivariate form of activation energy function. The f_1 and f_2 are coefficients that makes the activation energy convex combination of E_1 and E_2. However, the function $H^t(E_1,E_2)$ is to be defined explicitly so that the pyrolysis process can be examined qualitatively. Thus, there is a need of linking function that can correlate the activation energies, E_1 and E_2, in a consolidated form. In the subsequent sections, Equation 6.4 is examined within the scope of asymptotic limits imposed on activation energies.

6.1.2 Application of Archimedean Copulas

The concept of a "Copula" was first introduced in Sklar (1959), which was further examined by Deheuvels (1979) and Genest and Mackay (1986). Archimedean family of copulas includes Clayton, Frank, and Gumbel functions, which provides a richer dependence than the usual Gaussian assumptions. In the qualitative aspect of energy modeling, the function $H^t(E_1,E_2)$ is described as a joint probability density function of activation energies needed to initiate the primary as well as the secondary pyrolysis reactions. It follows the Gumbel-Hougaard copula. The concept of implementing copula pivots around the stage decomposition of biomaterial. In thermogravimetry, material exhibit a plateauing for a certain time unless it reaches a critical temperature to initiate the chemical reactions. The same phenomenon is explained in different perspective that the primary pyrolysis influences the autocatalytic reactions, but it does not participate in the secondary phase of thermal decomposition of biomass. Therefore, these two simultaneous steps are tantamount to mutually exclusive processes. Inability of the single-step process to interpret intrinsic behavior pyrolysis process requires some remedial solution; thus, the new guideline is necessary for reformation of the classical form of DAEM. Some other mathematical schemes are also proposed to solve modeling problem of pyrolysis (Burnham and Braun 1999; Güneş and Güneş 2002; Cai and Ji 2007; De Caprariis et al. 2012; Zhang et al. 2014). To establish relationship among different regimes of pyrolysis of biomass, there is a need to improve the form of initial distribution function. It can precisely be modeled by a MRM because it is necessary to involve all sets of reactions that may affect the releasing fraction of volatiles. Pyrolysis of biomass is significantly depending on the reaction scheme and the local temperature. However, the overall reaction pathway of pyrolysis is labyrinth of thousand numbers of concurrent reactions that cannot easily be recognized during experimentation. But the new hypothesis and simplification of the existing model in the correct manner may bring paradigm shift in the classical theory of kinetic modeling. Interdependence of various reactions on each other cannot be delineated by a single marginal function; hence, it is replaced by a bivariate function of molecular activation energies. The fundamental essence of this investigation can be comprehended by the fact that classical formulation is incapable to demarcate the different pyrolysis steps.

To demonstrate this approach, a copula function is represented as a cumulative probability function of activation energies (E_1; E_2).

Now, assume $E \in (E_1, E_2)$ and $h(E_1)$ and $h(E_2)$ are any two univariate distributions. It can be shown that

$$H(E_1,E_2) = C_\theta\left(h(E_1);h(E_2)\right) \tag{6.5}$$

where C is the copula introduced to link the univariate marginal probability functions $h(E_1)$ and $h(E_2)$.

Here, the marginal distribution functions $h(E_1) = 2\chi(E-\phi)exp\{-\chi(E-\phi)^2\}$ and $h(E_2) = \frac{\rho}{\eta}(E-\gamma/\eta)^{(\rho-1)}e^{-(E-\gamma/\eta)^\rho}$ are represented by the Rayleigh and the Weibull distribution functions, respectively. The expression for joint probability of activation energies is considered to follow the Gumbel-Hougaard Copula. χ and ρ are the shape parameters of Rayleigh and Weibull distribution function, respectively, and η is the scale factor of the Weibull distribution function. The threshold or location parameters of the Rayleigh and the Weibull distribution functions are represented by ϕ and γ, respectively. Then

$$C_\theta\left(f(E_1);f(E_2)\right) = e^{-\left[\left[-log\left\{2\chi(E-\phi)exp\left\{-\chi(E-\phi)^2\right\}\right\}\right]^{\frac{1}{\theta}} + \left[-log\left(\frac{\rho}{\eta}\left(\frac{E-\gamma}{\eta}\right)^{(\rho-1)} e^{-\left(\frac{E-\gamma}{\eta}\right)^\rho}\right)\right]^{\frac{1}{\theta}}\right]^\theta}, \forall\theta \in [1,\infty) \tag{6.6}$$

The parameter θ provides different kind of dependences. For $\theta = 1$, this becomes independent copula. But as value of θ tends to infinity, comonotonicity will be exhibited.

The functions $h(E_1)$ and $h(E_2)$ vary accordingly to the Rayleigh and the Weibull distributions, respectively. In other words, distribution pattern of activation energies of all reactions occurring during primary pyrolysis are represented by the density function of Rayleigh distribution; whereas the secondary pyrolysis reactions are demarcated by the Weibull distribution. Equation 6.4 has two terms: double exponential (DExp) and the distribution function of activation energies. DExp is an implicit function of time and temperature; although the second term is independent of time, it depends on the distribution characteristic of volatiles release during pyrolysis. Miura (1995) and Miura and Maki (1998) gave an approximation of double exponential term to determine the effective boundary between energies. He replaced the double exponential term with its proposed approximation equation form without keeping in the view error component associated with expression. Later, he evaluated A and E pairs by using Equation 6.7. Equation 6.7 is also referred as "temperature integral."

$$exp\left[\int_{T_0}^{T}\frac{A}{\beta}exp\left(\frac{-E}{RT}\right)dT\right]\sim exp\left[\frac{-0.545ART^2}{\delta E}\right] \tag{6.7}$$

Double exponential term has no analytical solution; therefore, it is approximated by using Laplace method of integral; thus this rapidly varying term is approximated by a piecewise linear function. It has three distinct regions: one where DExp is zero, one where it is equal to unity and one in domain of (0,1), where it rises linearly from zero to one. The width and location of the linear segment varies with time. To avoid the transient heat-up time characteristic of isothermal experiments, linear thermal history is chosen (Eq. 6.8).

$$T(K)=\beta l(s) \tag{6.8}$$

To demonstrate the scheme, the parameter $-E/RT$ is assumed to be large; therefore, the dominant contribution from the integral is when l is near t (at neighborhood of temperature range where maximum decomposition takes place). This provide the concrete asymptotic approximation to the function:

$$exp\left[\int_{0}^{t}\frac{A}{\beta}exp\left(\frac{-E}{R\beta l}\right)dT\right]\sim exp\left[\frac{-AR\delta t^2}{E}e^{\frac{-E}{R\delta l}}\right] \text{ as } \frac{E}{R\beta l}\to\infty \tag{6.9}$$

Equation 6.9 can also be rewritten in the form

$$\sim exp\left[-exp\left(\frac{E_s-E}{E_w}\right)\right] \tag{6.10}$$

It is clear from Equation 6.10 that the function switches swiftly from zero to one as E increase by size E_w around E_s.

Suppose

$$g(E)=\left(\frac{E_s-E}{E_w}\right)$$

Then, Equation 6.10 will be

$$\sim exp\left[-exp\left(g(E)\right)\right]$$

Here,

$$g(E)=\frac{-E}{R\beta t}+ln\left(\frac{AR\beta t^2}{E}\right) \tag{6.11}$$

As the nature of function $g(E)$ around central value, E_s, is of main interest; hence, it is expanded by the Taylor series expansion.

$$g(E) \sim g(E_s) + (E - E_s) g'(E_s) + \frac{(E - E_s)^2}{2} g''(E_s) + \ldots \tag{6.12}$$

The value of function $g(E)$ are chosen in such a way that

$$g(E_s) = 0 \text{ and } g'(E_s) = \frac{1}{E_w}$$

After solving Equations 6.11 and 6.12, we get

$$E_s = R\beta t W(At) \text{ and } E_w = \left(\frac{R\beta t E_s}{R\beta t + E_s} \right)$$

where $W(x)$ represents Lambert W function.

To incorporate the energy correction factor, "y," and time scale factor, "τ," the terms can be modified as

$$y_{s1} = \frac{f_1 R\beta t W(\tau)}{\phi} \text{ and } y_{s2} = \frac{f_2 R\beta t W(\tau)}{\gamma}$$

Similarly,

$$y_{w1} = \frac{y_{s1}}{1 + W(\tau)} \text{ and } y_{w2} = \frac{y_{s2}}{1 + W(\tau)}$$

Here subscripts "1" and "2" represent the energy correction factors for different marginal distribution functions.

The time factor is defined as $\tau = At$.

The central value of activation energies, E_s, and the activation energies spacing, E_w, are used throughout in the numerical solution. The notations ϕ and γ represent the location or threshold parameters of the Rayleigh and the Weibull distributions, respectively.

6.1.3 Transmuted Functions

Regression-related models are convincing and time saving, but whether they estimate the parametric values precisely and address the kinetic mechanism of pyrolysis properly or not is an important question. Therefore, it must be carefully examined. Iso-conversional and model-based schemes are some of regression supporting means of estimating the kinetic parameters. For instance, the fundamental basis of model-free scheme states that the same reactions occur in the same ratio at different heating rates, and it must be independent of temperature. However, the scheme is scuppered when a system of parallel independent reactions whose relative reactivity is proportional to temperature, and it changes because of the variation of activation energies (Golikeri and Luss 1972). There is also likelihood of existence of such a reaction pathway of having different activation energies changes at different temperature (Burnham and Dinh 2007). The other form of kinetic modeling, model-fitting, pivots on force-fitting of data to hypothetical reaction model. Determination of kinetic triplet by phenomenological model, which is already assumed. This scheme is incapable to distinguish the temperature dependence of rate constant and the conversion. Consequently. resulting a drastic variation and fallacious values of A and E. This kind of modeling is highly misleading for kinetic analysis (Brown et al. 2000). Implementation of conventional logistical distribution functions are also not only way to represent activation energies of parallel reactions, but there are also various means of solving DAEM, such as linear mixing of density function $H^t(E_1,E_2)$

$$H^t(E_1,E_2) = (1+\lambda)H(E_1,E_2) - \lambda\left(H(E_1,E_2)\right)^2, |\lambda| \le 1 \tag{6.13}$$

where E_1 and E_2 are activation energies of two stage decomposition reactions

A function $H(E_1,E_2)$ is said to be transmuted if it follows Equation 6.13. Although the same expression is also valid for cumulative distribution function of activation energy. Shaw and Buckley (2009) were the first leading researchers who introduced the distributional flexibility by imbuing a new parameter into an existing distribution. They observed that involvement of parameters did not make solution redundant but increased the reproducibility of the modeling of natural process (Shaw and Buckley 2009). The generated or transmuted family includes the parent distribution and imparts more dynamical characteristic to simulated results than the conventional logistical approach for various types of data. One of such investigations reported for modeling the metrological data, and it has been found that the transmuted Gumbel distribution has effectively analyzed the snowfall data of a region (Aryal and Tsokos 2009). In another study of wind speed data, the transmuted Frechet distribution has provided much better results than exponential Frechet distribution to gather the statistical properties for a wind energy-conversion system (Elbatal et al. 2014). In reliability analysis of component, it is seen that the behavior of transmuted function is useful for modeling reliability data in the manufacturing industries (Ashour and Eltehiwy 2013). Therefore, it is possible that transmuted model may provide better solution than the conventional modeling procedure. For asymptotic and shapes of transmuted function, the following limits are adopted (Bourguignon et al. 2016).

Proposition 1: The asymptotic of Equation 6.1, if $p(E) \to 0$, are

$$(i)\ p^t(E) \sim (1+\lambda)p(E)$$

$$(ii)\ G^t(E) \sim (1+\lambda)G(E)$$

Proposition 2: The asymptotic of Equation 6.1 as $E \to \infty$, are

$$(i)\ p^t(E) \sim (1-\tau)p(E)$$

$$(ii)\ G^t(E) \sim (1-\tau)G(E)$$

NOTE: Refer Appendix "A" for numerical solution of Gumbel-Hougaard Copula.

6.1.4 Experimental Set-up and Computational Algorithms

To analyze the pine needles, the chemical and thermogravimetry techniques are adopted to obtain the chemical and thermal data of pine needles. The pine wasted is collected from pest county of Hungary.

The chemical assessment of biowaste is done by CHNS-O analyzer (vario MACRO). The elemental composition (C%, N%, S%, H%, and O%) is derived on dry basis. Before performing chemical test, the furnace is preheated till 1473 K for 30 minutes. Once the apparatus is operative, the capsule form of sample wrapped in a silver foil is placed inside the rotating turret. The flow rate of oxygen gas maintains the catalytic combustion. Helium gas is used as carrier gas to carry away the product of combustion. The volatile gases thereafter separate at different reduction columns, where the gas mixture is resolved into its components through trap chromatography. These tubes are adjacent to each other, and they are retrofitted with thermal conductivity detector (TCD). The TCD generates the electrical signal, which is send to computer-based software. The combustion properties are analyzed by the bomb calorimeter at constant volume.

On the other hand, thermal characteristic of pine wasted is evaluated by the thermogravimetry (Diamond TG/DTA, Perkin Elmer, USA). The samples of pine waste are tested at different heating rate (5°C/min, 10°C/min, and 15°C/min). The temperature varies from 303 to 923 K in each experimental run. To avoid buoyancy effect, the horizontal differentia type balance is used for experimentation (Brown 1988). Thermocouple type "R" (Platinum-Rhodium- 13%/Platinum) is used to measure the temperature inside the furnace. The purge flow rate of noninteractive gas is measured 200 mL min^{-1}. Indium and Tin are used as reference material for differential thermograms. Table 6.1 lists the chemical characteristic of pine needles.

TABLE 6.1

Chemical Characteristic of Pine Waste

C%	H%	N%	O%	S%	Ash Content (%)	Higher Heating Value (MJ/kg)
54.28	6.55	0.61	35.50	0.11	2.92	22.07

MATLAB 2015a is used as an interface between chemical model and the numerical solution. Computational programming pivots on testing the solution at different heating rate (β). Each assigned value is examined in each loop if it satisfies the given tolerance level, it is accepted for use, else the loop terminated itself. To initiate the computational task, the inputs are varied. They are allowed to pass through the given end condition unless it converges. The numerical solution is obtained by the Laplace integral of the dynamical system. The two different limits are imposed on the numerical solution of MRM, so that the precise solution can be portrayed.

6.3 Results and Discussions

The numerical solutions of DAEM are obtained by replacing the univariate function with the Gumbel-Hougaard copula. The solutions are derived by imposing different limits ($E \to \infty$ and $E \to 0$) on the transmuted form of activation energies. It is assumed that the activation energies are divided into two mutually exclusive decomposition stages of biomass pyrolysis; thus, the plots are drawn at different heating rates. The conventional form of Laplace method of integration is adopted. To know the relative significance of temperature integral (double exponential function), the value of "y" gets varied around the central value (y_s) of the temperature integral.

The effect of transmuted form of DAEM for activation energy limit of $E \to \infty$ is illustrated in Figure 6.1a–d. The estimated values of numerical parameters corresponding to the limit $E \to \infty$ are given in Tables 6.2 and 6.3. It is clear from the thermo-analytical data obtained for different heating rates that the released fraction of volatile content $(1-\alpha)$ must be at neighborhood of one during initial stage of biomass pyrolysis. However, the remaining mass fraction for the predicted solution is less than one during the initial decomposition of biomass. Yet, it is able to simulate the primary devolatilization stage obtained after dehydration of biomass. The two-stage decomposition can be easily demarcated through the second plateau obtained between temperature range of 600 and 650 K. The plateauing effect during the initial stage of pyrolysis happened due to the stationary attribute of the moving maxima (Figure 6.1c). It indicates that the exponential functions $h(y_1^e)$ and $h(y_2^e)$ remains constant. But as the maxima points y_1^e and y_2^e increases with time, the remaining mass fraction of biomass decreases

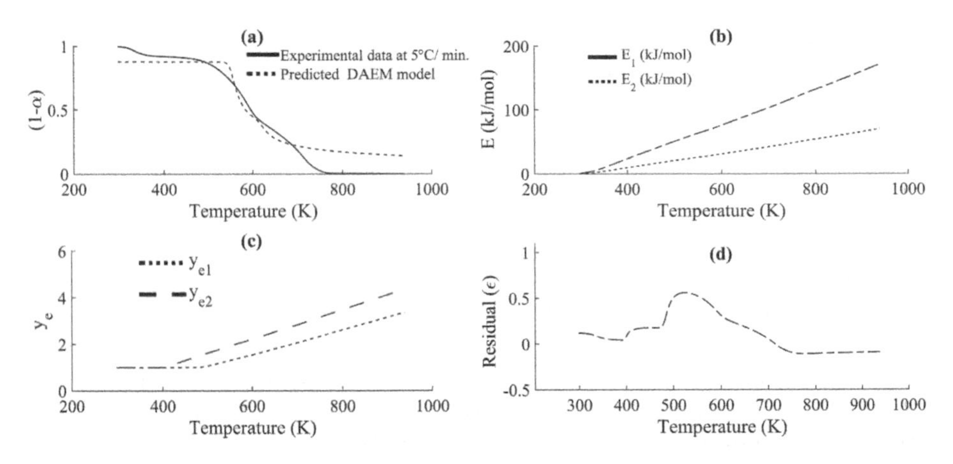

FIGURE 6.1
The effect of transmuted Gumbel-Hougaard on the activation energy and characteristic of numerical solution ($E \rightarrow \infty$, at 5°C/min). (a) Comparative sketch; (b) Activation energies for two-stage pyrolysis; (c) Location of moving maxima with respect to temperature; (d) Validation of result through the residual plot.

TABLE 6.2

Parametric Value of Kinetic Parameters at Different Heating Rates $(E \rightarrow \infty)$

Heating Rate β (°C/min)	Primary Pyrolysis $\overline{E_1}$ (kJ/mol)	Secondary Pyrolysis $\overline{E_2}$ (kJ/mol)	Primary Pyrolysis (kJ/mol) E_{10} (Min)	E_{11} (Max)	Secondary Pyrolysis (kJ/mol) E_{20} (Min)	E_{21} (Max)	A (min⁻¹)	Coefficient of Regression (R^2)	λ
5	84.66	34.57	0.0094	171.06	0.0038	69.87	10^{20}	0.93	0.1780
10	80.81	33.00	0.0177	163.39	0.0072	66.73	10^{19}	0.94	0.1780
15	80.05	32.85	0.0249	161.94	0.0102	66.46	10^{18}	0.95	0.1780

TABLE 6.3

Additional Parameters for Kinetic Analysis $(E \rightarrow \infty)$

E_1 a (kJ-min²/-°C-mol)	E_1 b (kJ-min/°C-mol)	E_1 c (kJ/mol)	E_2 a (kJ-min²/°C-mol)	E_2 b (kJ-min/-°C-mol)	E_2 c (kJ/mol)	β (°C/min)	Combination Coefficients f_1	f_2
0.062	−1.7	92	0.028	−0.74	38	5	0.71	0.29
						10	0.71	0.29
						15	0.70	0.30

drastically, and the pyrolysis process is finished with the residual mass of 0.9 mg. The activation energy domains $\{E_1, E_2\}$ for the two-stage pyrolysis at 5°C/min, 10°C/min, and 15°C/min are {[0.0094,171.06],[0.0038,69.37]}, {[0.0177,163.39],[0.0072,66.73]} and {[0.0249,161.94],[0.0102,66.46]}, respectively. The effect of activation energies on conversion (α) and temperature (T) is illustrated in Figures 6.2 and 6.3. It can be understood from the estimated values of activation energies for the primary and the secondary stages of pyrolysis that the autocatalytic reactions occur at the higher temperature regime with the relatively low range of activation energies (Figures 6.2 and 6.3). On the other hand, the primary pyrolysis requires

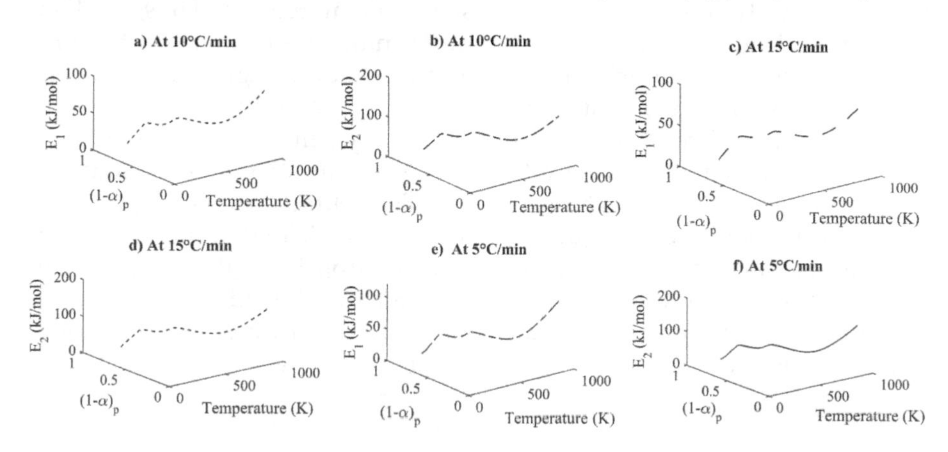

FIGURE 6.2
The variation of activation energies (E_1 and E_2) with conversion and temperature (T) (Proposition 1). (a, b: primary and secondary stages, respectively, at 10°C/min; c, d: primary and secondary stages, respectively, at 15°C/min; e, f: primary and secondary stages, respectively, at 5 °C/min).

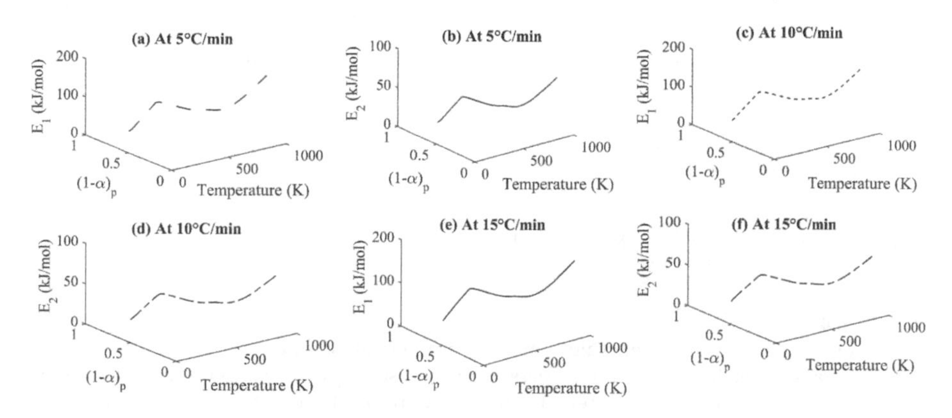

FIGURE 6.3
The variation of activation energies (E_1 and E_2) with conversion and temperature (T) (Proposition 2). (a, b: primary and secondary stages, respectively, at 5°C/min; c, d: primary and secondary stages, respectively, at 10°C/min; e, f: primary and secondary stages, respectively, at 15 °C/min).

an appreciable range of activation energy to initiate the chemical reactions among various constituents of biomass. The fraction of the primary pyrolysis is found to be $f_1 = 0.71$, whereas it is $f_2 = 0.29$ for the secondary stage of pyrolysis. The overall estimated values of kinetic parameters are uniform and consistent to each other, and there is no anomality found with respect to time and temperature. The transmuted form of DAEM for limit $E \to \infty$ exhibits the good simulation results at the transient stage of pyrolysis. But its inability to converge with thermo-analytical data is observed in the temperature interval of 600 to 900 K. It happened because the relative decrease of slope of moving maxima (y_2^e) with respect to temperature which fastens the decomposition of biomass with respect to temperature. Thus, the temperature scale gets shifted up for the similar number of experimental runs. It can also be seen through graphical representation (Figure 6.1a) that the relative variation of slope at temperature range of 500 to 700 K. To validate the proposed scheme for activation energy limit $E \to \infty$, the tests are conducted at different heating rate, and the similar attributes are found at 10°C/min and 15°C/min. The schematic plots obtained at 10°C/min and 15°C/min are depicted in Figures 6.4a–d and 6.5a–d. To determine whether the computational algorithms provided the appropriate value of coefficient of regression, the residual graphs (Figures 6.1d, 6.4d, 6.5d, 6.6d, 6.7d, 6.8d) are plotted for the obtained predicted solutions at different heating rates. The effect of heating rate on the averaged value of activation energies is

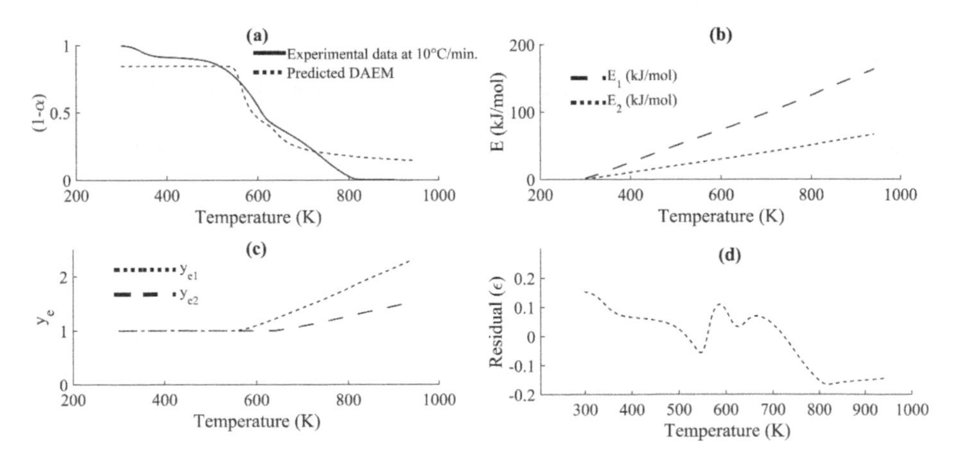

FIGURE 6.4
Characteristic of activation energies and the numerical solution ($E \to \infty$, at 10°C/min). (a) Comparative sketch; (b) Activation energies for two-stage pyrolysis; (c) Location of moving maxima with respect to temperature; (d) Validation of result through the residual plot.

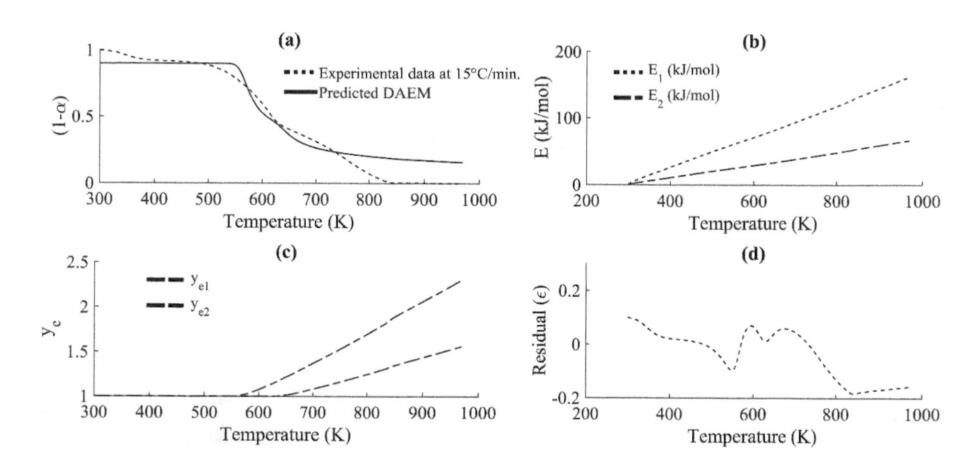

FIGURE 6.5
Characteristic of activation energies and the numerical solution ($E \rightarrow \infty$, at 15°C/min). (a) Comparative sketch; (b) Activation energies for two-stage pyrolysis; (c) Location of moving maxima with respect to temperature; (d) Validation of result through the residual plot.

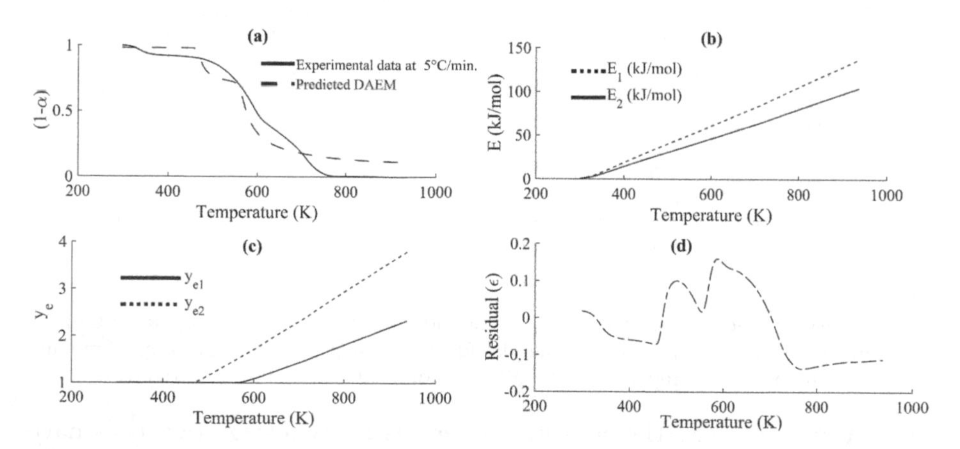

FIGURE 6.6
Characteristics of activation energies and the numerical solution ($E \rightarrow 0$, at 5°C/min). (a) Comparative sketch; (b) Activation energies for two-stage pyrolysis; (c) Location of moving maxima with respect to temperature; (d) Validation of result through the residual plot.

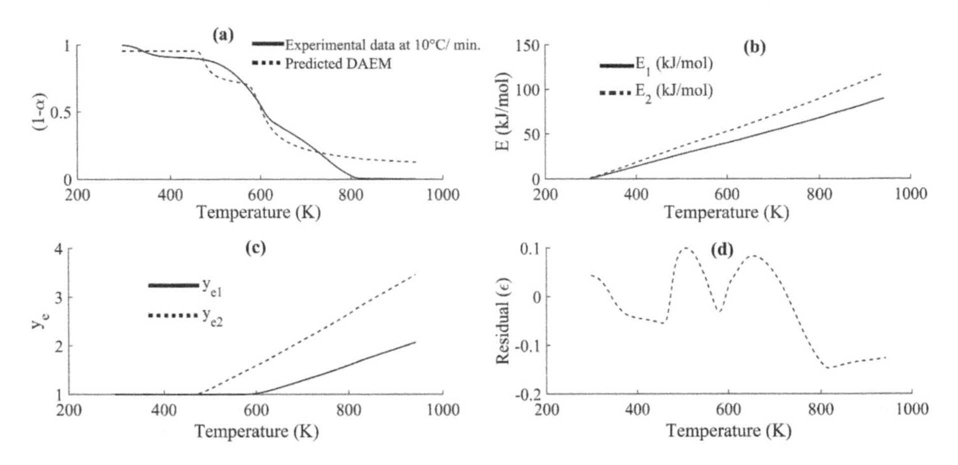

FIGURE 6.7
Characteristic of activation energies and the numerical solution ($E \to 0$, at 10°C/min). (a) Comparative sketch; (b) Activation energies for two-stage pyrolysis; (c) Location of moving maxima with respect to temperature; (d) Validation of result through the residual plot.

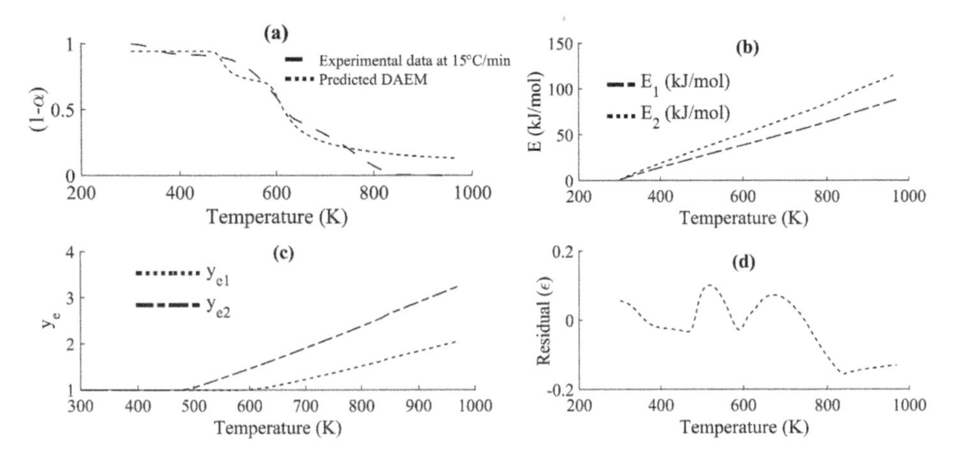

FIGURE 6.8
Characteristic of activation energies and the numerical solution ($E \to 0$, at 15°C/min). (a) Comparative sketch; (b) Activation energies for two-stage pyrolysis; (c) Location of moving maxima with respect to temperature; (d) Validation of result through the residual plot.

shown in Figure 6.9. The activation energies for two-stage pyrolysis have shown the similar trend with heating rate. But the activation energy (E_2) decreases faster than E_1 as heating rate increases from 5°C/min to 15°C/min. Furthermore, the domain of activation energies $\{E_1, E_2\}$ also get shrunk with the increasing heating rate. Therefore, on the basis of predicted model, it can be concluded that the activation energy associated with the secondary pyrolysis is highly affected by temperature and the residence time.

The effect of transmuted Gumbel-Hougaard on the activation energy for limit $E \to 0$ is illustrated at Figure 6.6a–d. Unlike the higher range of activation

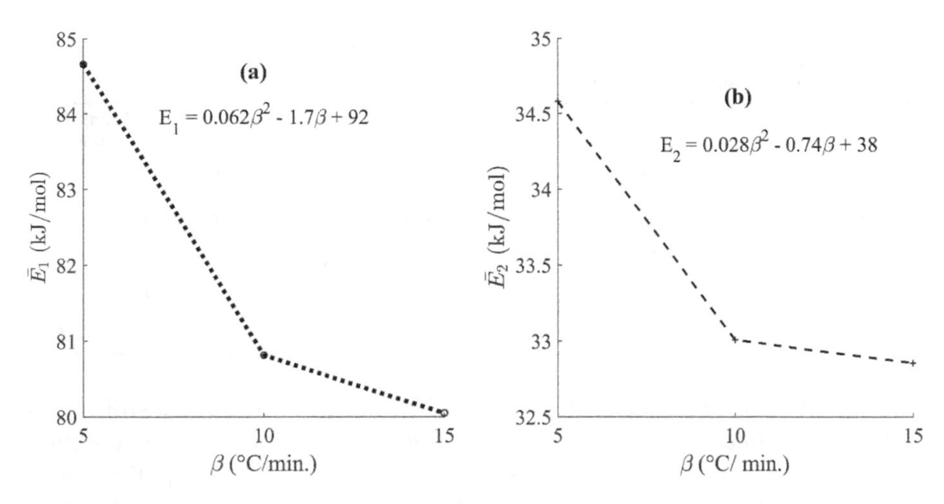

FIGURE 6.9
The effect of the heating rate, δ, on the averaged activation energies (E_1 and E_2) for biomass pyrolysis. (a, primary stage; b, secondary stage).

energies, the obtained solution, when the activation energy is tending to reduce with time and temperature, is found to be much promising at $E \to 0$. The overall predictability of the solution depends on the components of numerical solution. In case of Gumbel-Hougaard copula, the contribution of the marginal distribution functions in the final numerical solution predicts the qualitative improvement of the final solution. The plateauing effect as well as the variation of activation energies is relatively low for $E \to 0$. The obtained residual mass of biomass is 0.5 mg, which is 44% less than the residual mass derived for $E \to \infty$. The estimated values of numerical parameters are given in Tables 6.4 and 6.5. The activation energy domains associated with the primary and secondary stages of pyrolysis at heating rate of 5°C/min, 10°C/min, and 15°C/min are $\{[0.0057,104.08],[0.0075,136.85]\},\{[0.0095,89.42],[0.0124,117.57]\}$ and $\{[0.0132,88.16],[0.0173,115.91]\}$, respectively. Similar to the predicted solution obtained for $E \to \infty$, the second plateau can be demarcated at temperature interval of 550 to 600 K. The shifting of plateau happens because of the relative

TABLE 6.4
Parametric Value of Kinetic Parameters at Different Heating Rates $(E \to 0)$

Heating Rate β (°C/min)	Primary Pyrolysis $\bar{E}_1$ (kJ/mol)	Secondary Pyrolysis $\bar{E}_2$ (kJ/mol)	Primary Pyrolysis (kJ/mol) E_{10} (Min)	$E_{11.}$ (Max)	Secondary Pyrolysis (kJ/mol) E_{20} (Min)	E_{21} (Max)	A (min^{-1})	Coefficient of Regression (R^2)	λ
5	51.51	67.72	0.0057	104.08	0.0075	136.85	10^{20}	0.94	0.1780
10	44.17	58.08	0.0095	89.42	0.0124	117.57	10^{17}	0.96	0.1780
15	43.51	57.22	0.0132	88.16	0.0173	115.91	10^{16}	0.95	0.1780

TABLE 6.5

Additional Parameters for Kinetic Analysis ($E \rightarrow 0$)

E_1			E_2				Combination coefficients	
a (kJ-min²/-°C-mol)	b (kJ-min/°C-mol)	c (kJ/mol)	a (kJ-min²/°C-mol)	b (kJ-min/°C-mol)	c (kJ/mol)	β (°C/min)	f_1	f_2
0.013	−3.5	66	0.018	−4.6	86	5	0.43	0.57
						10	0.43	0.57
						15	0.43	0.57

variation of convex coefficients, f_1 and f_2 (Table 6.5). Moreover, the shifting of plateau also affects the configurational value of activation energies assigned for the primary and the secondary stages of pyrolysis. More the plateau shift to the lower temperature regime, the higher activation energy is needed. This effect makes the fact more concrete that the transmuted form for limit $E \rightarrow 0$ exhibit the characteristic required for dehydration and devolatilization. Thus, it can be concluded that the activation energy required for the initial of evaporation, or preliminary decomposition, is relatively low to the activation energy required for decomposition of cellulose and lignin parts of biomass. It may provide better insight about the stage decomposition if both the limits get merged into a single robust model. In this manner, the clarity of activation energy boundaries would be grasped to much better extent. The similar effect is produced by both the limits except the boundaries of activation energies required to trigger the primary and the secondary pyrolysis are altered. The effect of heating rate on the activation energies is illustrated in Figure 6.10. On the contrary, the results obtained at $E \rightarrow \infty$, E_1 decreases more quickly than

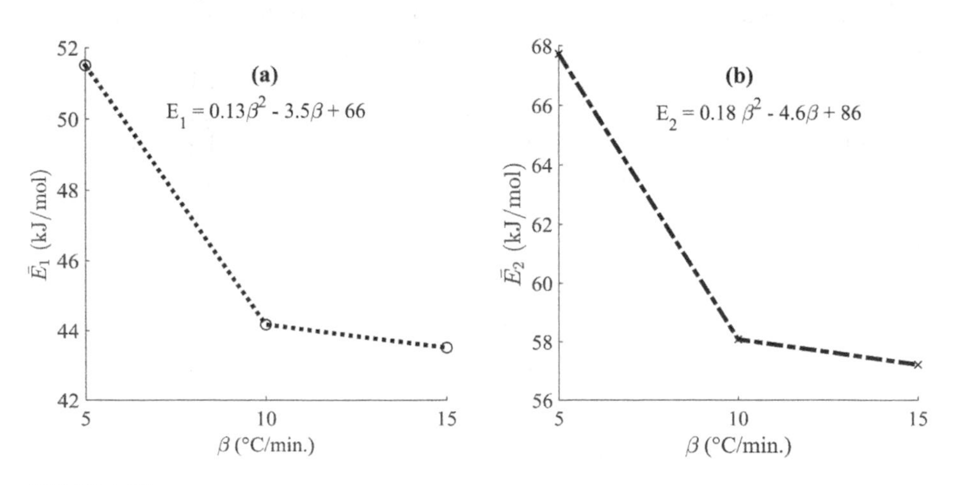

FIGURE 6.10
The effect of the heating rate, δ, on the averaged activation energies (E_1 and E_2) for biomass pyrolysis. (a, primary stage; b, secondary stage).

E_2 as the heating rate increases. The effect of variation of E_1 reduces the plateauing effect at the onset of pyrolysis of biomass. Moreover, it can also be explained through relative shifting of location of moving maxima y_e (Figures 6.6c, 6.7c, 6.8c) as compared to y_e (Figures 6.1c, 6.4c, 6.5c) as imposed limit on activation energy changes. The dormant state of y_e reduces by 100 K for the similar input sets of data. Therefore, it can be inferred that the initial distribution of activation energy affects the whole process. The overlapping of activation energy boundaries changes the rate of conversion with time ($d\alpha/dt$), albeit, the experimental curves do not show wide variation in ($d\alpha/dt$) with time or temperature at the initial stage of pyrolysis. With increasing heating rate, the remaining mass fraction curves shifted right, which is unanimously followed by the thermogravimetric (TG) curves as well. Comparative evaluation of transmutation for different limits is graphically illustrated in Figure 6.11. It is seen that the initial stage of pyrolysis is predicted differently at different range of activation energies. The plateauing effect is relatively low in the non-transmuted DAEM, yet it is not able to describe the initial and the final stages of decomposition for the similar condition. This is because the transmutation allows an additional degree of freedom that adds-up to flexibility in the final predicted solution.

Dhaundiyal et al. (2018b) used different member of Archimedean family, Frank copula, but they assumed the activation energies to be similar for each stage, which provided activation energy for pine needles to be 84 kJ/mol. They have not demarcated activation energy boundaries, which provided them a nontrivial solution for the whole pyrolysis. There is a fair possibility that the thermal decomposition can be represented by a model that imbibes the features of demarcating the different stages of pyrolysis. The obtained value of activation energies is consistent with the values obtained from other statistical-based models (Loy et al. 2018; Font et al. 2009).

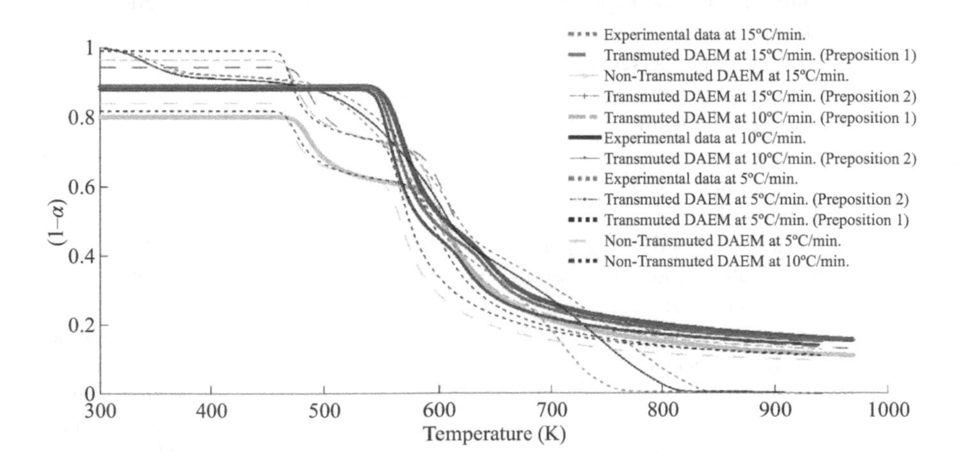

FIGURE 6.11
Comparative sketch of transmuted and nontransmuted form of distributed activation energy model (DAEM) with the experimental data.

6.4 Conclusion

The effect of activation energy is introduced through transmutation of the Gumbel-Hougaard Copula in DAEM. Pyrolysis is distinguished into the primary and the secondary pyrolysis. On the basis of the same, boundary of activation energies is estimated by using asymptotic limits on the activation energies. For the limit $E \rightarrow \infty$, it is found that the average activation energy required for initiating the primary stage of pyrolysis varies from 80.5 to 84.66 kJ/mol, whereas the secondary stage needs the activation energy of 32.85 to 34.57 kJ/mol to trigger autocatalytic reaction. On the other hand, it varies from 43.51 to 51.51 kJ/mol and 57.22 to 67.72 kJ/mol for the primary and the secondary pyrolysis respectively for the limit $E \rightarrow 0$. The frequency factor (A) min^{-1} range varies from 10^{16} to 10^{20} for both the limits. The convex coefficients $\{f_1, f_2\}$ of the primary and the secondary pyrolysis vary from {[0.43,0.71], [0.29,0.57]} for both the limits. The dormant state of moving maxima, y_e, varies as the imposed asymptotic limits changes from 0 to ∞. The more clarity of activation energy boundaries can be determined if both the limits are combined into a model. The results derived E_1 for the limit $E \rightarrow 0$ provided the activation energy boundary required for the preliminary stage of pyrolysis (i.e., dehydration and decomposition of hemicellulose). The proposed scheme can provide a better insight about activation energy boundary and their overlapping at different stages.

NOTATIONS

E_1	Activation energy for primary stage	kJ/mol
E_2	Activation energy for secondary stage	kJ/mol
χ	Threshold or location parameter of the Rayleigh distribution	kJ/mol
ϕ	Threshold or location parameter of the Weibull distribution	kJ/mol
β	Heating rate	°C/min
η	Scale factor of the Weibull distribution	kJ/mol
θ	Copula parameter of the Gumbel- Hougaard copula	dimensionless
δ	Shape parameter of the Weibull distribution	Dimension-less
ω_1 and ω_2	Constants	Dimension-less
R	Gas constant	kJ/mol-K
T	Temperature	K
E_s	Central value of double exponential function	kJ/mol
E_w	Step width of double exponential function	kJ/mol
A	Frequency factor	min^{-1}
y	Energy correction	Dimension-less
f_1 & f_2	Convex coefficient	Dimension-less
λ	Transmutation parameter	Dimension-less
α	Conversion	dimensionless
σ	Standard deviation	kJ/mol
p, s, w	the subscripts denoted predicted model and the values corresponding to the central as well as the step-width, respectively	—

Appendix: A.6

The Gumbel copula for two marginal distribution function $\left(f(E_1), f(E_2)\right)$ is given by expression

$$C\left(f(E_1), f(E_2)\right) = exp\left[-\left\{\left(1-\left(\ln\left(f(E_1)\right)+1\right)\right)^{\theta} + \left(1-\left(\ln\left(f(E_2)\right)+1\right)\right)^{\theta}\right\}^{\frac{1}{\theta}}\right] \quad (A.6.1)$$

Here, θ is copula parameter and E_1 and E_2 are two different classes of activation required for triggering two independent stages of pyrolysis.

Applying the binomial expansion on the Equation A.6.1, we get

$$C\left(f(E_1), f(E_2)\right) \sim exp\left[-\left\{\left(1-\theta\left(\ln\left(f(E_1)\right)+\ln\left(f(E_2)\right)+2\right)\right)+\ldots\right\}^{\frac{1}{\theta}}\right] \quad (A.6.2)$$

After further simplification, the Equation A.6.2 becomes

$$C\left(f(y_1), f(y_2)\right) \sim 2.718 exp\left[\left\{\left(\begin{matrix}\left(\ln\left(f(E_1)\right)+\ln\left(f(E_2)\right)\right)+0.125\left(\dfrac{\theta-1}{\theta}\right) \\ \left(\ln\left(f(E_1)\right)+\ln\left(f(E_2)\right)+2\right)^2\end{matrix}\right)+\ldots\right\}\right]$$

The simplified form of marginal functions with respect to energy correction factors (y_1) and (y_2) for two-stage pyrolysis are given by Equations A.6.3 and A.6.4.

$$f(y_1) = 2\frac{\omega_1}{\sigma}(y_1-1) exp\left\{-\omega_1(y_1-1)^2\right\} \quad (A.6.3)$$

$$f(y_2) = \frac{\delta}{\eta}\omega_2^{\delta-1}(y_2-1)^{\delta-1} e^{-\left(\left(\omega_2(y_2-1)\right)^{\delta}\right)} \quad (A.6.4)$$

where,

$$\omega_1 = \frac{\phi^2}{\sigma^2\left(1-\dfrac{\pi}{4}\right)} \text{ and } \omega_2 = \frac{\gamma}{\eta}$$

The mean, the variance and the scale parameter of the Rayleigh distribution are $E_{01}, \sigma^2{}_1$ and χ, respectively

$$E_{01} = \phi + \frac{\sqrt{\pi}}{2\chi} \text{ and } \sigma^2 = \frac{1}{\chi} - \frac{\pi}{4\chi}.$$

The mean, the scale parameter, and the variance of the Weibull distribution are E_{02}, η, and $\sigma^2{}_2$ respectively

$$E_{02} = \gamma + \eta\Gamma\left(\frac{1}{\delta}+1\right) \text{ and } \sigma_2^2 = \eta^2\left[\Gamma\left(\frac{2}{\delta}+1\right)-\left\{\Gamma\left(\frac{1}{\delta}+1\right)\right\}^2\right].$$

After incorporating the approximated value of double exponential function in Equation A.6.4, the expression becomes

$$\int_0^\infty exp\left(-exp\left(\frac{(y_{s1}+y_{s2})-(y_1+y_2)}{(y_{w1}+y_{w2})}\right)\right)\left(H(y_1,y_2)\right)\left(f_1\phi dy_1 + f_2\gamma dEy_2\right) \quad \text{(A.6.5)}$$

Put the values of $H(y_1,y_2)$ $f(y_1)$ and $f(y_2)$ from Equations 6.5, A.6.3, and A.6.4 into Equation A.6.5.

$$(1-\alpha) \sim 5.42\frac{\omega_1}{\phi}\frac{\delta}{\eta}\omega_2{}^{\delta-1}\int_0^\infty\left\{(y_1-1)(y_2-1)^{\delta-1}\right\}exp\left(-exp\left(\frac{(y_{s1}+y_{s2})-(y_1+y_2)}{(y_{w1}+y_{w2})}\right)\right)$$
$$exp\left[\left\{-\omega_1(y_1-1)^2-\left(\omega_2(y_2-1)^\delta\right)\right\}\right]d\left(f_1\phi y_1+f_2\gamma y_2\right) \quad \text{(A.6.6)}$$

Simplifying Equation A.6.6, we get

$$(1-\alpha)\sim 5.42\left[\begin{array}{l} f_1\omega_1\frac{\delta}{\eta}\omega_2{}^{\delta-1}\int_0^\infty\left\{(y_1-1)(y_2-1)^{\delta-1}\right\}exp\left(-exp\left(\frac{(y_{s1}+y_{s2})-(y_1+y_2)}{(y_{w1}+y_{w2})}\right)\right) \\ exp\left\{-\omega_1(y_1-1)^2-\left(\omega_2(y_2-1)^\delta\right)\right\}dy_1+\frac{\omega_1}{\phi}\delta\omega_2{}^\delta f_2\int_0^\infty\left\{(y_1-1)(y_2-1)^{\delta-1}\right\} \\ exp\left(-exp\left(\frac{(y_{s1}+y_{s2})-(y_1+y_2)}{(y_{w1}+y_{w2})}\right)\right) \\ exp\left\{-\omega_1(y_1-1)^2-\left(\omega_2(y_2-1)^\delta\right)\right\}dy_2 \end{array}\right] \quad \text{(A.6.7)}$$

Apply the Laplace method of integration and obtain the approximated solution of Equation A.6.7

$$h\left(y_1{}^e\right) = -exp\left(\frac{(y_{s1}+y_{s2})-(y_1+y_2)}{(y_{w1}+y_{w2})}\right)-\left\{\omega_1(y_1-1)^2+\left(\omega_2(y_2-1)\right)^\delta\right\} \quad \text{(A.6.8)}$$

$$h\left(y_2^e\right)=-exp\left(\frac{\left(y_{s1}+y_{s2}\right)-\left(y_1+y_2\right)}{\left(y_{w1}+y_{w2}\right)}\right)-\left\{\omega_1\left(y_1-1\right)^2+\left(\omega_2\left(y_2-1\right)\right)^{\delta}\right\} \quad \text{(A.6.9)}$$

Differentiating Equations A.6.8 and A.6.9 with respect to "y_1" and "y_2" respectively, we have

$$h\left(y_1^e\right)'=0 \text{ and } h\left(y_2^e\right)'=0$$

The location of maxima (y_1^e, y_2^e) for the functions $h(y_1)$ and $h(y_2)$ are given below

$$y_1^e=1+\left(y_{w1}+y_{w2}\right)W\left(\frac{1}{2\omega_1\left(y_{w1}+y_{w2}\right)^2}exp\left(\frac{\left(y_{s1}+y_{s2}-y_2\right)-1}{\left(y_{w1}+y_{w2}\right)}\right)\right)$$

$$y_2^e=1+\left(\delta-1\right)\left(y_{w1}+y_{w2}\right)W\left(\begin{array}{c}\frac{1}{\left(\delta-1\right)\left(y_{w1}+y_{w2}\right)}\left(\frac{1}{\delta\left(y_{w1}+y_{w2}\right)\omega_2^{\delta}}\right)^{\frac{1}{(\delta-1)}}\\ exp\left\{\frac{1}{\left(\delta-1\right)}\left(\frac{\left(y_{s1}+y_{s2}-y_1\right)-1}{\left(y_{w1}+y_{w2}\right)}\right)\right\}\end{array}\right)$$

The second derivative of function $h(y_1)$ and $h(y_2)$with respect to "y_1" and "y_2" respectively are:

$$h\left(y_1^e\right)''==-2\omega_1\left(\frac{\left(y_1-1\right)}{\left(y_{w1}+y_{w2}\right)}+1\right) \quad \text{(A.6.10)}$$

$$h\left(y_2^e\right)''=-\delta\left(\omega_2\left(y_1-1\right)\right)^{(\delta-2)}\left(\frac{\omega_2\left(y_1-1\right)}{\left(y_{w1}+y_{w2}\right)}+\left(\delta-1\right)\right) \quad \text{(A.6.11)}$$

Finding the maximum values of $h(y_1)$ and $h(y_2)$ at $y_1=y_1^e$ and $y_2=y_2^e$, respectively, we have

$$h\left(y_1^e\right)=-\left(\omega_1\left(y_1-1\right)^2+\left(\omega_2\left(y_2-1\right)\right)^{\delta}+2\omega_1\left(y_{w1}+y_{w2}\right)\left(y_1-1\right)\right) \quad \text{(A.6.12)}$$

$$h\left(y_2^e\right)=-\left(\omega_1\left(y_1-1\right)^2+\left(\omega_2\left(y_2-1\right)\right)^{\delta}+\delta\left(y_{w1}+y_{w2}\right)\left(\omega_1\left(y_1-1\right)\right)^{(\delta-1)}\right) \quad \text{(A.6.13)}$$

The leading behavior of $(1-\alpha)$ in Equations A.6.12 and A.6.13 is

$$(1-\alpha)\approx 5.42\sqrt{2\pi}\,(1+\lambda)\left[\begin{array}{l}(y_1^e-1)(y_2-1)^{\delta-1} f_1\omega_1\omega_2^{\delta-1}\dfrac{\delta}{\eta}\dfrac{exp(h(y_1^e))}{\sqrt{\left|h(y_1^e)''\right|}}\\ +(y_1-1)(y_2^e-1)^{\delta-1} f_2\,\delta\omega_1\omega_2^{\delta}\dfrac{exp(h(y_2^e))}{\sqrt{\left|h(y_2^e)''\right|}}\end{array}\right] \quad \text{(A.6.14)}$$

The predicted solutions of $(1-\alpha)$ for different limits of transmutation function $H^t(y_1, y_2)$ are

$$(1-\alpha)\sim 5.42(1+\lambda)\sqrt{2\pi}\left[\begin{array}{l}\omega_1\omega_2^{\delta-1}\dfrac{\delta}{\eta}(y_1^e-1)(y_2-1)^{\delta-1}\\ \left\{\dfrac{exp\left\{-\Phi\left(\omega_1(y_1^e-1)^2+(\omega_2(y_2-1))^{\delta}+2\omega_1(y_{w1}+y_{w2})(y_1^e-1)\right)\right\}}{\sqrt{\Phi\omega_1\left(\dfrac{(y_1^e-1)}{(y_{w1}+y_{w2})}+1\right)}}\right\}\\ +\delta\omega_1\omega_2^{\delta}(y_1-1)(y_2^e-1)^{\delta-1}\\ \left\{\dfrac{exp\left\{-\Phi\left(\omega_1(y_1-1)^2+(\omega_2(y_2^e-1))^{\delta}+\delta(y_{w1}+y_{w2})(\omega_2(y_2^e-1))^{(\delta-1)}\right)\right\}}{\sqrt{\delta\Phi(\omega_2(y_2^e-1))^{(\delta-2)}\left(\dfrac{\omega_2(y_2^e-1)}{(y_{w1}+y_{w2})}+(\delta-1)\right)}}\right\}\end{array}\right]$$

(A.6.15)

$$(1-\alpha)\sim 5.42(1-\lambda)\sqrt{2\pi}\left[\begin{array}{l}\omega_1\omega_2^{\delta-1}(y_1^e-1)(y_2-1)^{\delta-1}\dfrac{\delta}{\eta}\\ \left\{\dfrac{exp\left\{-\Phi\left(\omega_1(y_1^e-1)^2+(\omega_2(y_2-1))^{\delta}+2\omega_1(y_{w1}+y_{w2})(y_1^e-1)\right)\right\}}{\sqrt{\Phi\omega_1\left(\dfrac{(y_1^e-1)}{(y_{w1}+y_{w2})}+1\right)}}\right\}\\ +(y_1-1)(y_2^e-1)^{\delta-1}\delta\omega_1\omega_2^{\delta}\\ \left\{\dfrac{exp\left\{-\Phi\left(\omega_1(y_1-1)^2+(\omega_2(y_2^e-1))^{\delta}+\delta(y_{w1}+y_{w2})(\omega_2(y_2^e-1))^{(\delta-1)}\right)\right\}}{\sqrt{\delta\Phi(\omega_2(y_2^e-1))^{(\delta-2)}\left(\dfrac{\omega_2(y_2^e-1)}{(y_{w1}+y_{w2})}+(\delta-1)\right)}}\right\}\end{array}\right]$$

(A.6.16)

Equations A.6.15 and A.6.16 are the required expressions for two-stage pyrolysis problem.

References

Akita, K. 1959. "Studies on the Mechanism of Ignition of Wood." *Report of the Fire Research Institute of Japan* 9: 1–105.

Antal, M. J. 1983. "Biomass Pyrolysis: A Review of the Literature Part 1—Carbohydrate Pyrolysis." In *Advances in Solar Energy*, pp. 61–111. Boston, MA: Springer. doi:10.1007/978-1-4684-8992-7_3.

Anthony, D. B., and J. B. Howard. 1976. "Coal Devolatilization and Hydrogastification." *AIChE Journal* 22 (4): 625–656. doi:10.1002/aic.690220403.

Aryal, G. R., and C. P. Tsokos. 2009. "On the Transmuted Extreme Value Distribution with Application." *Nonlinear Analysis: Theory, Methods & Applications* 71 (12): e1401–e1407. doi:10.1016/j.na.2009.01.168.

Ashour, S. K., and M. A. Eltehiwy. 2013. "Transmuted Lomax Distribution." *American Journal of Applied Mathematics and Statistics* 1 (6): 121–27. doi:10.12691/ajams-1-6-3.

Bamford, C. H., J. Crank, and D. H. Malon. 1946. "The Combustion of Wood." In *Proceedings of the Cambridge Philosophical Society* 42: 166–182.

Bhoi, P. R., R. N. Singh, A. M. Sharma, and S. R. Patel. 2006. "Performance Evaluation of Open Core Gasifier on Multi-Fuels." *Biomass and Bioenergy* 30 (6): 575–579. doi:10.1016/j.biombioe.2005.12.006.

Blasi, C. D. 1993. "Modeling and Simulation of Combustion Processes of Charring and Non-Charring Solid Fuels." *Progress in Energy and Combustion Science* 19 (1): 71–104. doi:10.1016/0360-1285(93)90022-7.

Bourguignon, M., I. Ghosh, and G. M. Cordeiro. 2016. "General Results for the Transmuted Family of Distributions and New Models." *Journal of Probability and Statistics* 2016: 1–12. doi:10.1155/2016/7208425.

Bradbury, A. G. W., Y. Sakai, and F. Shafizadeh. 1979. "A Kinetic Model for Pyrolysis of Cellulose." *Journal of Applied Polymer Science* 23 (11): 3271–3280. doi:10.1002/app.1979.070231112.

Broido, A., and M. A. Nelson. 1975. "Char Yield on Pyrolysis of Cellulose." *Combustion and Flame* 24: 263–268. doi:10.1016/0010-2180(75)90156-X.

Brown, H. P., A. J. Panshin, and C. C. Forsaith. 1952. "The Physical, Mechanical, and Chemical Properties of the Commercial Woods of the United States." In *Textbook of Wood Technology*, Volume II. New York: McGraw-Hill.

Brown, M. E., M. Maciejewski, S. Vyazovkin, R. Nomen, J. Sempere, A. Burnham, J. Opfermann et al. 2000. "Computational Aspects of Kinetic Analysis: Part A: The ICTAC Kinetics Project-Data, Methods and Results." *Thermochimica Acta* 355 (1–2): 125–143. doi:10.1016/S0040-6031(00)00443-3.

Brown, M. E. 1988. "Thermogravimetry (TG)." In *Introduction to Thermal Analysis*, pp. 7–22. Dordrecht, the Netherlands: Springer. doi:10.1007/978-94-009-1219-9_3.

Burnham, A. K., and L. N. Dinh. 2007. "A Comparison of Isoconversional and Model-Fitting Approaches to Kinetic Parameter Estimation and Application Predictions." *Journal of Thermal Analysis and Calorimetry* 89 (2): 479–490. doi:10.1007/s10973-006-8486-1.

Burnham, A. K., and R. L. Braun. 1999. "Global Kinetic Analysis of Complex Materials." *Energy & Fuels* 13 (1): 1–22. doi:10.1021/ef9800765.

Cai, J., and L. Ji. 2007. "Pattern Search Method for Determination of DAEM Kinetic Parameters from Nonisothermal TGA Data of Biomass." *Journal of Mathematical Chemistry* 42 (3): 547–553. doi:10.1007/s10910-006-9130-9.

Cai, J., T. Li, and R. Liu. 2011. "A Critical Study of the Miura–Maki Integral Method for the Estimation of the Kinetic Parameters of the Distributed Activation Energy Model." *Bioresource Technology* 102 (4): 3894–3899. doi:10.1016/j.biortech.2010.11.110.

De Caprariis, B., P. De Filippis, C. Herce, and N. Verdone. 2012. "Double-Gaussian Distributed Activation Energy Model for Coal Devolatilization." *Energy & Fuels* 26 (10): 6153–6159. doi:10.1021/ef301092r.

Cho, J., J. M. Davis, and G. W. Huber. 2010. "The Intrinsic Kinetics and Heats of Reactions for Cellulose Pyrolysis and Char Formation." *ChemSusChem* 3 (10): 1162–1165. doi:10.1002/cssc.201000119.

Constable, F. H. 1925. "The Mechanism of Catalytic Decomposition." *Proceedings of the Royal Society A: Mathematical, Physical and Engineering Sciences* 108 (746): 355–378. doi:10.1098/rspa.1925.0081.

Deheuvels, P. 1979. "La Fonction de Dependence Empirique et Ses Propiet´es. Un Test Non Parametrique d'independence." *Académie Royale de Belgique. Bulletin de La Classe Des Sciences* 5 (65): 274–292.

Dhaundiyal, A., and S. B. Singh. 2017a. "Mathematical Insight to Non-Isothermal Pyrolysis of Pine Needles for Different Probability Distribution Functions." *Biofuels* 9 (5): 647–658. doi:10.1080/17597269.2017.1329495.

Dhaundiyal, A., and S. B. Singh. 2017b. "Study of Distributed Activation Energy Model Using Various Probability Distribution Functions for the Isothermal Pyrolysis Problem." *Rudarsko-Geološko-Naftni Zbornik* 32 (4): 1–14. doi:10.17794/rgn.2017.4.1.

Dhaundiyal, A., and P. C. Tewari. 2015. "Comparative Analysis of Pine Needles and Coal for Electricity Generation Using Carbon Taxation and Emission Reductions." *Acta Technologica Agriculturae* 18 (2): 29–35. doi:10.1515/ata-2015-0007.

Dhaundiyal, A., and V. K. Gupta. 2014. "The Analysis of Pine Needles as a Substrate for Gasification." *Hydro Nepal: Journal of Water, Energy and Environment* 15: 73–81. doi:10.3126/hn.v15i0.11299.

Dhaundiyal, A., and M. M. Hanon. 2018. "Calculation of Kinetic Parameters of the Thermal Decomposition of Residual Waste of Coniferous Species: Cedrus Deodara." *Acta Technologica Agriculturae* 21 (2): 76–81. doi:10.2478/ata-2018-0014.

Dhaundiyal, A., and S. B. Singh. 2017c. "Approximations to the Non-Isothermal Distributed Activation Energy Model for Biomass Pyrolysis Using the Rayleigh Distribution." *Acta Technologica Agriculturae* 20 (3): 78–84. doi:10.1515/ata-2017-0016.

Dhaundiyal, A., S. B. Singh, and M. M. Hanon. 2018a. "Study of Distributed Activation Energy Model Using Bivariate Distribution Function, f (E 1, E 2)." *Thermal Science and Engineering Progress* 5: 388–404. doi:10.1016/j.tsep.2018.01.009.

Dhaundiyal, A., S. B. Singh, M. M. Hanon, and R. Rawat. 2018b. "Determination of Kinetic Parameters for the Thermal Decomposition of Parthenium Hysterophorus." *Environmental and Climate Technologies* 22 (1): 5–21. doi:10.1515/rtuect-2018-0001.

Dhaundiyal, A., and P. Tewari. 2017. "Kinetic Parameters for the Thermal Decomposition of Forest Waste Using Distributed Activation Energy Model (DAEM)." *Environmental and Climate Technologies* 19 (1): 15–32. doi:10.1515/rtuect-2017-0002.

Dhaundiyal, A., and P. C. Tewari. 2016. "Performance Evaluation of Throatless Gasifier Using Pine Needles as a Feedstock for Power Generation." *Acta Technologica Agriculturae* 19 (1): 10–18. doi:10.1515/ata-2016-0003.

Elbatal, I., G. Asha, and A. V. Raja. 2014. "Transmuted Exponentiated Fr^echet Distribution: Properties and Applications." *Journal of Statistics Applications & Probability* 3 (3): 379–394. doi:10.12785/jsap/030309.

Fan, L. T., L.-S. Fan, K. Miyanami, T. Y. Chen, and W. P. Walawender. 1977. "A Mathematical Model for Pyrolysis of a Solid Particle: Effects of the Lewis Number." *The Canadian Journal of Chemical Engineering* 55 (1): 47–53. doi:10.1002/cjce.5450550109.

Ferdous, D., A. K. Dalai, S. K. Bej, and R. W. Thring. 2002. "Pyrolysis of Lignins: Experimental and Kinetics Studies." *Energy & Fuels* 16 (6): 1405–12. doi:10.1021/ef0200323.

Font, R., J. A. Conesa, J. Moltó, and M. Muñoz. 2009. "Kinetics of Pyrolysis and Combustion of Pine Needles and Cones." *Journal of Analytical and Applied Pyrolysis* 85 (1–2): 276–286. doi:10.1016/j.jaap.2008.11.015.

Friedman, H. L. 1964. "Kinetics of Thermal Degradation of Char-Forming Plastics from Thermogravimetry. Application to a Phenolic Plastic." *Journal of Polymer Science Part C: Polymer Symposia* 6 (1): 183–195. doi:10.1002/polc.5070060121.

Galgano, A., and C. Di Blasi. 2003. "Modeling Wood Degradation by the Unreacted-Core-Shrinking Approximation." *Industrial & Engineering Chemistry Research* 42 (10): 2101–2111. doi:10.1021/ie020939o.

Genest, C., and J. Mackay. 1986. "The Joy of Copulas: Bivariate Distributions with Uniform Marginals." *American Statistician* 40 (4): 280–283. doi:10.1080/00031305.1986.10475414.

Golikeri, S. V., and D. Luss. 1972. "Analysis of Activation Energy of Grouped Parallel Reactions." *AIChE Journal* 18 (2): 277–282. doi:10.1002/aic.690180205.

Güneş, M., and S. Güneş. 2002. "A Direct Search Method for Determination of DAEM Kinetic Parameters from Nonisothermal TGA Data (Note)." *Applied Mathematics and Computation* 130 (2–3): 619–628. doi:10.1016/S0096-3003(01)00124-2.

Hanbaba, P. 1967. "Reaktionkinetische Untersuchungen Sur Kohlenwas serstoffenbindung Aus Steinkohlen Bie Niedregen Aufheizgeschwindigkeiten." University of Aachen, Aachen, Germany.

Ibbett, R., S. Gaddipati, S. Davies, S. Hill, and G. Tucker. 2011. "The Mechanisms of Hydrothermal Deconstruction of Lignocellulose: New Insights from Thermal–analytical and Complementary Studies." *Bioresource Technology* 102 (19): 9272–9278. doi:10.1016/j.biortech.2011.06.044.

Jalan, R. K., and V. K. Srivastava. 1999. "Studies on Pyrolysis of a Single Biomass Cylindrical Pellet—kinetic and Heat Transfer Effects." *Energy Conversion and Management* 40 (5): 467–494. doi:10.1016/S0196-8904(98)00099-5.

Kansa, E. J., H. E. Perlee, and R. F. Chaiken. 1977. "Mathematical Model of Wood Pyrolysis Including Internal Forced Convection." *Combustion and Flame* 29: 311–324. doi:10.1016/0010-2180(77)90121-3.

Koufopanos, C. A., N. Papayannakos, G. Maschio, and A. Lucchesi. 1991. "Modelling of the Pyrolysis of Biomass Particles. Studies on Kinetics, Thermal and Heat Transfer Effects." *The Canadian Journal of Chemical Engineering* 69 (4): 907–915. doi:10.1002/cjce.5450690413.

Lanzetta, M., and C. Di Blasi. 1998. "Pyrolysis Kinetics of Wheat and Corn Straw." *Journal of Analytical and Applied Pyrolysis* 44 (2): 181–192. doi:10.1016/S0165-2370(97)00079-X.

Lee, C. K., R. F. Chaiken, and J. M. Singer. 1977. "Charring Pyrolysis of Wood in Fires by Laser Simulation." *Symposium (International) on Combustion* 16 (1): 1459–1470. doi:10.1016/S0082-0784(77)80428-1.

Loy, A. C. M., D. K. W. Gan, S. Yusup, B. L. F. Chin, M. K. Lam, M. Shahbaz, P. Unrean, M. N. Acda, and E. Rianawati. 2018. "Thermogravimetric Kinetic Modelling of In-Situ Catalytic Pyrolytic Conversion of Rice Husk to Bioenergy Using Rice Hull Ash Catalyst." *Bioresource Technology* 261: 213–222. doi:10.1016/j.biortech.2018.04.020.

Maa, P. S., and R. C. Bailie. 1973. "Influence of Particle Sizes and Environmental Conditions on High Temperature Pyrolysis of Cellulosic Material—I (Theoretical)." *Combustion Science and Technology* 7 (6): 257–269. doi:10.1080/00102207308952366.

Matsumoto, T., T. Fujiwara, and J. Kondo. 1969. "12th Symposium on Combustion." In *12th Symposium on Combustion*, pp, 515–531. Pittsburgh, PA: The Combustion Institute.

Miura, K. 1995. "A New and Simple Method to Estimate f(E) and K0(E) in the Distributed Activation Energy Model from Three Sets of Experimental Data." *Energy & Fuels* 9 (2): 302–307. doi:10.1021/ef00050a014.

Miura, K., and T. Maki. 1998. "A Simple Method for Estimating f (E) and k 0 (E) in the Distributed Activation Energy Model." *Energy & Fuels* 12 (5): 864–869. doi:10.1021/ef970212q.

Molten, P. M., T. F. Demmitt, J. M. Donovan, R. K. Miller. 1983. "Mechanism of Conversion of Cellulose Wastes to Liquid in Alkaline Solution." In *Energy from Biomass and Wastes III*, Klass, DL, p. 293. Chicago, IL: Institute of Gas Technology.

Murty, K. A., and P. L. Blackshear. 1967. "Pyrolysis Effects in the Transfer of Heat and Mass in Thermally Decomposing Organic Solids." *Symposium (International) on Combustion* 11 (1): 517–523. doi:10.1016/S0082-0784(67)80176-0.

Ozawa, T. 1965. "A New Method of Analyzing Thermogravimetric Data." *Bulletin of the Chemical Society of Japan* 38 (11): 1881–1886. doi:10.1246/bcsj.38.1881.

Panton, R. L., and J. G. Rittmann. 1971. "Pyrolysis of a Slab of Porous Material." *Symposium (International) on Combustion* 13 (1): 881–891. doi:10.1016/S0082-0784(71)80089-9.

Pitt, G. J. 1962. "The Kinetics of the Evolution of Volatile Products from Coal." *Fuel*, 41: 267–274.

Roberts, A. F. 1971. "The Heat of Reaction during the Pyrolysis of Wood." *Combustion and Flame* 17 (1): 79–86. doi:10.1016/S0010-2180(71)80141-4.

Roberts, A. F., and G. Clough. 1963. "Thermal Decomposition of Wood in an Inert Atmosphere." *Symposium (International) on Combustion* 9 (1): 158–166. doi:10.1016/S0082-0784(63)80022-3.

Shaw, W. T., and I. R. C. Buckley. 2009. "The Alchemy of Probability Distributions: Beyond Gram-Charlier Expansions, and a Skew-Kurtotic-Normal Distribution from a Rank Transmutation Map." arXiv:0901.0434.

Sinha, S., A. Jhalani, M. R. Ravi, and A. Ray. 2000. "Modelling of Pyrolysis in Wood: A Review." *Solar Energy Society of India* 10 (1): 41–62.

Sklar, M. 1959. "Fonctions de Repartition an Dimensions et Leurs Marges." *Publications de l'Institut de Statistique* 8: 229–231.

Talib, A., D. Bienstock, J. Goss, and V. Flanigan. 1986. "U.S./India Co-Operative Biomass Conversion Program." *Energy* 11: 1401–1410. doi:10.1016/0360-5442(86)90076-9.

Ungerer, P., and R. Pelet. 1987. "Extrapolation of the Kinetics of Oil and Gas Formation from Laboratory Experiments to Sedimentary Basins." *Nature* 327 (6117): 52–54. doi:10.1038/327052a0.

Vand, V. 1943. "A Theory of the Irreversible Electrical Resistance Changes of Metallic Films Evaporated in Vacuum." *Proceedings of the Physical Society* 55 (3): 222–246. doi:10.1088/0959-5309/55/3/308.

Zaror, D. L., and C. A. Pyle. 1984. "Models for Low Temperature Pyrolysis of Wood." In *Thermochemical Processing of Biomass*. London, UK: Butterworths & Co Publishers.

Zhang, J., T. Chen, J. Wu, and J. Wu. 2014. "Multi-Gaussian-DAEM-Reaction Model for Thermal Decompositions of Cellulose, Hemicellulose and Lignin: Comparison of N_2 and CO_2 Atmosphere." *Bioresource Technology* 166: 87–95. doi:10.1016/j.biortech.2014.05.030.

Zhang, Z., M. Zhu, and D. Zhang. 2018. "A Thermogravimetric Study of the Characteristics of Pyrolysis of Cellulose Isolated from Selected Biomass." *Applied Energy* 220: 87–93. doi:10.1016/j.apenergy.2018.03.057.

List of Figures

7

Differential Quadrature Method: A Robust Technique to Solve Differential Equations

G. Arora and M. Bashir

CONTENTS

7.1 Introduction

Partial differential equations (PDEs), along with a set of initial or boundary conditions, play an important role in finding the solution of engineering problems. Most of the differential equations arise as a result of mathematical modeling of the various phenomenon of science. The solution of the PDEs is always required to obtain the values of the related parameters in the domain of interest. For example, the well-known Fisher's equation is used to study the mutation rate of genes to analyze the spread of creature or plant population and to analyze the advancement of neutron population in an atomic reactor. Similarly, the solution of Burgers' equation plays a significant role in the investigation of turbulence revealed by the association of the convection and dispersion phenomenon. Because it is not always possible to find a closed form solution of the equation using an analytical approach, there exists a need for a numerical approach.

Numerical methods are in the prime focus of researchers because these methods are easy to implement to solve various forms of differential equations in a programmable approach. With the advancement of technology, there is a wide availability of mathematical software that assists researchers in finding the numerical solution of differential equations by programming of the considered algorithm. Among the numerous available numerical approaches, finite-difference and finite-element methods are the well-known approaches because of their ease of applicability for solving various forms of differential equations. A lot of advancement has occurred in the field of numerical analysis, but there is no such numerical method that is capable of solving all the types of differential equations. Therefore, researchers are continuously putting their efforts into the development of new numerical schemes. As a result, various numerical schemes have been proposed by researchers that include collocation method, wavelet-based methods, Galerkin approach, radial basis function approach, pseudospectral methods, and differential quadrature method (DQM).

This chapter is a concise summary of the basic theory of the DQM and its modified forms. One of the key features of the DQM is the basis functions, on which the value of weighting coefficients can be calculated. These factors are required to approximate the derivatives of functions. Various approaches, their formulation, and use for calculating weighting coefficients is detailed. A generalized equal width equation is used for illustrating the methodology of the DQM. In the presented approach, the B-spline basis functions are used as a hybrid of its standard and trigonometric form to obtain the derivative approximation. Numerical examples are solved to illustrate the methodology with discussion of the results presented in form of figures and tables.

7.2 Origin of Differential Quadrature Method

Numerical methods are broadly categorized as methods of higher order and of lower order. The methods such as finite difference and finite element that fall under the category of lower-order method require large number of domain partitions. But in practical applications, the solution of the differential equation is not always required at all the node points of the domain. Hence, there is a need for the development of higher-order methods that provides accurate results with less number of grid points by discretization of the domain. This requirement resulted in the origin of the DQM proposed by Bellman and Casti [1] and based on the idea of integral quadrature. Quadrature rule is a well-known numerical integration method suitable for a programmable approach. The integration formula for a given set of data values is given by:

$$\int_a^b f(x)dx = \sum_{i=1}^{N} w_i f(x_i) \tag{7.1}$$

where N is the total number of points in domain $[a,b]$ and w_i are the corresponding weights. The value of these weights gives rise to the different form of integration quadrature formulas. One of the famous integration rules is the trapezoidal rule given by formula:

$$\int_a^b f(x)dx = \frac{h}{2}\left(f(x_0)+2f(x_1)+\ldots+2f(x_{n-1})+f(x_n)\right) \tag{7.2}$$

This formula follows with the calculation of $w_i's$ given as:

$$w_i = \begin{cases} h/2, & i=1,n \\ h, & i=2,\ldots,n-1 \end{cases}$$

As proposed by Bellman et al., if $f(x)$ is a sufficiently smooth continuous function, then the integration formula can be extended to calculate the approximate value of the derivative given by:

$$\left.\frac{df}{dx}\right)_{x_i} = \sum_{j=1}^{N} \omega_{ij} f(x_j), \quad i=1 \text{ to } N \tag{7.3}$$

The DQM has been developed as a numerical technique capable of finding the numerical solution of initial and boundary value problems, including ordinary and partial differential equations in uniform and nonuniform domain. It is also referred to as a polynomial-based collocation method and has been applied successfully to find numerical solutions of differential equations with less computational cost and with high efficiency as well. This method has been used in the study of phenomenon in such areas such as biosciences, nuclear phenomenon, fluid mechanics, transport phenomenon, and chemical processes. Hence, this method has proven its potential compared to the traditional finite-element and finite-difference approaches.

The DQM has gone through a lot of development in terms of the degree of the polynomial, different forms of basis functions, implementation of boundary conditions, and the choice of grid points. Because of various developments in DQM, it is a now well-known numerical method for its admirable quality of computational efficiency and fast convergence. The initial phases of advancement of the DQM and its use in various applications to solve differential equations are reported by Shu [2]. The utmost requirement of DQM is the calculation of weighting coefficient whose formulation was further improved by Quan and Chang [3,4]. One of the key features of the DQM is the basis functions. The efficiency of the obtained numerical solution also depends on the choice of the basis function. Various kinds of basis functions, such as spline functions, Lagrange interpolation polynomials, and sinc function [5,6], for example, are successfully implemented to determine the weighting coefficients.

7.3 Developments in the DQM

During the early stages of development, this method is implemented to solve partial differential equations using a higher-degree polynomial as basis function and become famous as polynomial-based differential quadrature method (PDQM).

In PDQM, the derivative of a function for a given knot point is estimated by a linear summation of the values of the polynomial function calculated along a grid line. In 1972, Bellman et al. [7] proposed two different methodologies to calculate the weighting coefficients that resulted in an algebraic system of the equation.

7.3.1 Bellman's First Approach

This methodology is based on the test function taken as $\phi_k(x) = x^k$, $k = 0$ to $N-1$ given as $\{1, x, x^2, \dots, x^{N-1}\}$. On substituting the value of derivative of N test functions on the left-hand side of Equation 7.3 and the functional value on the right-hand side of the equation results in the following system of equation for each value of i:

$$\begin{cases} 0 = \sum_{j=1}^{N} \omega_{1j} \\ 1 = \sum_{j=1}^{N} \omega_{2j} x_j \\ 2x_i = \sum_{j=1}^{N} \omega_{2j} x_j^2 \\ kx_i^{k-1} = \sum_{j=1}^{N} \omega_{ij} x_j^k, k = 2, \ldots, N-1 \end{cases}$$

The obtained system of equation is an $N \times N$ matrix system to be solved for unknown weighting coefficients. But this approach has a disadvantage of generating an ill-conditioned matrix system as the number of grid points is increased.

7.3.2 Bellman's Second Approach

This methodology is based on the test function framed using the Legendre polynomial of degree N, $L_N(x)$ and is given as $\phi_k(x) = \frac{L_N(x)}{(x-x_k)L_N'(x)}$, $k = 1$ to N.

Here, the node points are the roots of the shifted Legendre polynomial. Using this approach, the weighting coefficients are calculated as:

$$\omega_{ij} = \begin{cases} \dfrac{L_N'(x_i)}{(x_i - x_j)L_N'(x_j)}, & \text{for } i \neq j \\ \dfrac{1-2x_i}{2x_i(x_i-1)}, & \text{for } i = j \end{cases}$$

In the first approach given by Bellman, the existence of an ill-conditioned matrix system with more grid points restricted its use in the application in which more grid points are required. But the first approach is preferred over the second because in the second approach the node points are the roots of shifted Legendre polynomial, which is impractical to obtain more number of node points and also leads to complicated calculations.

7.3.3 Other Famous Approaches

To overcome these drawbacks, researchers devised many amendments. Among them, the approach of Shu [2] gets the credit in resolving this issue. He proposed the calculation of weighting coefficients by an algebraic formulation independent of the number of grid points. To calculate the higher-order derivatives, he obtained the weighting coefficients using the recurrence relation starting with weighting coefficients of the first-order derivative (discussed in the next section). Because of the dependency of the weighting coefficients on the spatial grid spacing, consider n grid points on the real axis distributed uniformly, $a = x_1 < x_2 < ... < x_{n-1} < x_n = b$ with $x_{j+1} - x_j = h$. The approximate values of derivatives can be computed as follows:

$$\begin{aligned} u_x &= \sum_{j=1}^{N} a_{ij} u(x_j); \quad u_{xx} = \sum_{j=1}^{N} b_{ij} u(x_j); \\ u_{xxx} &= \sum_{j=1}^{N} c_{ij} u(x_j); \quad u_{xxxx} = \sum_{j=1}^{N} d_{ij} u(x_j) \end{aligned} \tag{7.4}$$

Shu further extended this approach for calculating the weighting coefficients by using Fourier series expansion. This approach of the DQM became known as the Fourier expansion-based differential quadrature method (FDQM) [2]. Another approach for calculating the weighting coefficients was based on harmonic functions; hence, the DQM is named a harmonic differential quadrature (HDQ). Civalek [8] compared the procedures to calculate the weighting coefficients of the DQ and HDQ.

7.4 Approaches to Obtain Higher-order Derivatives

After calculating the first-order derivative, the procedure of calculating the value of higher-order derivatives is next objective. There are various approaches to calculate the higher-order derivatives. One of the most well-known approaches was given by Shu [2] that uses the different values of weighting coefficients to calculate the derivatives. The formula to calculate the weighting coefficients is given in two forms; one is based on polynomial function and another is based on nonpolynomial functions given as follows:

$$a_{ij}^{m} = \begin{cases} m\left(a_{ij}^{1} a_{ij}^{m-1} - \dfrac{a_{ij}^{m-1}}{x_i - x_j}\right), & \text{for } i \neq j \\ -\displaystyle\sum_{i=1, i\neq j}^{N} a_{ij}^{m}, & \text{for } i = j \end{cases} \tag{7.5}$$

Here, $m \geq 2$, $a_{ij}^2 = b_{ij}$; $a_{ij}^3 = c_{ij}$; $a_{ij}^4 = d_{ij}$.

For nonpolynomial basis, the following expression can be used to calculate the weighting coefficients:

$$b_{ij} = \begin{cases} a_{ij}\left(\dfrac{2\sin(x_i)}{\cos(x_i - x_j)}\right) + 2a_{ii} + \cot(x_i), & \text{for } i \neq j \\ s_i^2(x_i) + 2\cot(x_i)s_i^1(x_i) - 1, & \text{for } i = j \end{cases} \tag{7.6}$$

where, $s_i^1(x_i) = \dfrac{a_{ij}}{\left(\sin(x_i)/\sin(x_j)\right)}; s_i^2(x_i) = \dfrac{b_{ij} - 2a_{ij}\cot(x_i)}{\left(\sin(x_i)/\sin(x_j)\right)}$

In another approach, Wang [9] proposed the calculation of a second-order derivative by using the weighting coefficients calculated for the first-order derivative. The approximation for second-order derivative is then given by:

$$\begin{aligned} \frac{\partial^2 u}{\partial x^2} &= \frac{\partial}{\partial x}\left(\frac{\partial u}{\partial x}\right) = \sum_{k=1}^{N} a_{ik}\left(\frac{\partial u}{\partial x}\right)_{x=x_k} \\ &= \sum_{k=1}^{N} a_{ik}\left(\sum_{j=1}^{N} a_{kj}u(x_j)\right) = \sum_{j=1}^{N} b_{ij}u(x_j),\ i = 1 \text{ to } N \end{aligned} \tag{7.7}$$

Similarly, the third- and higher-order derivatives can be obtained as:

$$\begin{aligned} \frac{\partial^3 u}{\partial x^3} &= \frac{\partial}{\partial x}\left(\frac{\partial^2 u}{\partial x^2}\right) = \sum_{k=1}^{N} a_{ik}\left(\frac{\partial^2 u}{\partial x^2}\right)_{x=x_k} \\ &= \sum_{k=1}^{N} a_{ik}\left(\sum_{j=1}^{N} b_{kj}u(x_j)\right) = \sum_{j=1}^{N} c_{ij}u(x_j),\ i = 1 \text{ to } N \end{aligned} \tag{7.8}$$

7.5 Introduction to Generalized Equal Width Equation

Study of solitary waves plays an important role in knowing the equivalent processes of particle physics. The generalized equal width (GEW) equation is a PDE that results in the secondary solitary waves solutions. This nonlinear wave equation was first expressed by Peregrine [10] and Benjamin et al. [11] during the mathematical modeling of small amplitude long waves on the surface of the water in a channel. In applications related to the generation of waves, this equation plays a significant role. For example, it is used to model the unidirectional transmitting of waves in the water channel and in the study of long-crested waves that generate near the seashore.

The GEW wave equation is given by

$$u_t + \varepsilon u^p u_x - \mu u_{xxt} = 0 \tag{7.9}$$

with $u(a,t) = u(b,t) = 0$ has boundary conditions with parameters, ε, μ, and p, which is a positive constant on domain $[a,b]$.

For $p = 1$, this equation is reduced to equal width (EW) equation. The EW equation has been solved by many numerical and exact solution methods including Galerkin's method [12], Petrov-Galerkin method [13], least-squares technique [14], and collocation method [15,16] to find the accurate and capable numerical solution of the EW equation.

For $p = 2$, the GEW equation results in a modified equal width (MEW) equation. This equation is solved by Esen [17] using the lumped-Galerkin method based on B-spline function; Saka [18] solved the equation using the collocation method; and Zaki [19] has used the Petrov-Galerkin method based on cubic B-spline for the numerical solution of MEW equation.

Thus, the GEW equation has a foundation on the EW equation and is also connected to the wave equations that result in solitary solutions such as generalized regularized long wave equation and the generalized Korteweg-de-Vries equation. Until now, the GEW wave equation has been solved by many analytical and numerical solution techniques, such as Panahipour [20], who solved the equation using a radial basis function (RBF) approach. A moving least-squares collocation (MLSC) method was implemented by Kaplan and Derel [21] to get the numerical solution of GEW equation. Karakoc and Zeybek [22] have applied the collocation method with two different techniques for linearization. For the numerical simulations of the GEW and GRLW equation, Battal and Zeybek [23] used a lumped-Galerkin approach using B-spline functions of the third degree. Mohammadi [24] used the collocation method with an exponential B-spline basis to solve the GRLW equation. The collocation method with quadratic B-spline was further studied for the GEW wave equation by Evans and Raslan [25].

7.6 Hybrid Basis Functions DQM

Various arrangements of PDEs have been solved by the application of DQM that include Fisher's equation, advection equation, hyperbolic telegraph equation, Burger's equation, nonlinear wave equation, and sine Gordon in two and three dimensions.

To present the application of DQM for solving the PDE, the GEW equation is solved with a linear combination of the standard and trigonometric form of B-spline as basis function in the DQM. Application of this method results in a matrix system to be solved recursively by the implementation of the numerical method. The resulted system is thus solved by the matrix inversion method to obtain the numerical solution of GEW equation. The details about B-spline basis functions and the methodology of DQM are explained in the next sections.

7.7 What Is a B-spline?

A B-spline is an important tool in computer graphics and is successfully implemented in finding the solution of differential equations arising in various areas of engineering and sciences. A B-spline is a name given to basic type of spline. Most other spline functions can be written as the linear combinations of B-splines. This basis function is used to solve many PDEs with finite-difference method, collocation method, Galerkin method, weighted residual method, and DQM. Here, the two forms of B-spline basis function (i.e., standard and trigonometric forms) are used in linear combination with DQM.

7.7.1 Linear Combinations of Basis Functions

To employ the distinct properties of the standard and trigonometric form of B-spline basis functions, a linear combination is proposed as basis function to be used in DQM. Let the linear combination be given as:

$$l_m(x) = \gamma CT_m(x) + (1-\gamma)MB_m(x)$$

Taking $(1-\gamma)=\delta$, we have the following form of basis function:

$$l_m(x) = \gamma CT_m(x) + \delta MB_m(x) \tag{7.10}$$

where, $CT_m(x)$ represents the modified trigonometric basis function [26], and $MB_m(x)$ represents the modified B-spline basis function [27] of degree three. The modified form of both standard and trigonometric B-spline basis function is applied to deal with the ill-conditioning of the matrix system, which exists with the use of a general form of these basis functions.

The definition of standard B-spline basis function is given by:

$$B_m(x) = \begin{cases} 1, x = x_{m-1} \text{ or } x_{m+1} \\ 4, x = x_m \\ 0, \text{otherwise} \end{cases} ; \; B'_m(x) = \begin{cases} 3/h, x = x_{m-1} \\ -3/h, x = x_{m+1} \\ 0, \text{otherwise} \end{cases}$$

$$B''_m(x) = \begin{cases} 6/h^2, x = x_{m-1} \text{ or } x_{m+1} \\ -6/h^2, x = x_m \\ 0, \text{otherwise} \end{cases} \tag{7.11}$$

The definition of trigonometric B-spline basis function is given by:

$$CTB_m(x) = \begin{cases} d_1, x = x_{m-1} \text{ or } x_{m+1} \\ d_2, x = x_m \\ 0, \text{otherwise} \end{cases} ; \; CTB'_m(x) = \begin{cases} d_3, x = x_{m-1} \\ d_4, x = x_{m+1} \\ 0, \text{otherwise} \end{cases} ;$$

$$CTB''_m(x) = \begin{cases} d_5, x = x_{m-1} \text{ or } x_{m+1} \\ d_6, x = x_m \\ 0, \text{otherwise} \end{cases} \tag{7.12}$$

with $d_i's$ given by

$$d_1 = \frac{\sin^2(\frac{h}{2})}{\sin(h)\sin(\frac{3h}{2})}, \quad d_2 = \frac{2}{1+2\cos(h)}, \quad d_3 = \frac{-3}{4\sin(\frac{3h}{2})}$$

$$d_4 = \frac{3}{4\sin(\frac{3h}{2})}, \quad d_5 = \frac{3(1+3\cos(h))}{16\sin^2(\frac{h}{2})2\cos(h)\cos(\frac{3h}{2})}, \quad d_6 = \frac{-3\cos^2(\frac{h}{2})}{\sin^2(\frac{h}{2})(2+4\cos(h))}$$

7.8 Determination of the Weighting Coefficients

After finalizing the basis functions to be used in DQM, the next step is to obtain the weighting coefficients that can be used to approximate the derivatives.

To approximate the first-order derivative, expanding the derivatives as a linear summation of basis functions one can obtain:

$$l'_m(x_i) = \sum_{j=1}^{N} a_{ij} l_m(x_j), \quad i = 1,\ldots,N; \; m = 1,\ldots,N \tag{7.13}$$

For the first knot point, x_1, the approximation can be given as

$$l'_m(x_1) = \sum_{j=1}^{N} a_{1j} l_m(x_j), \quad m = 1,\ldots,N \tag{7.14}$$

On substituting the values of $l'_m(x_1)$ and $l_m(x_j)$, for $j = 1,\ldots,N$, from the definition of basis function, results in a tridiagonal system of equations as AX = B where,

$$A = \begin{bmatrix} \gamma(2d_1+d_2)+6\delta & \gamma d_1+\delta & & & & & \\ 0 & \gamma d_2+4\delta & \gamma d_1+\delta & & & & \\ & \gamma d_1+\delta & \gamma d_2+4\delta & \gamma d_1+\delta & & & \\ & & \ddots & \ddots & \ddots & & \\ & & & \gamma d_1+\delta & \gamma d_2+4\delta & \gamma d_1+\delta & \\ & & & & \gamma d_1+\delta & \gamma d_2+4\delta & 0 \\ & & & & & \gamma d_1+\delta & \gamma(2d_1+d_2)+6\delta \end{bmatrix}$$

$$X = \begin{bmatrix} a_{11} \\ a_{12} \\ \cdot \\ \cdot \\ \cdot \\ a_{1N-1} \\ a_{1N} \end{bmatrix} \quad B = \begin{bmatrix} -2\gamma d_4 - \frac{6}{h}\delta \\ \frac{\gamma(d_3-d_4)}{2} + \frac{6}{h}\delta \\ 0 \\ 0 \\ \vdots \\ \vdots \\ 0 \end{bmatrix}$$

Here, $d_i's$ are the values of the trigonometric B-splines summarized as given in the last section. The obtained tridiagonal system of equations results in the weighting coefficients $a_{11}, a_{12}, \ldots, a_{1N}$.

Similarly, the approximation for the second knot x_2 can be determined as follows:

$$l'_m(x_2) = \sum_{j=1}^{N} a_{2j} l_m(x_j), \quad m = 1, \ldots, N \tag{7.15}$$

The tridiagonal system of equations for the second knot is again of the form $AX = B$ with same coefficient matrix A and the following values of X and B:

$$X = \begin{bmatrix} a_{21}, & a_{22}, & \ldots & . & \ldots & a_{2N-1}, & a_{2N} \end{bmatrix}^T$$

$$B = \begin{bmatrix} \gamma d_4 - \frac{3}{h}\delta, & 0, & \gamma d_4 + \frac{3}{h}\delta, & 0 & \ldots. & \ldots & 0 \end{bmatrix}^T$$

The solution of this system yields the weighting coefficients $a_{21}, a_{22}, \ldots, a_{2N}$. Similarly, one can proceed up to second last knot point x_{N-1}, with values of X and B given as:

$$X = \begin{bmatrix} a_{N-1,1}, & a_{N-1,2}, & \ldots & . & \ldots & a_{N-1,N-1}, & a_{N-1,N} \end{bmatrix}^T$$

$$B = \begin{bmatrix} 0, & 0, & \ldots & \ldots & \gamma d_4 - \frac{3}{h}\delta, & 0 & \gamma d_3 + \frac{3}{h}\delta \end{bmatrix}^T$$

For the last knot point, x_N, the tridiagonal system of equations produces the following values of X and B with the same value of coefficient matrix A:

$$X = \begin{bmatrix} a_{N,1}, & a_{N,2}, & \ldots & . & \ldots & a_{N,N-1}, & a_{N,N} \end{bmatrix}^T$$

$$B = \begin{bmatrix} 0, & 0, & \ldots & \ldots & 0, & (d_4-d_3)\gamma - \frac{6}{h}\delta, & 2\gamma d_3 + \frac{6}{h}\delta \end{bmatrix}^T$$

Thus, the solution of this system yields the weighting coefficients to approximate $a_{ij}'s$. To approximate the second derivative, the weighting coefficients $b_{ij}'s$ can be calculated by using the coefficients of a_{ij} by the relation given as follows:

$$b_{ij} = \begin{cases} 2a_{ij} - \left(\dfrac{1}{x_i - x_j} \right) & \text{for } i \neq j \\ -\sum_{i=1,}^{N} b_{ij}, & \text{for } i = j \end{cases}$$

7.9 Method Implementation

Let us discuss the approach of solving the GEW equation with this method. On discretizing the time derivatives in the GEW equation by forward difference scheme, and applying the Crank Nicolson scheme for the nonlinear term, the equation resulted in:

$$\frac{u^{n+1} - u^n}{\Delta t} + \varepsilon \left(\frac{(u^p u_x)^{n+1} + (u^p u_x)^n}{2} \right) - \mu \left(\frac{u_{xx}^{n+1} - u_{xx}^n}{\Delta t} \right) = 0 \tag{7.16}$$

Using quasi-linearization technique for linearizing the nonlinear $u^p u_x$, the term leads to expression:

$$\left(u^p u_x \right)^{n+1} = (u^p)^{n+1} (u_x)^n + (u^p)^n (u_x)^{n+1} - (u^p u_x)^n$$

Substituting the linearized form and separating the terms at different time levels leads to following simplified form:

$$u^{n+1} + \lambda \left(u_x^{n+1} (u^p)^n + u_x^n (u^p)^{n+1} \right) - \mu u_{xx}^{n+1} = u^n - \mu u_{xx}^n \tag{7.17}$$

where $\lambda = \frac{\varepsilon \Delta t}{2}$

To find the solution of the equation at a particular level, substituting the values of derivatives by the approximations as defined in the previous section generates a system of equations that can be solved by any numerical approach. Then the scheme can be recursively used to obtain the solution at the later time level.

7.10 Results

The GEW wave Equation (7.9) has the analytical solution of the form

$$u(x,t) = \left(\frac{v(p+1)(p+2)}{2\varepsilon} \sec^2 h\{k(x - vt - x_0)\} \right)^{1/p}$$

where v is constant wave velocity, $\left(\frac{v(p+1)(p+2)}{2\varepsilon} \right)^{1/p}$ is the amplitude, and $k = \frac{p}{2\sqrt{\mu}}$ is the wave width, respectively.

Moreover, the solution of the GEW equation also yields three conservative laws given as:

$$I_1 = \int_a^b u\,dx; \; I_2 = \int_a^b (u^2 + \mu u_x^2)dx; \; I_3 = \int_a^b u^{p+2}\,dx,$$

which are equivalent to conservation of mass, conservation of momentum, and energy, respectively. In this section, we have presented the solution of GEW equation as the motion of a single solitary with two different set of parameters using the proposed algorithm. To illustrate the applicability and effectiveness of the numerical solution, errors are the most common way. Here we have used L_2 and L_∞ error norms to validate the obtained results.

Example 7.1

In this first example, we have taken $p = 1$ with $x_0 = 10, h = 0.15, \varepsilon = 1$ and $\mu = 1$ as the parameters for our computational work with $\Delta t = 0.05$ over an interval [0, 30]. The generated initial condition obtained from the exact solution can be written as:

$$u(x,t) = 3v \sec^2 h[0.5(x - vt - x_0)]$$

Results are calculated for two choices of wave velocity as $v = 0.1$ and $v = 0.3$. The obtained solution depicts the behavior of a solitary wave marching forward with 0.3 and 0.09 unit of amplitude, respectively, and a fixed width of 0.5 with zero values at the boundaries.

The solutions are calculated for $v = 0.1$ at different time levels in Table 7.1 for different values of γ. From the obtained results it can be concluded that there is no much effect on the solution with respect to the different values of γ as the basis function in standard form, trigonometric form, or a linear combination of both. Table 7.2 presents the solution calculated at $v = 0.3$ only for $\gamma = 0.5$ at different time

TABLE 7.1
Results Obtained for Wave Velocity 0.1 for Different Time Levels for Example 7.1

		$T = 10$	$T = 20$	$T = 30$	$T = 40$	$T = 80$
$\gamma = 0.5$	L_2	0.1529×10^{-4}	0.5815×10^{-4}	0.2985×10^{-4}	0.2874×10^{-4}	0.7451×10^{-4}
	L_∞	0.2004×10^{-4}	0.7373×10^{-4}	0.2712×10^{-4}	0.1736×10^{-4}	0.7373×10^{-4}
$\gamma = 1$	L_2	0.1527×10^{-5}	0.5640×10^{-5}	0.2144×10^{-5}	0.09866×10^{-5}	0.8088×10^{-5}
	L_∞	0.2004×10^{-5}	0.7373×10^{-5}	0.2712×10^{-5}	0.09978×10^{-5}	0.7373×10^{-5}
$\gamma = 0$	L_2	0.1529×10^{-5}	0.5880×10^{-5}	0.2144×10^{-5}	0.3337×10^{-5}	0.5679×10^{-5}
	L_∞	0.2004×10^{-5}	0.7373×10^{-5}	0.2712×10^{-5}	0.2006×10^{-5}	0.7373×10^{-5}

TABLE 7.2
Results Obtained for Wave Velocity 0.3 for Different Time Levels for Example 7.1

	$T = 10$	$T = 20$	$T = 30$	$T = 40$	$T = 80$
L_2	0.0922×10^{-4}	0.6832×10^{-5}	0.5066×10^{-5}	0.3751×10^{-5}	0.1254×10^{-5}
L_∞	0.1211×10^{-4}	0.8969×10^{-5}	0.6645×10^{-5}	0.4923×10^{-5}	0.1483×10^{-5}
I_1	2.3998	2.3999	2.3999	2.3999	2.3999
I_2	0.1728	0.1728	0.1728	0.1728	0.1728
I_3	0.0104	0.0104	0.0104	0.0104	0.0104

levels in range $t = 10$ to $t = 80$ along with the values of the three conservation laws as specified previously. The initial solution of the equation is presented in Figure 7.1, followed by the comparison of solution at different values of γ as shown in Figure 7.2 for time levels $t = 10$ to $t = 40$.

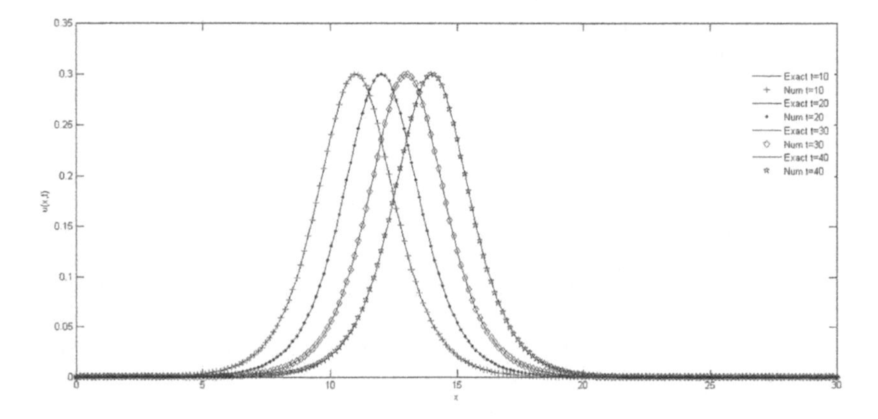

FIGURE 7.1
Traveling wave solution of GEW equation for amplitude 0.3 with comparisons of numerical and exact solution at $t = 10$ to $t = 40$.

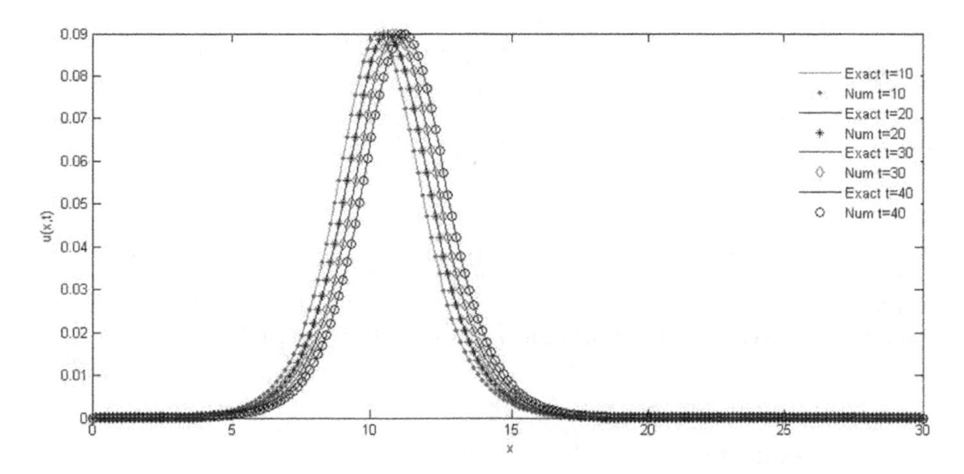

FIGURE 7.2
Traveling wave solution of GEW equation for amplitude 0.09 with comparisons of numerical and exact solution at $t = 10$ to $t = 40$.

Example 7.2

Taking the value of $p = 2$ with $x_0 = 30, h = 0.15, \varepsilon = 3, \mu = 1$ and $v = \frac{1}{32}$ as the considered parameters, the solution is obtained for $\Delta t = 0.05$ on the domain [0, 80]. The solution is obtained for the exact solution given by:

$$u(x,t) = \sqrt{2v\, \mathrm{sec}h^2(x - vt - x_0)}$$

Results are obtained and presented in Table 7.3 for the time ranging from $t = 5$ to $t = 20$ at a fixed value of parameter $\gamma = 0.5$. The obtained numerical results are also presented in Figure 7.3. This figure represents the behavior solution at different time levels for limited domain [20, 40] because the solution difference is not easy to visualize with domain [0, 80]. It can be verified from the tables that the results are in accordance with the results reported in the literature. The values of the three conservation constants are also calculated and depicted in Table 7.4 for parameter $\gamma = 0.5$.

TABLE 7.3
Results Obtained for Wave Velocity 1/32 for Different Time Levels for Example 7.2

	$T = 5$	$T = 10$	$T = 15$	$T = 20$
L_2	0.2659×10^{-4}	0.5981×10^{-4}	0.9126×10^{-4}	0.1245×10^{-3}
L_∞	0.1836×10^{-4}	0.3999×10^{-4}	0.6463×10^{-4}	0.0924×10^{3}

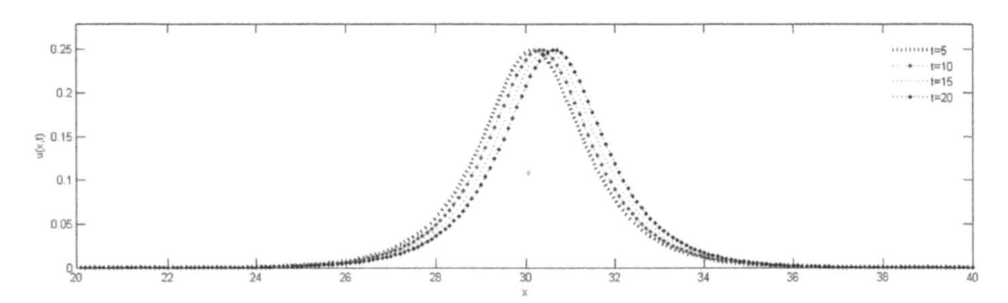

FIGURE 7.3
Traveling wave solution of GEW equation for $t = 5$ to $t = 20$ for example 7.2.

TABLE 7.4
The Values of Three Conservative Constants Obtained for Example 7.2 at Different Time Levels

	$T = 5$	$T = 10$	$T = 15$	$T = 20$
I_1	7.8533	7.8525	7.8518	7.8511
I_2	1.6664	1.6662	1.6659	1.6657
I_3	0.0521	0.0521	0.0520	0.0520

7.11 Summary

This chapter briefly summarizes the developments in DQM with emphasis on the different approaches available for calculating the weighting coefficients. The method has gone through various modifications in terms of basis functions and the methods to obtain the weighting coefficients. The approximation of derivatives calculated by the DQM can be easily applied along with finite-difference and finite-element approaches. To investigate the methodology of DQM, the GEW equation is solved numerically using hybrid basis functions with DQM. The proposed hybrid basis function is a linear combination of modified-standard and modified-trigonometric B-spline basis functions. The solution to the equation demonstrates the traveling wave solutions. Furthermore, to access the capability of the discussed approach, a solution of the GEW equation is calculated for solitary waves under different parameters. This method has undergone a lot of advancement and modifications and is used with other well-known legacy methods for numerical solutions of equations.

References

1. R. E. Bellman, and J. Casti, Differential quadrature and long-term integration. *J. Math. Anal. Appl.* 34 (1971), 235–238.
2. C. Shu, *Differential Quadrature and Its Application in Engineering*. Springer, London, UK, 2000.
3. J. R. Quan, and C. T. Chang, New insights in solving distributed system equations by the quadrature methods-I. *Comput. Chem. Eng.* 13 (1989), 779–788.
4. J. R. Quan, and C. T. Chang, New insights in solving distributed system equations by the quadrature methods-II. *Comput. Chem. Eng.* 13 (1989), 1017–1024.
5. A. Korkmaz, and I. Dag, Shock wave simulations using sinc differential quadrature method. *Eng. Comput.* 28 (6) (2011), 654–674.
6. X. Wu, and Y. Ren, Differential quadrature method based on the highest derivative and its applications. *J. Comput. Appl. Math.* 205 (1) (2007), 239–250.
7. R. Bellman, B. G. Kasher, and J. Casti, Differential quadrature: A technique for the rapid solution of nonlinear partial differential equation. *J. Comput. Phys.* 10 (1972), 40–52.
8. Ö. Civalek, Application of differential quadrature (DQ) and harmonic differential quadrature (HDQ) for buckling analysis of thin isotropic plates and elastic columns. *Eng. Struct.* 26 (2) (2004), 171–186.
9. X. Wang, *Differential Quadrature and Differential Quadrature Based Element Methods*. Butterworth-Heinemann publications, Amsterdam, the Netherlands, 2015.
10. D. H. Peregrine, Long waves on a beach. *J. Fluid Mech.* 27 (1967), 815–827.
11. T. B. Benjamin, J. L. Bona, and J. J. Mahony, Model equations for long waves in non-linear dispersive systems. *Philos. Trans. Royal Soc. A* 272 (1972), 47–78.
12. L. R. T. Gardner, and G. A. Gardner, Solitary waves of the equal width wave equation. *Comput. Phys.* 101 (1992), 218–223.
13. L. R. T. Gardner, G. A. Gardner, F. A. Ayoub, and N. K. Amein, Simulations of the EWE undular bore. *Commun. Num. Methods Eng.* 13 (1997), 583–592.
14. S. I. Zaki, A least-squares finite element scheme for the EW equation. *Comp. Methods Appl. Mech. Eng.* 189 (2000), 587–594.
15. I. Dag, and B. Saka, A cubic B-spline collocation method for the EW equation. *Math. Comput. Appl.* 9 (2004), 381–392.
16. F. Fazal-I Haq, I. A. Shah, and S. Ahmad, Septic B-spline collocation method for numerical solution of the equal width wave (EW) equation. *Life Sci. J.* 10 (2013), 253–260.
17. A. Esen, A lumped Galerkin method for the numerical solution of the modified equal-width wave equation using quadratic B-splines. *Int. J. Comput. Math.* 83 (2006), 449–459.
18. B. Saka, Algorithms for numerical solution of the modified equal width wave equation using collocation method. *Math. Comput. Model.* 45 (2007), 1096–1117.
19. S. I. Zaki, Solitary wave interactions for the modified equal width equation. *Comput. Phys.* 126 (2000), 219–231.
20. H. Panahipour, Numerical simulation of GEW equation using RBF collocation method. *Commun. Numer. Anal.* 59 (2012), 28.

21. A. G. Kaplan, and Y. Derel, Numerical solutions of the GEW equation using MLS collocation method. *Int. J. Mod. Phys.* 28 (2017), 23.
22. S. B. G. Karakoc, and H. Zeybek, A septic B-spline collocation method for solving the generalized equal width wave equation. *Kuwait J. Sci.* 43 (3) (2016), 20–31.
23. S. B. G. Karakoc, and H. Zeybek, A cubic B-spline Galerkin approach for the numerical simulation of the GEW equation. *Stat. Optim. Inf. Comput.* 4 (2016), 30–41.
24. R. Mohammadi, Exponential B-spline collocation method for numerical solution of the generalized regularized long wave equation. *Chin. Phys. B* 24 (5) (2015) 050206.
25. D. J. Evans, and K. R. Raslan, Solitary waves for the generalized equal width (GEW) equation. *Int. J. Comput. Math.* 82 (2005), 445–455.
26. G. Arora, and V. Joshi, Comparison of numerical solution of 1D hyperbolic telegraph equation using B-spline and trigonometric B-spline by differential quadrature method. *Indian J. Sci. Technol.* 9 (2016) 12–14.
27. G. Arora, and B. K. Singh, Numerical solution of Burgers' equation with modified cubic B-spline differential quadrature method. *Appl. Math. Comput.* 224 (2013), 166–177.

List of Figures

8

The TSCSTFCF Procedure for Reliability Demonstration

Amos E. Gera

CONTENTS

8.1 Introduction

Start-up demonstration tests are set up for proving the reliability of different kinds of equipment such as lawn mowers, batteries, and power generators. A whole bunch of procedures have been developed during recent decades for this purpose. In general, the results of these tests lead to the decision whether to accept or to reject the unit that is tested. The underlying theory is based on the theory of runs. Runs are sets of consecutive successes or failures, and they are related to the commonly known consecutive k_c-out-of- n systems. A survey of the theory may be found, for instance, in Kuo and Zuo [1] and in Eryilmaz [2].

There are several ways of handling start-up demonstration tests. These include using generating functions, the Markov Chain Embedding technique (MCE), and the combinatorial approach. An extensive survey of the various start-up procedures has been presented by Balakrishnan, Koutras and Milienos [3]. At first, the consecutive successes (CS) procedure was introduced by Hahn and Gage [4]. Accordingly, the tested unit is accepted if there is a consecutive set of k_{cs} successes. Thereafter, Gera [5] presented the total successes consecutive successes (TSCS) model. Further on, adding the number of failures into the procedure, the consecutive successes total failures (CSTF) procedure evolved. Accordingly, the equipment is accepted if there exists a consecutive set of k_{cs} successes, and it is rejected if k_f failures appeared before (Balakrishnan and Chan [6], Martin [7,8], Smith and Griffith [9–11]). A generalization of these models to the total successes consecutive successes total failures (TSCSTF) and Total successes consecutive failures total failures consecutive failures (TSCSTFCF) procedures has been carried out by Gera [12,13]. Accordingly, accept the tested unit if either k_s successes or a run of length k_{cs} successes are encountered before the counting of k_f failures and the appearance of a run of k_{cf} consecutive failures. Otherwise, the unit is rejected. The models herein include only simple i.i.d. binary tests (success or failure). Thereafter, previous sum-dependent tests have been handled in which the probability of the success of each test depends on the total number of previous ones [14].

This single-dimensional concept may be generalized to the two-dimensional case [15]. This involves testing M units in parallel. Work in this direction has been presented by Zhao, Cui, and Xie [16] for the case of $M = 2$ units. The extended procedure works as follows. The equipment is accepted if either there is a total of k_s successes or if all M units have successes along the same consecutive k_{cs} tests (so that we have a rectangular grid of $M \times k_{cs}$ successes). It is rejected if before fulfilling these criteria for success, either there is a total of k_f failures or if all units fail at the occurrence of the same consecutive k_{cf} tests (thus, creating a rectangular grid consisting of $M \times k_{cf}$ failures).

For the purpose of designing a set of demonstration tests, two quantities are of interest. The basic one is the number of tests (N) that will be involved, and it is represented by their expected number $E\{N\}$. In addition, we are interested in the probability of success and of accepting the tested unit (P_a).

In this chapter a combinatorial approach to evaluate the quantities of interest is used. It consists of defining auxiliary probability functions through which the parameters of interest are calculated. Further on, an optimization problem of minimizing the expected number of required tests subject to some confidence level constraints is solved. The present technique also proved to be useful for solving nearby problems; see Gera [12–15,17].

8.2 Basic Quantities

A run of successes is a set of consecutive successes (CS): "ss…ss." The number of successes is the length of the run. Thus, short runs may be included within a longer one. Meanwhile, we are dealing with a single unit that must undergo as set of demonstration tests.

Let N be the random variable representing the total number of tests till it stops because of some stopping criterion yielding acceptance or rejection. $E\{N\}$ is the expected number of tests, and P_a is the probability of accepting the tested unit. Let p and q be the probabilities of success and failure for each test. Practically, their values are specified after performing some estimation procedure [13]. Here they are assumed to be identical for each of them. This restriction may be released, including the case of dependence between the tests [14].

Considering first the CS model, the point probability of N is given by the following difference equation [18]:

$$P\{N=n\}=\left\{\begin{array}{cc} 0 & n<k_{cs} \\ p^{k_{cs}} & n=k_{cs} \\ qp^{k_{cs}} & 2k_{cs}\ge n>k_{cs} \\ qp^{k_{cs}}\cdot\left[1-\sum_{i=1}^{n-2k_{cs}}P\{N=k_{cs}+i-1\}\right] & n>2k_{cs} \end{array}\right\} \quad (8.1)$$

Proof: Let

$$X_n=\left\{\begin{array}{ll} 1 & \text{success at } n \\ 0 & \text{failure at } n \end{array}\right\}$$

Then, for $n>2k_{cs}$,

$$\begin{aligned} P\{N=n\}=P\{&X_n=X_{n-1}=\ldots=X_{n-k_{cs}+1}=1, X_{n-k_{cs}}=0, \\ &\text{no run of } k_{cs} \text{ successes within } [1,n-k_{cs}-1]\} \end{aligned} \quad (8.2)$$

A more complicated model is consecutive successes consecutive failures (CSCF) for which termination of the set of tests occurs if we meet either a run of k_{cs} successes or a run of k_{cf} failures. Accordingly, the tested unit is accepted or rejected.

The direct calculation of E{N} and P_a for a special case is now presented:

Example A: CSCF with $k_{cs}=k_{cf}=2$.

For evaluating P{$N=n$}, the following sets of results lead to the termination of the tests at the n-th stage. Let m represent a natural number.

a. 'sf…(no 'ss' or 'ff' strings)…fss', even $n=2m$, $P=p^2(qp)^{m-1}$
b. 'fs…(no 'ss' or 'ff' strings)…fsff', even $n=2m$, $P=q^2(qp)^{m-1}$
c. 'fs…(no 'ss' or 'ff' strings)…fsfss', odd $n=2m+1$, $P=qp^2(qp)^{m-1}$
d. 'sf…(no 'ss' or 'ff' strings)…sfsff', odd $n=2m+1$, $P=pq^2(qp)^{m-1}$

Therefore, the probability for termination at the n'th test is, for $n>1$:

$$P\{N=n\}=\left\{\begin{array}{cc} \left(p^2+q^2\right)\cdot\left(pq\right)^{m-1} & n=2m \\ \left(qp\right)^{m} & n=2m+1 \end{array}\right\} \quad (8.3)$$

Now we wish to find the expected number (E{N}) of tests that are required to stop the testing.

In terms of Equation 8.3,

$$E\{N\} = \sum_{n=1}^{\infty} n \cdot P\{N = n\} = \sum_{m=1}^{\infty} (2m) \cdot (p^2 + q^2) \cdot (pq)^{m-1} + \sum_{m=1}^{\infty} (2m+1) \cdot (qp)^m \tag{8.4}$$

Using the known summation formulas:

$$\sum_{m=1}^{\infty} r^m = \sum_{m=0}^{\infty} r^m - 1 = \frac{1}{1-r} - 1 \tag{8.5}$$

and

$$\sum_{m=1}^{\infty} m r^{m-1} = \frac{1}{(1-r)^2} \tag{8.6}$$

We arrive at

$$E\{N_r\} = 2\left(p^2 + q^2\right) \cdot \frac{1}{(1-pq)^2} + 2qp \cdot \cdot \frac{1}{(1-pq)^2} + \frac{1}{1-qp} - 1 \tag{8.7}$$

After some algebraic manipulation, the following closed form expression is derived:

$$E\{N\} = \frac{\left(1-p^2\right)\left(1-q^2\right)}{pq\left[1-(1-p)(1-q)\right]} = \frac{(1+p)(1+q)}{1-pq} \tag{8.8}$$

This result fits the formula for E{N} presented in [3,11,19] for general CSCF problems (k_{cs}, k_{cf}),

$$E\{N\} = \frac{\left(1-p^{k_{cs}}\right)\left(1-q^{k_{cf}}\right)}{pq\left[1-\left(1-p^{k_{cs}-1}\right)\left(1-q^{k_{cf}-1}\right)\right]} \tag{8.9}$$

Another value of interest is the probability of acceptance (P_a) of the tested unit when performing all the tests (i.e., if at any stage we meet a set of two consecutive successes).

For any n, either answer a or c contributes to the acceptance. Thus, till $n = 2m$, summing up on all even numbered tests:

$$P_a(n) = p^2 \cdot \sum_{m=1}^{\frac{n-2}{2}} (pq)^m = p^3 q \cdot \frac{1-(pq)^{\frac{n-2}{2}}}{1-pq} \quad (n \geq 4) \tag{8.10}$$

and till $n = 2m + 1$, for all odd numbered ones:

$$P_a(n) = p^2 q \cdot \sum_{m=1}^{\frac{n-3}{2}} (pq)^m = p^3 q^2 \cdot \frac{1-(pq)^{\frac{n-3}{2}}}{1-pq} \quad (n \geq 5) \tag{8.11}$$

Obviously,

$$P_a(2) = p^2$$

$$P_a(3) = qp^2$$

When considering a large number of tests, the all in all probability of acceptance will tend to:

$$P_a(n) \to p^2 + qp^2 + \frac{p^3 q}{1-pq} + \frac{p^3 q^2}{1-pq} \quad \text{as} \quad n \to \infty \tag{8.12}$$

Some algebraic manipulation shows that the term simplifies to:

$$P_a = \frac{p^2(1+q)}{1-pq} \tag{8.13}$$

Again, this may be compared to the expression given in [3,11,19] for general CSCF procedure.

$$P_a = \frac{p^{k_{cs}-1}\left(1-q^{k_{cf}}\right)}{1-\left(1-p^{k_{cs}-1}\right)\left(1-q^{k_{cf}-1}\right)} \tag{8.14}$$

Consider a furthermore generalization of the previous procedure to the total successes consecutive successes consecutive failures (TSCSCF) procedure [12]. Accordingly, termination also occurs if there are k_s successes (yielding acceptance). The same $k_{cs} = k_{cf} = 2$ are assumed.

The following cases should be considered for yielding stopping of the process:

a. 'sf…(no 'ss', 'ff' strings, no $k_s$xs, $k_f$xf)…sfs', odd $n = 2m + 1$, $k_s - 1 = 0.5(n - 1) < k_f, P = p(qp)^{0.5(n-1)}$
b. 'sf…(no 'ss', 'ff' strings, no $k_s$xs, $k_f$xf)…sf', even $n = 2m$, $k_f = 0.5n < k_s$, $P = (qp)^{0.5n}$
c. 'fs…(no 'ss', 'ff' strings, no $k_s$xs, $k_f$xf)…fsf', odd $n = 2m + 1$, $k_f - 1 = 0.5(n-1) < k_s$, $P = q(qp)^{0.5(n-1)}$

d. 'fs…(no 'ss', 'ff' strings, no $k_s$xs, $k_f$xf)…fs', even $n = 2m$, $k_s = 0.5n < k_f$, $P = (qp)^{0.5n}$
e. 'sf…(no 'ss', 'ff' strings, no $k_s$xs, $k_f$xf)…ss', even $n = 2m$, $k_s - 2 = 0.5(n-2) < k_f$, $P = p^2(qp)^{0.5(n-2)}$
f. 'fs…(no 'ss', 'ff' strings, no $k_s$xs, $k_f$xf)…fss', odd $n = 2m + 1$, $k_s - 1 = 0.5(n-1) < k_f$, $P = p(qp)^{0.5(n-1)}$
g. 'sf…(no 'ss', 'ff' strings, no $k_s$xs, $k_f$xf)…sff', odd $n = 2m + 1$, $k_f - 1 = 0.5(n-1) < k_s$, $P = q(qp)^{0.5(n-1)}$
h. 'fs…(no 'ss', 'ff' strings, no $k_s$xs, $k_f$xf)…fsff' even $n = 2m$, $k_f - 2 = 0.5(n-2) < k_s$, $P = q^2(qp)^{0.5(n-2)}$

This is a more complicated model. Referring to the various cases (a–h):

$$P\{N=n\} = \left\{\begin{array}{ccc} q(pq)^{k_s-1} & n = 2k_s & (a),(h) \\ (1+p)(pq)^{k_s-1} & n = 2k_s - 1 & (d),(f),(g) \\ (p^2+q^2)(pq)^{m-1} & n \le 2k_s - 2 & n = 2m,\ (e),(h) \\ (pq)^m & n < 2k_s - 2 & n = 2m+1,(f),(g) \\ 0 & n > 2k_s & \end{array}\right\} \tag{8.15}$$

Like before, we need to use the finite summation formulas:

$$\sum_{m=1}^{M} r^m = r \cdot \frac{1-r^M}{1-r} \tag{8.16}$$

$$\sum_{m=1}^{M} m r^{m-1} = \frac{1-(M+1)r^M}{1-r} + \frac{r(1-r^M)}{(1-r)^2} \tag{8.17}$$

It is then observed that even for the relative simple case that is considered, the direct approach for deriving the expressions for $E\{N\}$ and P_a become cumbersome so that we need to resort to numerical techniques.

8.3 Auxiliary Functions

The functions of interest may be calculated with aid of some auxiliary functions. In the following, we consider at first the CSCF system where any run of k_{cs} successes or k_{cf} failures yields the termination of the set of tests. Like before, it is assumed that each trial may result in a success or a failure and that the probability of success is identical for each test (binary i.i.d.). $T_{n,s}$ denotes the total number of successes till n-th test, and likewise for $T_{n,f}$. $L_{n,s}$ is the length of the longest run of successes till n and likewise for $L_{n,f}$.

Define the auxiliary functions:

$$f_r\left(i,k_{cs},k_{cf},n\right)=P\left\{T_{n,s}=i,L_{n,s}<k_{cs},L_{n,f}<k_{cf},X_n=r\right\}\quad r=0,1 \tag{8.18}$$

Some interconnecting relations are observed to be valid [13].

For $n>i\geq 1$:

$$f_0(i,k_{cs},k_{cf},n)=\sum_{a=1}^{\min(k_{cf},n)-1} q^a\cdot f_1\left(i,k_{cs},k_{cf},n-a\right) \tag{8.19}$$

$$f_1(i,k_{cs},k_{cf},n)=\sum_{b=1}^{\min(k_{cs}-1,i,n-1)} p^b\cdot f_0(i-b,k_{cs},k_{cf},n-b) \tag{8.20}$$

The boundary conditions are:

$$f_0\left(0,k_{cs},k_{cf},n\right)=q^n\cdot u\left[\left(k_{cf}-1\right)-n\ \right]$$

$$f_1\left(0,k_{cs},k_{cf},n\right)=0$$

For $i>n$ and also for $i=n\geq k_{cs}$:

$$f_0\left(i,k_{cs},k_{cf},n\right)=0$$

$$f_1\left(i,k_{cs},k_{cf},n\right)=0$$

Proof of Equations 8.19 and 8.20. Use is made of the law of total probability. Summing up on the various runs of failures that do not yield termination of the testing,

$$f_0\left(i,k_{cs},k_{cf},n\right)=\sum_{a=1}^{\min(k_{cf}-1,n-1)} q^a\cdot P\left\{T_{n-a,s}=i,L_{n-a,s}<k_{cs},L_{n-a,f}<k_{cf},X_{n-a}=1\right\} \tag{8.21}$$

Referring to the definition for the f_1 function (8.18), we arrive at Equation 8.19.

The same goes for Equation 8.20.

It may be noticed that for $k_{cs}>n\geq i>0$, the following is valid:

$$f_1\left(i,k_{cs},k_{cf},n\right)=\binom{n}{i-1}p^i q^{n-i}\cdot u\left[\left(k_{cf}-1\right)-(n-i)\right] \tag{8.22}$$

Illustration: $k_{cs} = k_{cf} = 2$

$$f_0(2,2,2,5) = p^3q^2 \text{ case of 'fsfsf'}$$

$$f_1(3,2,2,5) = p^3q^2 \text{ case of 'sfsfs'}$$

The system of difference equations is normally solved numerically. However, some insight into the analytical form of solution may be gained through the use of generating functions.

8.4 Solution by Generating Functions

A closed-form solution to the system of Equations 8.19 and 8.20 may be provided using the method of generating functions. We will illustrate this technique for the special case of the CSCF procedure with $k_{cs} = k_{cf} = 2$. Then, for $i > 0$, $n > 1$:

$$\begin{aligned} f_0(i,n) &= q \cdot f_1(i,n-1) \\ f_1(i,n) &= p \cdot f_0(i-1,n-1) \end{aligned} \tag{8.23}$$

These equations may be mingled together into a single one, for instance (for $n > 2$)

$$f_0(i,n) = pq \cdot f_0(i-1,n-2) \tag{8.24}$$

Appropriate boundary conditions must be added:

$$f_0(0,n) = q \cdot \delta[n-1]$$

$$f_0(1,n) = \begin{Bmatrix} 0 & n=1 \\ pq & n=2 \\ pq^2 & n=3 \\ 0 & n>3 \end{Bmatrix}$$

$$f_0(i,1) = q \cdot \delta[i]$$

$$f_0(i,2) = pq \cdot \delta[i-1]$$

The generating function $F_0(x, y)$ (formal power series) for $f_0(i, n)$ is:

$$F_0(x,y) = \sum_{i=0}^{\infty}\sum_{n=1}^{\infty} f_0(i,n)x^iy^n \tag{8.25}$$

It may be split up as follows:

$$F_0(x,y) = \sum_{n=1}^{\infty} f_0(0,n)y^n + \sum_{n=1}^{\infty} f_0(1,n)\, xy^n + \sum_{i=2}^{\infty} f_0(i,1)x^i y$$

$$+\sum_{i=2}^{\infty} f_0(i,2)x^i y^2 + \sum_{i=2}^{\infty}\sum_{n=3}^{\infty} f_0(i,n)x^i y^n \tag{8.26}$$

Referring to the given boundary conditions,

$$\sum_{n=1}^{\infty} f_0(0,n)\, y^n = qy$$

$$\sum_{n=1}^{\infty} f_0(1,n)\, xy^n = x\left(pqy^2 + pq^2y^3\right)$$

$$\sum_{i=2}^{\infty} f_0(i,1)\, x^i y = \sum_{i=2}^{\infty} f_0(i,2)\, x^i y^2 = 0$$

Regarding the last term in Equation 8.26 and referring to our Equation 8.24:

$$\begin{aligned}
\sum_{i=2}^{\infty}\sum_{n=3}^{\infty} f_0(i,n)\, x^i y^n &= pq\cdot\sum_{i=2}^{\infty}\sum_{n=3}^{\infty} f_0(i-1,n-2)\, x^i y^n \\
&= pqxy^2\cdot\sum_{i=2}^{\infty}\sum_{n=3}^{\infty} f_0(i-1,n-2)\, x^{i-1}y^{n-2} \\
&= pqxy^2\cdot\sum_{i=1}^{\infty}\sum_{n=1}^{\infty} f_0(i,n)\, x^i y^n \\
&= pqxy^2\cdot\left\{\sum_{i=0}^{\infty}\sum_{n=1}^{\infty} f_0(i,n)\, x^i y^n - \sum_{n=1}^{\infty} f_0(0,n)\, y^n\right\} \\
&= pqxy^2\cdot\{F_0(x,y) - qy\}
\end{aligned} \tag{8.27}$$

Therefore, the following equation evolves:

$$F_0(x,y) = qy + pqxy^2 + pq^2xy^3 + pqxy^2\cdot F_0(x,y) - pq^2xy^3 \tag{8.28}$$

so that

$$\left(1 - pqxy^2\right)\cdot F_0(x,y) = qy + pqxy^2 \tag{8.29}$$

Using formal geometric series expansion:

$$\begin{aligned} F_0(x,y) &= \left(1-pqxy^2\right)^{-1}\cdot\left(qy+pqxy^2\right) = \left(qy+pqxy^2\right)\cdot\sum_{r=0}^{\infty}(pq)^r x^r y^{2r} \\ &= \sum_{r=0}^{\infty} q(pq)^r x^r y^{2r+1} + \sum_{r=0}^{\infty}(pq)^{r+1}x^{r+1}y^{2r+2} \\ &= \sum_{r=0}^{\infty} q(pq)^r x^r y^{2r+1} + \sum_{r=1}^{\infty}(pq)^r x^r y^{2r} \end{aligned} \tag{8.30}$$

We, thus, learn that

$$\begin{aligned} f_0(i,2i) &= (pq)^i \quad (i>0) \\ f_0(i,2i+1) &= q(pq)^i \end{aligned} \tag{8.31}$$

The same procedure may be applied to $f_1(i, n)$ with the result:

$$\begin{aligned} f_1(i,2i) &= (pq)^i \quad (i>0) \\ f_1(i,2i+1) &= p(pq)^i \end{aligned} \tag{8.32}$$

Equivalent expressions are derived via a given "n":

$$\begin{aligned} f_0(i,n) &= \begin{cases} q(pq)^{\frac{n-1}{2}} & odd\ n=2m+1, i=\dfrac{n-1}{2} \\ (pq)^{\frac{n}{2}} & even\ n=2m, i=\dfrac{n}{2} \end{cases} \\ f_1(i,n) &= \begin{cases} p(pq)^{\frac{n-1}{2}} & odd\ n=2m+1, i=\dfrac{n+1}{2} \\ (pq)^{\frac{n}{2}} & even\ n=2m, i=\dfrac{n}{2} \end{cases} \end{aligned} \tag{8.33}$$

Then, for odd "n":

$$P\{N>n\} = f_0\left(\frac{n-1}{2},n\right) + f_1\left(\frac{n+1}{2},n\right) = (pq)^{\frac{n-1}{2}} \tag{8.34}$$

and for even "n":

$$P\{N>n\} = f_0\left(\frac{n}{2},n\right) + f_1\left(\frac{n}{2},n\right) = 2(pq)^{\frac{n}{2}} \tag{8.35}$$

Because

$$P\{N=n\}=P\{N>n-1\}-P\{N>n\} \quad (n>1) \tag{8.36}$$

for odd n:

$$\begin{aligned} P\{N=\text{odd } n\} &= P\{N>\text{even } n-1\}-P\{N>\text{odd } n\} \\ &= 2(pq)^{\frac{n-1}{2}}-(pq)^{\frac{n-1}{2}}=(pq)^{\frac{n-1}{2}} \end{aligned} \tag{8.37}$$

and for even n:

$$\begin{aligned} P\{N=\text{even } n\} &= P\{N>\text{odd } n-1\}-P\{N>\text{even } n\} \\ &= (pq)^{\frac{n-2}{2}}-2(pq)^{\frac{n}{2}}=(pq)^{\frac{n-2}{2}}(1-2pq) \end{aligned} \tag{8.38}$$

Thus,

$$\begin{aligned} P\{n=2m+1\} &= (pq)^m \\ P\{n=2m\} &= (1-2pq)\cdot(pq)^{m-1} \end{aligned} \tag{8.39}$$

So that the expected number of tests will be given by:

$$E\{N\}=(1-2pq)\cdot\sum_{m=1}^{\infty}2m\cdot(pq)^{m-1}+\sum_{m=1}^{\infty}(2m+1)\cdot(pq)^m \tag{8.40}$$

Using the known closed-form expression for the summation, we arrive at (see Equation 8.8)

$$E\{N\}=\frac{2+pq}{1-pq} \tag{8.41}$$

We turn now to evaluate the probability of accepting the tested unit. This is achieved when we meet at any stage a run of two consecutive successes subject to not having before any two consecutive successes or failures. Thus, for any n,

$$P_{a,2,n}=p^2\cdot\delta[n-2]+qp^2\cdot\delta[n-3]+p^2\cdot\sum_{i=2}^{n-2}f_0(i-2,n-2) \tag{8.42}$$

Use is made of the explicit expression for $f_0(i, n)$ (8.33). Performing a transfer of variable from n to m (like before), we sum up on all odd and even stages n:

$$P_a=(1+q)p^2+\sum_{m=1}^{\infty}p^2q(pq)^m+\sum_{m=1}^{\infty}p^2(pq)^m \tag{8.43}$$

Thus, we get also a closed-form for the probability of acceptance:

$$P_a = (1+q)p^2 + p^2(1+q)\cdot\left[\frac{1}{1-pq} - 1\right] = \frac{p\left(1-q^2\right)}{1-pq} \tag{8.44}$$

This obviously should be identical to the previously derived expression (8.13).

8.5 Statistical Sampling

A comparison of the previously calculated results for $E\{N\}$ and P_a for the CSCF procedure with $k_{cs} = k_{cf} = 2$ to those derived via a statistical sampling process is now carried out. Using a random-number generator and assuming $p = 0.5$, we recorded the results for some samples as shown in Table 8.1.

Thus, for these 10 samples, $E\{N\} = 2\cdot 0.6 + 3\cdot 0.1 + 4\cdot 0.2 + 6\cdot 0.1 = 2.9$ and counting the number of samples for which there is acceptance, $P_a = 0.5$. These values may be compared to those obtained from Equations 8.41 and 8.44: $E\{N\} = 3$, $P_a = 0.5$.

Perform the same for $p = 0.8$; the results are presented in Table 8.2.

In this case, $E\{N\} = 2\cdot 0.7 + 3\cdot 0.2 + 4\cdot 0.1 = 2.4$ and $P_a = 1$.

This is compared to $E\{N\} = 2.57$ from Equation 8.41 and $P_a = 0.91$ from Equation 8.44. Thus, a much larger amount of samples is required for getting close to the true values of $E\{N\}$. The same goes with P_a.

A comparison between the resultant values is given in Table 8.3.

TABLE 8.1

Results of Some Samples, CSCF, $p = 0.5$

Sample No.	Sequence	$N = n$ for Termination	$N = n$ for Acceptance
1	fsfsffssss	6	0
2	sssfssfsfs	2	2
3	ssssffsfsf	2	2
4	sssssffssf	2	2
5	sffffffsff	3	0
6	fffsfffssff	2	0
7	sssssfssff	2	2
8	fsffsfssff	4	0
9	fsfffsfsfs	4	0
10	ssffffssss	2	2

TABLE 8.2

Results of Some Samples, CSCF, $p = 0.8$

Sample No.	Sequence	$N = n$ for Termination	$N = n$ for Acceptance
1	ssffsssfff	2	2
2	ssssffsssf	2	2
3	ssfsfsssss	2	2
4	sssffsssfs	2	2
5	ssssfsssss	2	2
6	sfssssssss	4	4
7	ssfssfssss	2	2
8	sssssssffs	2	2
9	fssssssfss	3	3
10	fssfffssfs	3	3

TABLE 8.3

Results for $E\{N\}$, P_a $k_{cs} = k_{cf} = 2$

	CSCF, $p = 0.5$	CSCF, $p = 0.8$
$E\{N\}$ (41)	3	2.57
$E\{N\}$ simul.	2.9	2.4
P_a (44)	0.5	0.91
P_a simul.	0.5	1

8.6 The TSCSTFCF Procedure

Consider the more general TSCSTFCF procedure with specific values of k_s, k_f.

Thus, the set of tests is terminated if either we meet a run of k_{cs} successes or a total number of k_s successes before the occurrence of a run of k_{cf} failures or a total number of k_f failures. The acceptance or rejection of the tested unit is obviously determined through the way of termination.

Resorting to our previous example with $k_{cs} = k_{cf} = 2$, the more general TSCSTFCF formula for calculating the expected number of tests will be given by finite summations as follows:

Let $a = min(k_s, k_f)$. Then,

$$\begin{aligned} E\{N\} &= \sum_{m=1}^{a-1}(2m+1)(pq)^m + (1-2pq)\cdot\sum_{m=1}^{a} 2m(pq)^{m-1} = \\ &= 2pq\cdot\sum_{m=1}^{a-1} m\,(pq)^{m-1}pq\cdot\sum_{m=1}^{a-1}(pq)^{m-1} + 2\cdot\sum_{m=1}^{a} m(pq)^{m-1} \\ &\quad - 4pq\cdot\sum_{m=1}^{a-1} m\,(pq)^{m-1} \end{aligned} \tag{8.45}$$

For our finite summations, use

$$\sum_{m=0}^{M} r^m = \frac{1-r^{M+1}}{1-r} \tag{8.46}$$

and

$$\sum_{m=1}^{M} mr^{m-1} = \frac{1-r^M}{(1-r)^2} - \frac{Mr^M}{1-r} \tag{8.47}$$

Therefore,

$$E\{N\} = 2\left[\frac{1-(pq)^a}{(1-pq)^2} - \frac{a(pq)^a}{1-pq}\right] - 2pq\cdot\left[\frac{1-(pq)^{a-1}}{(1-pq)^2} - \frac{(a-1)(pq)^{a-1}}{1-pq}\right]$$
$$+ pq\cdot\frac{1-(pq)^{a-1}}{1-pq} \tag{8.48}$$

The more compact result is:

$$E\{N\} = \frac{(2+pq)-3(pq)^a}{1-pq} \tag{8.49}$$

Comparing this to Equation 8.41, the effect of shortening the number of required tests due to introducing k_s, k_f is obvious.

Considering the TSCSTFCF procedure with specific values of k_s, k_f, it may be observed that within the first summand in Equation 8.43, the following should be valid:

$$m + 2 \le k_s$$

$$m + 1 < k_f$$

and within the second summand:

$$m + 2 \le k_s$$

$$m < k_f$$

Thus, the infinite summations in ($P_{a,2}$ for CSCF) are replaced by finite ones:

$$P_{a,2} = (1+q)p^2 + \sum_{m=1}^{\min(k_s+k_f)-2+1} p^2q(pq)^m + \sum_{m=1}^{\min(k_s-2,k_f-1)} p^2(pq)^m \tag{8.50}$$

and the following modification takes place:

$$P_{a,2} = \frac{p^2(1+q) - p^2 q \cdot (pq)^{\min(k_s,k_f)-1} - p^2 \cdot (pq)^{\min(k_s-2,k_f-1)+1}}{1-pq} \tag{8.51}$$

Acceptance of the tested unit may also be achieved because of reaching k_s successes. In this specific case ($k_{cs} = k_{cf} = 2$), the probability $P_{a,1}$ for reaching this situation is given by

$$P_{a,1} = \begin{Bmatrix} p \cdot (pq)^{k_s-1} & k_f = k_s \\ p(1+q)(pq)^{k_s-1} & k_f > k_s \end{Bmatrix} \tag{8.52}$$

Finally, the probability of acceptance may be a result of any one of them because the probabilities $P_{a,1}$, $P_{a,2}$ are for disjoint events:

$$P_a = P_{a,1} + P_{a,2} \tag{8.53}$$

In general for the TSCSTFCF procedure, turning to the auxiliary functions f_0, f_1, the following is derived:

$$P\{N \geq n\} = \sum_{i=\max(n-k_f+1,0)}^{k_s-1} \left[f_0(i,k_{cs},k_{cf},n) + f_1(i,k_{cs},k_{cf},n) \right] \tag{8.54}$$

$$\begin{aligned} P_{a,1} &= \sum_{n=k_s}^{\infty} P\{T_{n,s} = k_s, L_{n,s} < k_{cs}, T_{n,f} < k_f, L_{n,f} < k_{cf}, X_n = 1\} \\ &= \sum_{n=k_s}^{k_s+k_f-1} f_1(k_s,k_{cs},k_{cf},n) \end{aligned} \tag{8.55}$$

Let

$$P_{a,2,i,n} = P\{T_{n,s} = i, L_{n-1,s} < k_{cs}, L_{n,s} = k_{cs}, T_{n,f} < k_f, L_{n,f} < k_{cf}\} \tag{8.56}$$

Then, for $\min(k_s, n) \geq i > \max(k_{cs}, n - k_f)$

$$P_{a,2,i,n} = p^{k_{cs}} \cdot P \begin{Bmatrix} T_{n-k_{cs},s} = i - k_{cs}, T_{n-1,s} < k_s, L_{n-k_{cs},s} < k_{cs}, T_{n-k_{cs},f} \\ < k_f, L_{n-k_{cs},f} < k_{cf}, X_{n-k_{cs}} = 0 \end{Bmatrix} \tag{8.57}$$

Thus,

$$P_{a,2,i,n} = p^{k_{cs}} \cdot f_0\left(i - k_{cs}, n - k_{cs}\right) \tag{8.58}$$

and

$$P_{a,2} = \sum_{n=k_{cs}}^{\infty} \sum_{i=\max\left(k_{cs}, n-k_f+1\right)}^{\min(k_s, n)} P_{a,2,i,n} \tag{8.59}$$

As stated previously, the total probability of acceptance is then obtained using Equation 8.53.

8.7 Planar Set of Tests

We can test two or more units simultaneously instead of a single one. Obviously, this will yield a much shorter time of testing. In essence, the same approach as before may be applied here. Consider the case of testing two units ($M = 2$) with $k_{cs} = k_{cf} = 2$ (CSCF).

Thus, we consider squares $\begin{smallmatrix} s & s \\ s & s \end{smallmatrix}$ compared with rectangles $\begin{smallmatrix} f & f \\ f & f \end{smallmatrix}$. If the square of successes is met first, then we accept the tested unit. Otherwise it is rejected.

Designate by "1" the result of success ("s") and by "0" the result of failure ("f"). We create at each stage of testing a column vector of length 2 with ones and zeros. This is a binary form of the possible results [15].

Let

$$V = \begin{pmatrix} 0 & 0 & 1 & 1 \\ 0 & 1 & 0 & 1 \end{pmatrix} \tag{8.60}$$

The columns of this matrix $V(m, j)$ represent the various possible results.

V_j will represent here the jth column vector and the previous variables X_n become vectors. Instead of considering $L_{n,s}$, $L_{n,f}$, we consider here squares of successes and of failures and $R_{n,s}$, $R_{n,f}$ designate their maximal area till stage n.

For the planar TSCSTFCF procedure, the following auxiliary function is used:

$$f(i, j, n) = P\left\{T_{n,s} = i, R_{n,s} < 2k_{cs}, R_{n,f} < 2k_{cf}, X_n = \underline{V_j}\right\} \tag{8.61}$$

The relevant difference equations are:

$i \geq 0, n > 1$:

$$f(i, 1, n) = \sum_{a=1}^{k_{cf}-1} q^{2a} \cdot \sum_{j'=2}^{4} f(i, j', n-a) \tag{8.62}$$

$i > 0, n > 1$:

$$j = 2,3:$$

$$f(i,j,n) = pq \cdot \sum_{j'=1}^{4} f(i-1,j',n-1) \tag{8.63}$$

$i > 1, n > 1$:

$$f(i,4,n) = \sum_{b=1}^{k_{cs}-1} p^{2b} \cdot \sum_{j'=1}^{3} f(i-2b,j',n-b) \tag{8.64}$$

The appropriate boundary conditions are:

$n = 1$: $f(i,1,1) = q^2\,\delta[i]$; $f(i,4,1) = p^2\,\delta[i-2]$; for $j = 2,3$ $f(i,j,1) = pq\,\delta[i-1]$

$i = 0$: for $n < k_{cf}$ $f(0,1,n) = q^{2n}$; for $j > 1$ $f(0,j,n) = 0$

Proof: For $j = 1$:
Split the event into a summation of independent events as follows:

$$f(i,1,n) = \sum_{a=1}^{k_{cf}-1} P\left\{X_n = \begin{pmatrix}0\\0\end{pmatrix}, \dots, X_{n-(a-1)} = \begin{pmatrix}0\\0\end{pmatrix}\right\} \cdot$$
$$P\left\{T_{n-a,s} = i, R_{n-a,s} < 2k_{cs}, R_{n-a,f} < 2k_{cf}, X_{n-a} \neq \begin{pmatrix}0\\0\end{pmatrix}\right\} \tag{8.65}$$

and using the definition in Equation 8.61, Equation 8.62 results.
Likewise, for $j = 4$:

$$f(i,4,n) = \sum_{b=1}^{k_{cs}-1} P\left\{X_n = \begin{pmatrix}1\\1\end{pmatrix}, \dots, X_{n-(b-1)} = \begin{pmatrix}1\\1\end{pmatrix}\right\} \cdot$$
$$P\left\{T_{n-b,s} = i-2b, R_{n-b,s} < 2k_{cs}, R_{n-b,f} < 2k_{cf}, X_{n-b} \neq \underline{V_4}\right\} \tag{8.66}$$

For $j = 2,3$:

$$f(i,j,n) = P\left\{T_{n-1,s} = i-1, R_{n-1,s} < 2k_{cs}, R_{n-1,f} < 2k_{cf}\right\} \cdot P\left\{X_n = \begin{pmatrix}0\\1\end{pmatrix} \text{or} \begin{pmatrix}1\\0\end{pmatrix}\right\} \tag{8.67}$$

and summing up on all possibilities results in Equation 8.63.

The tail distributions, expected number of stages of tests, and the probability of acceptance are calculated as for the linear case.

Example: Consider testing two units simultaneously using the CSCF procedure with $k_{cs} = k_{cf} = 2$. Assume the probability of success of each trial to be $p = 0.5$.

Solving the system of Equations 8.62 through 8.64, it turns out that $E\{N\} = 10$.

By performing a simulation on 50 samples, like in the linear case, we get $E\{N\} = 9.6$. Thus, even for a low number of samples, results are quite close (4%).

8.8 The Number of Runs

The defined auxiliary functions (8.18) may also be used for a different type of problem related to the number of runs of successes or failures within a set of tests.

Let $N^{(1)}(i, k_{cs}, n)$ be the number of ways to accommodate i successful tests within a set of n tests such that no k_{cs} or more successful tests are consecutive [1].

$$N^{(1)}(i,k_{cs},n) = \sum_{j=0}^{\min\left(\left[\frac{i}{k_{cs}}\right],n-i+1\right)} (-1)^j \binom{n-i+1}{j}\binom{n-j\cdot k_{cs}}{n-i} \tag{8.68}$$

Conclusion: The number of ways to accommodate i successful tests within a set of n tests so that they will include at least one set of k_{cs} or more consecutive successes is given by:

$$M^{(1)}(i,k_{cs},n) = \binom{n}{i} - N^{(1)}(i,k_{cs},n)) \tag{8.69}$$

Example: Take $k_{cs} = 2$, $i = 3$, $n = 5$. The only way to have no run of length $k_{cs} = 2$ is:

"sfsfs" so that $N^{(1)}(3,2,5) = 1$. Thus, $M^{(1)}(3,2,5) = 9$.

The $N^{(1)}(i, k_{cs}, n)$ term may be extended as follows. Instead of considering only the runs of successes, we add also the runs of failures. Thus, let $N^{(2)}(i, k_{cs}, k_{cf}, n)$ be the number of ways to have i successful tests within a set of n tests such that no k_{cs} or more successful tests are consecutive, and no k_{cf} or more failed tests are consecutive. Gera [13] presented some results regarding this direction.

Referring to the defined auxiliary functions, it is clear that for any p

$$N^{(2)}(i,k_{cs},k_{cf},n) = \frac{f_0(i,k_{cs},k_{cf},n) + f_1(i,k_{cs},k_{cf},n)}{p^i q^{n-i}} \tag{8.70}$$

Illustration: Use $k_{cs} = k_{cf} = 2$. It is easily observed that $f_0(3,2,2,6) = f_1(3,2,2,6) = p^3q^3$.

Thus, $N^{(2)}(3,2,2,6) = 2$. The two combinations are: "sfsfsf" and "fsfsfs."

8.9 An Optimization Problem

As mentioned previously, two important functions of interest are the expected number of required tests ($E\{N\}$) and the probability of acceptance of the tested equipment (P_a). Actually, we choose two values of probability, upper and lower p_U values. Then, the tested unit is accepted if the probability of success p of each individual test is higher than p_U, and it is rejected if it is lower than p_L. This sets up a confidence level on the acceptance of the tested equipment, and we consider here the two types of error for acceptance: Type I (rejection when $p > p_U$) and Type II (acceptance when $p < p_L$) [11,13].

Consider the TSCSTFCF procedure. The problem at stake is the minimization of the number of required tests (their expected number) subject to the confidence limits on the probability of acceptance. Given α and β, it is thus required to determine the values of k_s, k_{cs}, k_f, k_{cf} that will minimize $E\{N\}$ subject to satisfying the inequalities (8.71) on the probability of acceptance for the two specified probabilities:

$$\begin{aligned} &P\{a/p = p_U\} > 1 - \beta \\ &P\{a/p = p_L\} < \alpha \end{aligned} \tag{8.71}$$

Generally, we consider minimizing $E\{N\}$ at $p = p_U$.

It is problematic to provide a closed-form solution even in the simplest cases of total successes total failures (TSTF) or consecutive successes consecutive failures (CSCF) procedures. This is because the unknowns appear as exponents within the inequalities and normally do not lend simple closed-form expressions for their solution.

Some insight into the search of an optimum may be gained from observing the CSCF and CSTF procedures. There exist closed-form expressions for the expected number of tests and for the probability of acceptance in these cases [19].

Illustration: Consider the case of CSCF with $p_U = 0.9$, $p_L = 0.6$, $\alpha = \beta = 0.05$. We wish to find the values of the optimal k_{cs}, k_{cf}. Use an initially chosen value of $k_{cf} = 3$.

Let $x_U = 0.9^{k_{cs}}, x_L = 0.6^{k_{cs}}$. Then it turns out that we should minimize the function:

$$E\{N\} = 11.1 \cdot \frac{1 - x_U}{0.01 + 1.1x_U}$$

subject to the following inequality constraints:

$$P_{aU} = 1.11 \cdot \frac{x_U}{0.01 + 1.1x_U} > 0.95$$

$$P_{aL} = 1.56 \cdot \frac{x_L}{0.16 + 1.4x_L} < 0.05$$

It may be observed that $E\{N\}$ is a decreasing function in x_U, so that it is best to take an x_U value as large as possible. Because P_{aU} is increasing in x_U, this is in the right direction. However, the inequality involving P_{aL} limits the values of the permissible x_L. It yields the permissible region of $x_L < 0.01$ and, thus, the optimal value of $k_{cs} = 9$. This is a kind of suboptimum when choosing $k_{cf} = 3$. Other values should be checked as well.

Illustration: CSTF procedure for which there exist closed-form expressions for the expected number of tests and for the probability of acceptance. Take $p_U = 0.99$, $p_L = 0.70$, $k_f = 2$, $\alpha = \beta = 0.05$.

Let $x_U = 0.99^{k_{cs}}, x_L = 0.7^{k_{cs}}$. The function to be minimized is:

$$E\{N\} = 100 \cdot \frac{1 - 2x_U^2 + x_U^3}{x_U}$$

with the constraints:

$$P_{aU} = 2x_U - x_U^2 > 0.95$$

$$P_{aL} = 2x_L - x_L^2 < 0.05$$

Again, $E\{N\}$ is a decreasing function in x_U. Like before, the bound on the minimum is dictated by the inequality on x_L. It turns out that the optimal value is $k_{cs} = 11$.

A general optimization algorithm has been set up. Like in many optimization procedures, some reasonable initial guess is required. Thereafter, cut and try steps are used to arrive at the optimum. The process has been seen to be rather simple and easy to implement for purpose of calculations. At first this technique has been applied to the case of i.i.d. binary tests. It has been further extended to the two-dimensional TSCSTFCF procedure that was presented already.

8.10 Numerical Results

Various examples for the results of optimization have been given by Gera [13,15,19]. Although the TSCSTFCF procedure is more complicated, it is superior to simpler procedures like CSTF procedures. Using the upper and lower probability values of $p_U = 0.9$, $p_L = 0.6$, and specifying different confidence bounds of $\alpha = 0.05$, $\beta = 0.05, 0.15, 0.25$, the value of the expected number of required tests is presented in Table 8.4.

Likewise, for the two-dimensional case with two tested units ($M = 2$), we have the following optimal values of the variables in Table 8.5.

Further on testing $M = 5$ units in parallel, the following results are obtained (Table 8.6).

It might be that some of the results may be improved when using a systematic program of optimization, which does not exist as yet. Meanwhile, some cut-and-dry method has been used.

TABLE 8.4

Optimal Values of $E\{N\}$, CSTF, TSCSTFCF

α	β	$E\{N\}$, CSTF	$E\{N\}$, TSCSTFCF	Optimal Parameter Values k_s, k_{cs}, k_f, k_{cf}
0.05	0.05	17.75	15.69	20,10,6,3
0.05	0.15	12.36	11.62	14,9,4,2
0.05	0.25	10.46	10.06	13,8,3,2

TABLE 8.5

Optimal Values of $E\{N\}$, Planar $M = 2$, TSCSTFCF

α	β	$E\{N\}$, TSCSTFCF	Optimal Parameter Values k_s, k_{cs}, k_f, k_{cf}
0.05	0.05	8.14	20,5,6,2
0.05	0.15	6.79	14,5,4,2
0.05	0.25	6.36	14,5,3,2

TABLE 8.6

Optimal Values of $E\{N\}$, Planar $M = 5$, TSCSTFCF

α	β	$E\{N\}$, TSCSTFCF	Optimal Parameter Values k_s, k_{cs}, k_f, k_{cf}
0.05	0.05	3.54	20,2,7,∞
0.05	0.15	2.76	13,2,4,2
0.05	0.25	2.57	13,2,3,2

8.11 Concluding Remarks

A set of reliability demonstration procedures has been presented: CS, CSCF, CSTF, and TSCSTFCF. A direct approach of analysis has been first given for a specific simple case. Also the method of generating functions was seen to be of help in this instance. However, in general, we need to define auxiliary functions, set up a set of difference equations, and solve them numerically. The expected number of required tests and the probability of accepting the tested units are derived using the set of these functions. A constrained optimization problem is solved for minimizing the number of required tests subject to some confidence-level requirements. The variables for this optimization include the total number of successes, failures, and the maximal lengths of runs of successes and failures.

The results of this work are applicable to the case of i.i.d. tests (where all tests are assumed to have the same probability for success and there are only two possible results-to-success or -failure ratio). Extensions of this paper to tests with more than two results and to nonidentical probabilities of successes have been presented in various articles. Also the case of tests, which are considered as more or less important according to their sequential order, were handled (weighted tests).

The extension of the theory to the case of testing more than a single unit at a time has been considered. The direct approach and the method of generating functions seem to be too cumbersome and thus we must use the regular numerical approach. Obviously, running in parallel more than one unit shortens drastically the time of testing.

It has been observed that the TSCSTFCF procedure yields the shortest time of testing subject to the constraints. Hence, although it is a bit more complicated from the point of view of calculations, it seems preferable to use this model for reliability demonstration.

It has already been previously stated that the alternative Markov Chain Embedding approach might involve the inversion of large-scale matrices. There exist examples for which we could not apply that technique, and it was needed to use the present combinatorial approach. In some cases, it has been observed that the results were achieved in much shorter time than using MCE.

Some further questions may be handled. We might wish to consider more general patterns instead of runs of successes or failures. The amount of permissible uncertainty of the optimal parameters should be determined, and it should be considered. Computational run times and memory storage ought to be considered.

NOTATIONS

N	the total number of start-ups (trials) until termination of the experiment
$E\{N\}$	expected number of N
SD	the standard deviation for N
P_a	the probability of acceptance of the unit
X_i	outcome of i-th start-up test (=1 for success, =0 for failure)
u[.]	the unit step function
δ[.]	the delta function

References

1. Kuo, W., and Zuo, M.J. (2003). *Optimal Reliability Modeling, Principles and Applications*. New York: John Wiley & Sons.
2. Eryilmaz, S. (2010). Review of recent advances in reliability of consecutive k-out-of-n and related systems. *Proceedings of the Institution of Mechanical Engineers Part* O, 224(3), 225–236.
3. Balakrishnan, N., Koutras, M.V., and Milienos, F.S. (2014). Start-up demonstration test: Models, methods and applications, with some unifications. *Applied Stochastic Models in Business and Industry*, 30(4), 373–413.
4. Hahn, G.J., and Gage, J.B. (1983). Evaluation of a start-up demonstration test. *Journal of Quality Technology*, 15(3), 103–106.
5. Gera, A.E. (2004). Combined k-out-of-n:G and consecutive k_c-out-of-n:G systems. *IEEE Transactions on Reliability*, 53(4), 523–531.
6. Balakrishnan, N., and Chan, P.S. (2000). Start-up demonstration tests with rejection of units upon observing *d* failures. *Annals of the Institute of Statistical Mathematics*, 52(1), 184–196.
7. Martin, D.E.K. (2004). Markovian start-up demonstration tests with rejection of units upon observing *d* failures. *European Journal of Operational Research*, 155(2), 474–486.
8. Martin, D.E.K. (2008). Application of auxiliary Markov chains to start-up demonstration tests. *European Journal of Operational Research*, 184(2), 574–582.
9. Smith, M.L., and Griffith, W.S. (2011). Multi-state start-up demonstration tests. *International Journal of Reliability Quality and Safety Engineering*, 18(2), 99–117.
10. Smith, M.L., and Griffith, W.S. (2005). Start-up demonstration tests based on consecutive successes and total failures. *Journal of Quality Technology*, 37(3), 186–198.
11. Smith, M.L., and Griffith, W.S. (2008). The analysis and comparison of start-up demonstration tests. *European Journal of Operation Research*, 186(3), 1029–1044.

12. Gera, A.E. (2010). A new start-up demonstration test. *IEEE Transactions on Reliability*, 59(1), 128–131.
13. Gera, A.E. (2011). A general model for start-up demonstration tests. *IEEE Transactions on Reliability*, 60(1), 295–304.
14. Gera, A.E. (2013). A start-up demonstration procedure involving dependent tests. *Statistics and Probability Letters*, 83(10), 2191–2196.
15. Gera, A.E. (2015). Start-up demonstration tests involving a two-dimensional TSCSTFCF procedure. *International Journal of Reliability, Quality and Safety Engineering*, 22(1).
16. Zhao, X., Cui, L.R., and Xie, W.J. (2010). On parallel start-up demonstration test. *Proceedings of the 4-th Asia-Pacific International Symposium on Advanced Reliability and Maintainability*, pp. 851–857.
17. Gera, A.E. (2014). Discussion paper: Start-up demonstration tests, method and application with some unification. *Applied Stochastic Models in Business and Industry*, 30, 417–419. doi:10.1002/asmb.2048.
18. Viveros, R., and Balakrishnan, N. (1993). Statistical inference from start-up demonstration test data. *Journal of Quality Technology*, 25, 119–130.
19. Gera, A.E. (2018). A comparison of start-up demonstration test procedures based on a combinatorial approach. *International Journal of Mathematical, Engineering and Management Sciences*, 3(3), 195–219.

9

Reliability Evaluation of a MANET Incorporating Copula

Nisha Nautiyal and S. B. Singh

CONTENTS

9.1 Introduction

The rapid enhancement of wireless devices like laptops and mobile phones has given rise to portable computing. Mobile ad-hoc network (MANET) has become one of the most important topic of recent research because of the increasing demand of mobile devices. It is a kind of ad-hoc network that can change its locations and be organized by itself anytime. There are different types of MANETs; some are constrained to a local area of wireless devices like cluster of laptops, whereas others may be linked to the Internet. A vehicular ad hoc network (VANET) is a kind of MANET that allows automobiles to be in touch with roadside tools. Automobiles may not have direct Internet connections, but wireless roadside tools allow data from automobiles to be sent over the Internet. The data from the automobile can be used to compute traffic situations and to keep track of them. MANETs are generally not as secure as other networks because of its vibrant nature, so it is essential to be cautious what type of information should be sent over a MANET.

In wireless networks, radio frequencies in the air are used to obtain the data other than using physical cables and broadcasting them. Routers and hosts create a wireless network. The router upholds the packets in the network and the host can be source or terminal of the flow in the data in a wireless network. The communication in the components of the network is the basic difference between wired and wireless networks (Meena and Vasanthi 2016). For the transmission of data, a wired network depends upon the physical cables. The contact between the components of diverse network can be wired or wireless in a wireless network. There is no restriction on physical cables in a wireless network, which allows the host/router to be moved; this proves to be the advantage of the wireless network.

Some of the previous works on ad-hoc network are: Dimitar et al. (2004) who demonstrated a link reliability model for two hop ad-hoc networks, whereas Egeland and Engelstad (2009) presented a method to predict the use of redundant nodes for enhancing the reliability and availability of the wireless multihop networks. The probabilistic model of wireless multihop networks showing its fading nature has been presented by Ekpenyong and Isabona (2009). The end-to-end reliability model has been investigated by the authors by using the proposed probabilistic model to a multihop network. They introduced an idea to adjust the power transmitted for improving the end-to-end reliability by assuring the quality of service. Chowdhury and Neogy (2011) considered that the mobile agent system includes the concurrently functioning independent groups of mobile agents and evaluated reliability using Monte Carlo simulation technique. Ahmad and Mishra (2012) evaluated the reliability of the large-scale mobile ad-hoc networks, by recognizing the critical link within the network and showed that the MANETs having more than one critical link would divide the network into many subnetworks, thus reducing the complexity in calculation and also the reliability can be recursively calculated. Chaturvedi and Padmavathy (2013) presented an approach based on the Monte Carlo simulation for computing the reliability of the network. The algorithm has been used in the MATLAB® software. Bakshi et al. (2013) studied the significance of a mobile ad-hoc network. They analyzed how wireless networks allow certain freedom for the host because in wireless communication there are no physical cables. Choudhary et al. (2014) studied the behavior of MANETs under the momentary change in the structure of MANETs and also performed a gap analysis and considered some strategies for enhancing the reliability and performance of the MANET. Kaur (2014) introduced different techniques for reducing energy consumption. The research field of the reliability of the wireless sensor network has been discussed. Padmavathy and Chaturvedi (2015) evaluated the reliability of the mobile ad-hoc network using the Monte Carlo simulation and also analyzed that when mobility was considered, no significant impact on reliability was obtained. And when the mobility was not considered, the complexities in computation were reduced. Meena and Vasanthi (2016) evaluated the reliability of MANET using universal generating function technique (UGFT). The authors analyzed that both the proposed node and link UGFT were different from the existing algorithms, which were based on UGFT. For finding the final reliability of MANET, they used UGF for denoting the subpaths and merging their SDPs through a formally introduced composition operator.

Singh (1994) developed a new algorithm for calculating the minimal cutsets as a result of the node failure only. Sing concluded that the failure in the link of a network is only the complimentary representation, whereas the network fails because of the node failure only. Yeh (2001) presented a method for elucidating all the minimal cuts (MC), which are attained by modifying the original network and compared with the existing best-known algorithms. Malinowski (2016) presented a new algorithm—applicable for undirected, directed, and partly directed networks—for computing all *s*–*t* minimal cuts of a graph-networked model having nodes and link failure.

Reliability analysis of the systems is traditionally done with the help of probability distributions. Usually a single distribution is used in failure/repair analysis, but if two different distributions are to be applied simultaneously in the repair/failure, then a copula is used. The copula is a function that joins or couples a multivariate distribution function to its one-dimensional marginal distribution functions. There are some important families of copulas having their own characteristics. Using Gumbel-Hougaard family of copula, the availability, mean time to failure (MTTF), and cost of complex systems has been determined by Ram and Singh (2010), under preemptive repeat repair discipline. Singh et al. (2011) studied a system having three components out of which one was controlled by the controller and the other two were independent and computed reliability, availability, and MTTF of the system. Nailwal and Singh (2012) investigated the reliability characteristics of a complex system in which subsystems are arranged in the form of a matrix and have a configuration *k*-out-of-*n*, using the supplementary variable technique and copula methodology. They also analyzed different types of power failures that lead to failure of the system. Kumar and Singh (2013) analyzed the reliability of a complex system having two repairable components, which are connected in series, out of which one is *k*-out-of-*n*:G system and the other is consecutive 2-out-of-3:F system using the supplementary variable, Laplace transformation, and the Gumbel-Hougaard copula. Munjal and Singh (2014) considered the complex repairable system consisting of 2-out-of-3:G subsystem connected in parallel for finding the reliability characteristics using Gumbel-Hougaard family of copula. Li (2016) established a method for evaluating reliability characteristics of dormant *k*-out-of-*n* redundant systems. Amrutkar and Kamalja (2017) gave a general idea of importance measures of components of reliability system and also explained them with examples. Bisht and Singh (2018) obtained the reliability characteristics of complex bridge networks using different algorithms by the help of universal generating function (UGF). Also the signature reliability has been obtained using Owen's method. Kumar and Ram (2019) computed the interval-valued reliability using UGF and evaluated the reliability of a sliding-window system with the help of an algorithm. Kumar and Singh (2019) computed the signature reliability of a sliding-window-coherent system having multiple failures. The system consists of G parallel elements in A-within-B-from-D/G multistates.

From this discussion, it is clear that much research has been done in the field of MANET and other systems/networks, not only evaluating the reliability of the network using supplementary variable technique but also evaluating the reliability. Additionally, the characteristics of a MANET with the help of minimal cuts using supplementary variable technique have never been taken into consideration thus far.

We combine the concepts of minimal cuts (MC) and supplementary variable technique to evaluate the reliability of a MANET in this chapter. In the considered network, there are three possible states, working, partially failed, and completely failed. The network is said to be in a partially failed state if any of the edge fails, even though the sink node receives the signal, whereas if all the edges fail and there is no transmission of signal from the source node to the sink node, then the network is said to be in a completely failed state. Figure 9.1 shows the representation of the MANET taken in this chapter.

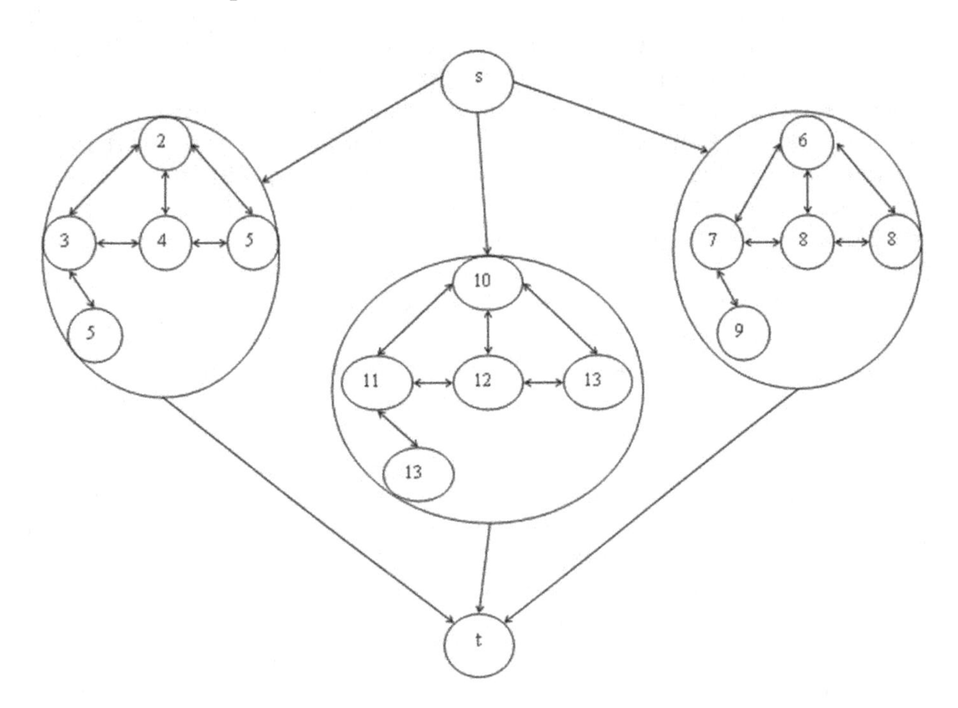

FIGURE 9.1
Diagram of proposed mobile ad-hoc network (MANET).

9.2 Materials and Methods

The chapter considers a MANET, which depicts the environment of the battlefield. The proposed MANET consists of main headquarters (source), three Indian forces (sub-MANETs), and the terrorist base camp (sink). Their aim is to destroy the base camp of the terrorists. Some strategies are made in the main branch/headquarters, which are then passed to the three armed forces through signals. These armed forces, themselves acts as a sub-MANET and passes information to its submembers. They work on these strategies and reach the base camp of the terrorists and to destroy them. The terrorist base camp acts as a sink and the headquarters as the source of the MANET.

In this chapter, the authors emphasized the repair of the failed nodes and evaluated the reliability measures of the MANET using the Gumbel-Hougaard copula, which has never before been considered. Different transition state probabilities of the network are evaluated with the help of supplementary variable technique and Laplace transformations. The failed nodes in the network are been repaired by copula methodology. The reliability of the network has been evaluated using the concept of MCs. First, the MCs are obtained for the proposed network, and then its transition state diagram is obtained by these MCs. By using the application of Gumbel-Hougaard copula, the reliability, MTTF, and sensitivity of the network have been evaluated. The nomenclature and the state specification of the model are given in the Tables 9.1 and 9.2, respectively, and the Figure 9.2 shows the different transition states of the network.

9.2.1 Assumptions

- Initially the MANET is in good state (i.e., each sub-MANET is fully operational). Transmission flow is good for each node.
- At time $t = 0$, all the team members are ready to work, and at $t > 0$, they start operating or working on their strategies.

TABLE 9.1

Nomenclature

S_i	Transition state of the network, where $i = 0$ to 27
$p_i(t)$	Probability of the network in S_i state at time, t, where $i = 0$ to 27
$\phi(x)$	Repair rate of each completely failed state

Let $u_1 = e^x$ and $u_2 = \phi(x)$ then the formula for the joint probability distribution as per Gumbel-Hougarrd family of copula where e^x is exponential repair rate and $\phi(x)$ is general repair rate, is given as

$$\phi(x) = \exp[(x^\theta + \log(\phi(x))^\theta]^{1/\theta}$$

TABLE 9.2

State Specification of the MANET

State	Network State	State	Network State	State	Network State
S_0	W	S_{11}	P	S_{22}	P
S_1	P	S_{12}	C	S_{23}	P
S_2	P	S_{13}	P	S_{24}	C
S_3	C	S_{14}	P	S_{25}	P
S_4	P	S_{15}	C	S_{26}	P
S_5	P	S_{16}	P	S_{27}	C
S_6	C	S_{17}	P	S_{28}	P
S_7	P	S_{18}	C	S_{29}	P
S_8	P	S_{19}	P	S_{30}	C
S_9	C	S_{20}	P		
S_{10}	P	S_{21}	C		

Notes: C, completely failed state; P, partially failed state; W, working/good state.

- The network consists of one main headquarters (source), three Indian forces (three sub-MANET), and terrorist base camp (sink), which is to be destroyed.
- When the soldiers do not receive signals from their team leaders, they are considered to be failed.
- All the nodes/team members are receiving same type of signals.
- Each sub-MANET contains other team members also, which one member heads.
- Repair of the nodes is done when the network fails completely.
- The repair of the failed nodes is perfect. After repair, each node is as good as new.
- The joint probability distribution of repairs when the failed nodes are under repair is computed by Gumbel-Hougaard copula.

The following MCs are obtained for the proposed network:

$$\{e_{s2}, e_{s10}, e_{s6}\}$$

$$\{e_{2t}, e_{10t}, e_{6t}\}$$

$$\{e_{23}, e_{24}, e_{25}\}$$

$$\{e_{67}, e_{68}, e_{69}\}$$

$$\{e_{1011}, e_{1012}, e_{1013}\}$$

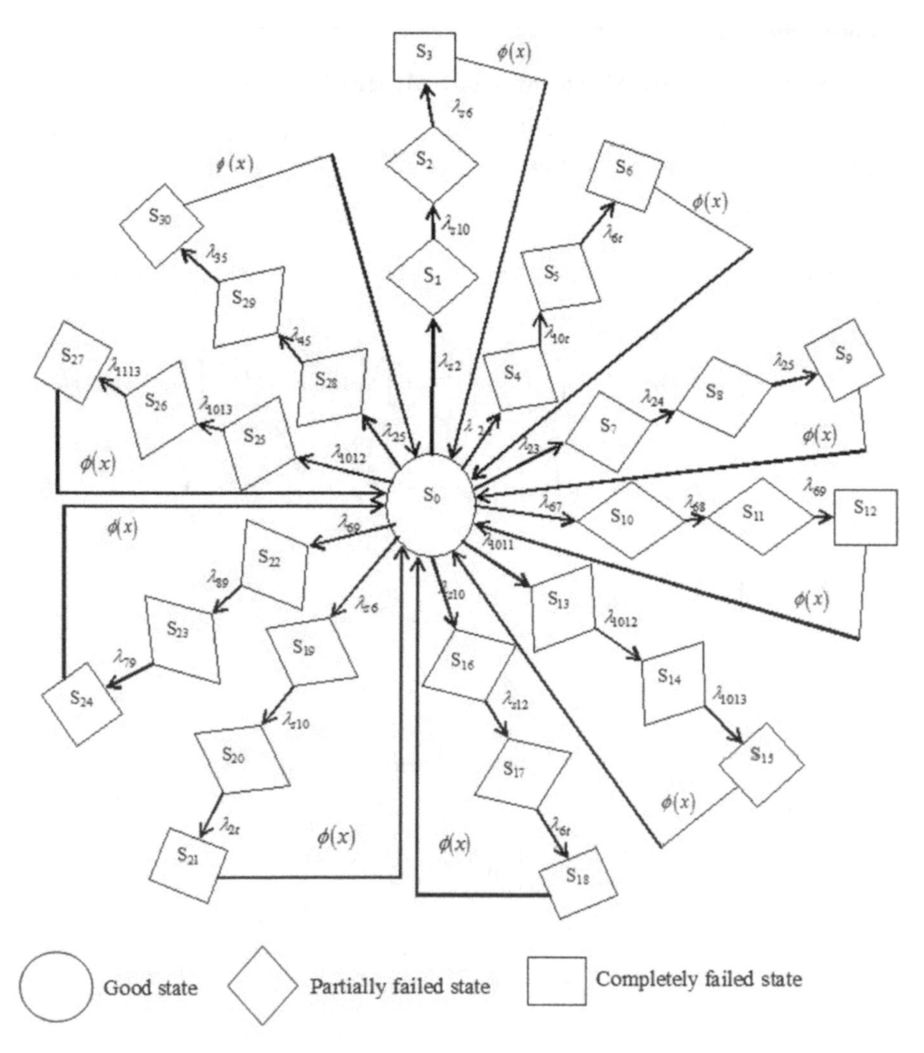

FIGURE 9.2
Transition diagram of the investigated network.

$$\{e_{s2}, e_{s10}\}, \{e_{6t}\}$$

$$\{e_{s10}, e_{s6}\}, \{e_{2t}\}$$

$$\{e_{25}, e_{45}\}, \{e_{35}\}$$

$$\{e_{69}, e_{89}\}, \{e_{79}\}$$

$$\{e_{1012}, e_{1013}\}, \{e_{1113}\}$$

9.3 Formulation of Mathematical Model

Using the supplementary variable technique, the following set of difference-differential equations associated with the model are obtained:

$$\left[\frac{d}{dt}+\lambda_{25}+\lambda_{69}+\lambda_{1012}+\lambda_{s6}+\lambda_{s10}+\lambda_{1011}+\lambda_{67}+\lambda_{23}+\lambda_{2t}+\lambda_{s2}\right]p_0(t)=\int_0^\infty p_3(x,t)\phi(x)dx$$
$$+\int_0^\infty p_6(x,t)\phi(x)dx+\int_0^\infty p_9(x,t)\phi(x)dx+\int_0^\infty p_{12}(x,t)\phi(x)dx+\int_0^\infty p_{15}(x,t)\phi(x)dx$$
$$+\int_0^\infty p_{18}(x,t)\phi(x)dx+\int_0^\infty p_{21}(x,t)\phi(x)dx+\int_0^\infty p_{24}(x,t)\phi(x)dx+\int_0^\infty p_{27}(x,t)\phi(x)dx$$
$$+\int_0^\infty p_{30}(x,t)\phi(x)dx \tag{9.1}$$

$$\left[\frac{d}{dt}+\lambda_{s10}\right]p_1(t)=\lambda_{s2}p_0(t) \tag{9.2}$$

$$\left[\frac{d}{dt}+\lambda_{s6}\right]p_2(t)=\lambda_{s10}p_1(t) \tag{9.3}$$

$$\left[\frac{\partial}{\partial t}+\frac{\partial}{\partial x}+\phi(x)\right]p_3(x,t)=0 \tag{9.4}$$

$$\left[\frac{d}{dt}+\lambda_{10t}\right]p_4(t)=\lambda_{2t}p_0(t) \tag{9.5}$$

$$\left[\frac{d}{dt}+\lambda_{6t}\right]p_5(t)=\lambda_{10t}p_4(t) \tag{9.6}$$

$$\left[\frac{\partial}{\partial t}+\frac{\partial}{\partial x}+\phi(x)\right]p_6(x,t)=0 \tag{9.7}$$

$$\left[\frac{d}{dt}+\lambda_{24}\right]p_7(t)=\lambda_{23}p_0(t) \tag{9.8}$$

$$\left[\frac{d}{dt}+\lambda_{25}\right]p_8(t)=\lambda_{24}p_7(t) \tag{9.9}$$

$$\left[\frac{\partial}{\partial t}+\frac{\partial}{\partial x}+\phi(x)\right]p_9(x,t)=0 \tag{9.10}$$

$$\left[\frac{d}{dt}+\lambda_{68}\right]p_{10}(t)=\lambda_{67}p_0(t) \tag{9.11}$$

$$\left[\frac{d}{dt}+\lambda_{69}\right]p_{11}(t)=\lambda_{68}p_{10}(t) \tag{9.12}$$

$$\left[\frac{\partial}{\partial t}+\frac{\partial}{\partial x}+\phi(x)\right]p_{12}(x,t)=0 \tag{9.13}$$

$$\left[\frac{d}{dt}+\lambda_{1012}\right]p_{13}(t)=\lambda_{1011}p_0(t) \tag{9.14}$$

$$\left[\frac{d}{dt}+\lambda_{1013}\right]p_{14}(t)=\lambda_{1012}p_{13}(t) \tag{9.15}$$

$$\left[\frac{\partial}{\partial t}+\frac{\partial}{\partial x}+\phi(x)\right]p_{15}(x,t)=0 \tag{9.16}$$

$$\left[\frac{d}{dt}+\lambda_{s12}\right]p_{16}(t)=\lambda_{s10}p_0(t) \tag{9.17}$$

$$\left[\frac{d}{dt}+\lambda_{6t}\right]p_{17}(t)=\lambda_{s12}p_{16}(t) \tag{9.18}$$

$$\left[\frac{\partial}{\partial t}+\frac{\partial}{\partial x}+\phi(x)\right]p_{18}(x,t)=0 \tag{9.19}$$

$$\left[\frac{d}{dt}+\lambda_{s10}\right]p_{19}(t)=\lambda_{s6}p_0(t) \tag{9.20}$$

$$\left[\frac{d}{dt}+\lambda_{2t}\right]p_{20}(t)=\lambda_{s10}p_{19}(t) \tag{9.21}$$

$$\left[\frac{\partial}{\partial t}+\frac{\partial}{\partial x}+\phi(x)\right]p_{21}(x,t)=0 \tag{9.22}$$

$$\left[\frac{d}{dt}+\lambda_{89}\right]p_{22}(t)=\lambda_{69}p_{0}(t) \tag{9.23}$$

$$\left[\frac{d}{dt}+\lambda_{79}\right]p_{23}(t)=\lambda_{89}p_{22}(t) \tag{9.24}$$

$$\left[\frac{\partial}{\partial t}+\frac{\partial}{\partial x}+\phi(x)\right]p_{24}(x,t)=0 \tag{9.25}$$

$$\left[\frac{d}{dt}+\lambda_{1013}\right]p_{25}(t)=\lambda_{1012}p_{0}(t) \tag{9.26}$$

$$\left[\frac{d}{dt}+\lambda_{1113}\right]p_{26}(t)=\lambda_{1013}p_{25}(t) \tag{9.27}$$

$$\left[\frac{\partial}{\partial t}+\frac{\partial}{\partial x}+\phi(x)\right]p_{27}(x,t)=0 \tag{9.28}$$

$$\left[\frac{d}{dt}+\lambda_{45}\right]p_{28}(t)=\lambda_{25}p_{0}(t) \tag{9.29}$$

$$\left[\frac{d}{dt}+\lambda_{35}\right]p_{29}(t)=\lambda_{45}p_{28}(t) \tag{9.30}$$

$$\left[\frac{\partial}{\partial t}+\frac{\partial}{\partial x}+\phi(x)\right]p_{30}(x,t)=0 \tag{9.31}$$

Boundary conditions:

$$p_{3}(0,t)=\lambda_{s6}p_{2}(t) \tag{9.32}$$

$$p_{6}(0,t)=\lambda_{6t}p_{5}(t) \tag{9.33}$$

$$p_{9}(0,t)=\lambda_{25}p_{8}(t) \tag{9.34}$$

$$p_{12}(0,t)=\lambda_{69}p_{11}(t) \tag{9.35}$$

$$p_{15}(0,t)=\lambda_{1013}p_{14}(t) \tag{9.36}$$

$$p_{18}(0,t) = \lambda_{6t} p_{17}(t) \tag{9.37}$$

$$p_{21}(0,t) = \lambda_{2t} p_{20}(t) \tag{9.38}$$

$$p_{24}(0,t) = \lambda_{79} p_{23}(t) \tag{9.39}$$

$$p_{27}(0,t) = \lambda_{1113} p_{26}(t) \tag{9.40}$$

$$p_{30}(0,t) = \lambda_{35} p_{29}(t) \tag{9.41}$$

Initial conditions:

$p_0(t) = 1$ and other transition state probabilities are 0 at $t = 0$.

9.3.1 Transition State Probabilities

On taking Laplace transformation and solving the Equations 9.1 through 9.31 using boundary conditions, the following transition state probabilities are obtained:

$$p_1(s) = \frac{\lambda_{s1}}{(s+\lambda_2)} p_0(s) \tag{9.42}$$

$$p_2(s) = \frac{\lambda_{s1}\lambda_{s2}}{(s+\lambda_{s2})(s+\lambda_{s3})} p_0(s) \tag{9.43}$$

$$p_3(s) = \frac{\lambda_{s1}\lambda_{s2}\lambda_{s3}}{(s+\lambda_{s2})(s+\lambda_{s3})} \left[\frac{1-\bar{S}(s)}{s}\right] p_0(s) \tag{9.44}$$

$$p_4(s) = \frac{\lambda_{s2}}{(s+\lambda_{s3})} p_0(s) \tag{9.45}$$

$$p_5(s) = \frac{\lambda_{s2}\lambda_{s3}}{(s+\lambda_{s3})(s+\lambda_{16})} p_0(s) \tag{9.46}$$

$$p_6(s) = \frac{\lambda_{s2}\lambda_{s3}\lambda_{16}}{(s+\lambda_{s3})(s+\lambda_{16})} \left[\frac{1-\bar{S}(s)}{s}\right] p_0(s) \tag{9.47}$$

$$p_7(s) = \frac{\lambda_{23}}{(s+\lambda_{25})} p_0(s) \tag{9.48}$$

$$p_8(s) = \frac{\lambda_{23}\lambda_{25}}{(s+\lambda_{s1})(s+\lambda_{25})} p_0(s) \tag{9.49}$$

$$p_9(s) = \frac{\lambda_{23}\lambda_{25}\lambda_{s1}}{(s+\lambda_{s1})(s+\lambda_{25})}\left[\frac{1-\bar{S}(s)}{s}\right] p_0(s) \tag{9.50}$$

$$p_{10}(s) = \frac{\lambda_{s3}}{(s+\lambda_{23})} p_0(s) \tag{9.51}$$

$$p_{11}(s) = \frac{\lambda_{s3}\lambda_{23}}{(s+\lambda_{23})(s+\lambda_{25})} p_0(s) \tag{9.52}$$

$$p_{12}(s) = \frac{\lambda_{s3}\lambda_{23}}{(s+\lambda_{23})(s+\lambda_{25})}\left[\frac{1-\bar{S}(s)}{s}\right] p_0(s) \tag{9.53}$$

$$p_{13}(s) = \frac{\lambda_{16}}{(s+\lambda_{56})} p_0(s) \tag{9.54}$$

$$p_{14}(s) = \frac{\lambda_{16}\lambda_{56}}{(s+\lambda_{56})(s+\lambda_{34})} p_0(s) \tag{9.55}$$

$$p_{15}(s) = \frac{\lambda_{16}\lambda_{56}\lambda_{34}}{(s+\lambda_{45})(s+\lambda_{56})(s+\lambda_{34})} p_0(s) \tag{9.56}$$

$$p_{16}(s) = \frac{\lambda_{16}\lambda_{56}\lambda_{34}\lambda_{45}}{(s+\lambda_{45})(s+\lambda_{56})(s+\lambda_{34})}\left[\frac{1-\bar{S}(s)}{s}\right] p_0(s) \tag{9.57}$$

$$p_{17}(s) = \frac{\lambda_{4t}}{(s+\lambda_{5t})} p_0(s) \tag{9.58}$$

$$p_{18}(s) = \frac{\lambda_{4t}\lambda_{5t}}{(s+\lambda_{6t})(s+\lambda_{5t})} p_0(s) \tag{9.59}$$

$$p_{19}(s) = \frac{\lambda_{4t}\lambda_{5t}\lambda_{6t}}{(s+\lambda_{6t})(s+\lambda_{5t})}\left[\frac{1-\bar{S}(s)}{s}\right] p_0(s) \tag{9.60}$$

$$p_{20}(s)=\frac{\lambda_{12}}{(s+\lambda_{16})}p_0(s) \tag{9.61}$$

$$p_{21}(s)=\frac{\lambda_{12}\lambda_{16}}{(s+\lambda_{25})(s+\lambda_{16})}p_0(s) \tag{9.62}$$

$$p_{22}(s)=\frac{\lambda_{12}\lambda_{16}\lambda_{25}}{(s+\lambda_{25})(s+\lambda_{16})(s+\lambda_{34})}p_0(s) \tag{9.63}$$

$$p_{23}(s)=\frac{\lambda_{12}\lambda_{16}\lambda_{25}\lambda_{34}}{(s+\lambda_{25})(s+\lambda_{16})(s+\lambda_{34})}\left[\frac{1-\bar{S}(s)}{s}\right]p_0(s) \tag{9.64}$$

$$p_{24}(s)=\frac{\lambda_{6t}}{(s+\lambda_{56})}p_0(s) \tag{9.65}$$

$$p_{25}(s)=\frac{\lambda_{6t}\lambda_{56}}{(s+\lambda_{56})(s+\lambda_{34})}p_0(s) \tag{9.66}$$

$$p_{26}(s)=\frac{\lambda_{6t}\lambda_{56}\lambda_{34}}{(s+\lambda_{25})(s+\lambda_{56})(s+\lambda_{34})}p_0(s) \tag{9.67}$$

$$p_{27}(s)=\frac{\lambda_{56}\lambda_{34}\lambda_{6t}\lambda_{25}}{(s+\lambda_{25})(s+\lambda_{56})(s+\lambda_{34})}\left[\frac{1-\bar{S}(s)}{s}\right]p_0(s) \tag{9.68}$$

where,

$$\bar{S}(s)=\frac{\phi(x)}{s+\phi(x)}$$

Substituting all these transition state probabilities of different states, the transition state probability at initial state is obtained as:

$$p_0(s)=\frac{1}{F(s)}$$

where

$$F(s) = (s + \lambda_{s2} + \lambda_{s1} + \lambda_{s3} + \lambda_{23} + \lambda_{12} + \lambda_{6t} + \lambda_{16} + \lambda_{4t})$$
$$-\bar{S}(s)\left[\frac{\lambda_{s1}\lambda_{s2}\lambda_{s3}}{(s+\lambda_{s2})(s+\lambda_{s3})} + \frac{\lambda_{s2}\lambda_{s3}\lambda_{16}}{(s+\lambda_{s3})(s+\lambda_{16})} + \frac{\lambda_{23}\lambda_{25}\lambda_{s1}}{(s+\lambda_{25})(s+\lambda_{s1})} + \frac{\lambda_{s3}\lambda_{23}\lambda_{25}}{(s+\lambda_{23})(s+\lambda_{25})}\right.$$
$$+\frac{\lambda_{16}\lambda_{56}\lambda_{34}\lambda_{45}}{(s+\lambda_{56})(s+\lambda_{34})(s+\lambda_{45})} + \frac{\lambda_{4t}\lambda_{5t}\lambda_{6t}}{(s+\lambda_{6t})(s+\lambda_{5t})} + \frac{\lambda_{12}\lambda_{16}\lambda_{25}\lambda_{34}}{(s+\lambda_{25})(s+\lambda_{34})(s+\lambda_{56})}$$
$$\left.+\frac{\lambda_{6t}\lambda_{56}\lambda_{34}\lambda_{25}}{(s+\lambda_{25})(s+\lambda_{34})(s+\lambda_{56})}\right]$$

Transition state probability of the network in the up state is as follows:

$$p_{up}(s) = p_o(s)+p_1(s)+p_2(s)+p_4(s)+p_5(s)+p_7(s)+p_8(s)$$
$$+p_{10}(s)+p_{11}(s)+p_{13}(s)+p_{14}(s)+p_{15}(s)+p_{17}(s)$$
$$+p_{18}(s)+p_{20}(s)+p_{21}(s)+p_{22}(s)+p_{24}(s)+p_{25}(s)+p_{26}(s)$$
$$=\frac{1}{F(s)}\left[1+\frac{\lambda_{s1}}{(s+\lambda_{s2})} + \frac{\lambda_{s1}\lambda_{s2}}{(s+\lambda_{s2})(s+\lambda_{s3})} + \frac{\lambda_{s2}}{(s+\lambda_{s3})} + \frac{\lambda_{s2}\lambda_{s3}}{(s+\lambda_{s3})(s+\lambda_{16})}\right.$$
$$+\frac{\lambda_{23}}{(s+\lambda_{25})} + \frac{\lambda_{23}\lambda_{25}}{(s+\lambda_{s1})(s+\lambda_{25})} + \frac{\lambda_{s3}\lambda_{23}}{(s+\lambda_{23})(s+\lambda_{25})} + \frac{\lambda_{s3}}{(s+\lambda_{23})} + \frac{\lambda_{16}}{(s+\lambda_{56})}$$
$$+\frac{\lambda_{16}\lambda_{56}}{(s+\lambda_{56})(s+\lambda_{34})} + \frac{\lambda_{16}\lambda_{56}\lambda_{34}}{(s+\lambda_{45})(s+\lambda_{56})(s+\lambda_{34})} + \frac{\lambda_{4t}}{(s+\lambda_{5t})} + \frac{\lambda_{4t}\lambda_{5t}}{(s+\lambda_{6t})(s+\lambda_{5t})}$$
$$\left.+\frac{\lambda_{12}}{(s+\lambda_{16})} + \frac{\lambda_{12}\lambda_{16}}{(s+\lambda_{25})(s+\lambda_{16})} + \frac{\lambda_{12}\lambda_{16}\lambda_{25}}{(s+\lambda_{25})(s+\lambda_{16})}\right] \tag{9.69}$$

Transition state probability of the network in the down state is obtained as:

$$p_{\text{down}}(s) = p_3(s)+p_6(s)+p_9(s)+p_{12}(s)+p_{16}(s)+p_{19}(s)+p_{23}(s)+p_{27}(s)$$
$$=\left[\frac{1-\bar{S}(s)}{s}\right]\left[\frac{1}{F(s)}\right]\left[\frac{\lambda_{s1}\lambda_{s2}\lambda_{s3}}{(s+\lambda_{s2})(s+\lambda_{s3})} + \frac{\lambda_{s2}\lambda_{s3}\lambda_{16}}{(s+\lambda_{s3})(s+\lambda_{16})} + \frac{\lambda_{23}\lambda_{25}\lambda_{s1}}{(s+\lambda_{s1})(s+\lambda_{25})}\right.$$
$$+\frac{\lambda_{12}\lambda_{16}\lambda_{25}\lambda_{34}}{(s+\lambda_{25})(s+\lambda_{16})(s+\lambda_{34})} + \frac{\lambda_{56}\lambda_{6t}\lambda_{25}}{(s+\lambda_{25})(s+\lambda_{56})(s+\lambda_{34})}$$
$$\left.+\frac{\lambda_{s3}\lambda_{23}}{(s+\lambda_{23})(s+\lambda_{25})} + \frac{\lambda_{16}\lambda_{56}\lambda_{34}\lambda_{45}}{(s+\lambda_{45})(s+\lambda_{56})(s+\lambda_{34})} + \frac{\lambda_{4t}\lambda_{5t}\lambda_{6t}}{(s+\lambda_{6t})(s+\lambda_{5t})}\right] \tag{9.70}$$

On adding Equations 9.69 and 9.70, we get,

$$p_{\text{up}}(s)+p_{\text{down}}(s)=\frac{1}{s}$$

9.3.2 Results

- **Reliability analysis**: Let the failure rates be $\lambda_{35}=0.4$, $\lambda_{45}=0.4$, $\lambda_{25}=0.1$, $\lambda_{1013}=0.3$, $\lambda_{1113}=0.4$, $\lambda_{1012}=0.1$, $\lambda_{89}=0.4$, $\lambda_{79}=0.4$, $\lambda_{69}=0.1$, $\lambda_{s10}=0.1$, $\lambda_{2t}=0.2$, $\lambda_{s6}=0.1$, $\lambda_{s12}=0.3$, $\lambda_{6t}=0.3$, $\lambda_{1011}=0.1$, $\lambda_{67}=0.2$, $\lambda_{68}=0.3$, $\lambda_{23}=0.2$, $\lambda_{24}=0.3$, $\lambda_{10t}=0.2$, $\lambda_{s2}=0.2$. Now substituting all these values in Equation 9.69 and then taking its inverse Laplace transformation, one can obtain

  $P_{\text{up}}(t)$ = (5exp $(-t/5))/12$ − (7exp $(-(2t)/5))/50$ + (295exp $(-t/10))/338$ + (317221exp $(-(7t)/5))/6134700$−(39exp $(-(3t)/10))/242$ + (t exp $(-(2t)/5))/25$ + (t exp $(-t/10))/65$ + ($3t$ exp $(-(3t)/10))/110$

  Table 9.3 and Figure 9.3 show the behavior of the reliability with respect to time.

- **MTTF analysis**: MTTF is given by:

$$\text{MTTF}=\lim_{s\to 0}P_{\text{up}}(s)$$

TABLE 9.3

Time w.r.t Reliability

Time	Reliability $R(t)$
0	1
1	0.991319
2	0.954713
3	0.896026
4	0.826288
5	0.753597
6	0.682786
7	0.616391
8	0.555516
9	0.500435
10	0.450975

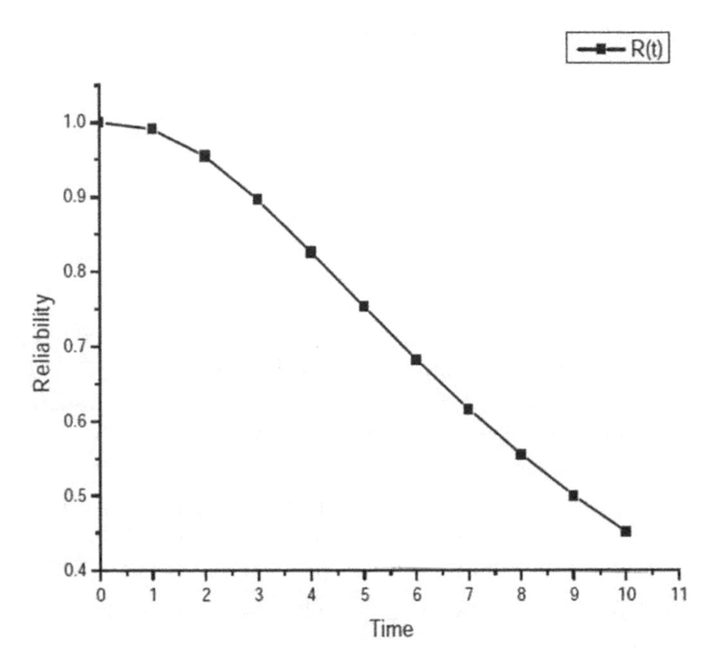

FIGURE 9.3
Time vs. Reliability.

Consider the failure rates to be $\lambda_{35} = 0.4$, $\lambda_{45} = 0.4$, $\lambda_{25} = 0.1$, $\lambda_{1013} = 0.3$, $\lambda_{1113} = 0.4$, $\lambda_{1012} = 0.1$, $\lambda_{89} = 0.4$, $\lambda_{79} = 0.4$, $\lambda_{69} = 0.1$, $\lambda_{s10} = 0.1$, $\lambda_{2t} = 0.2$, $\lambda_{s6} = 0.1$, $\lambda_{s12} = 0.3$, $\lambda_{6t} = 0.3$, $\lambda_{1011} = 0.1$, $\lambda_{67} = 0.2$, $\lambda_{68} = 0.3$, $\lambda_{23} = 0.2$, $\lambda_{24} = 0.3$, $\lambda_{10t} = 0.2$, $\lambda_{s2} = 0.2$.

- Now varying one parameter at a time (λ_{35}, λ_{45}, λ_{25}, λ_{1013}, λ_{1113}, λ_{1012}, λ_{89}, λ_{79}, λ_{69}, λ_{s10},λ_{2t}, λ_{s6}, λ_{s12}, λ_{6t}, λ_{1011}, λ_{67}, λ_{68}, λ_{23}, λ_{24}, λ_{10t} and λ_{s2}) from 0.1 to 1 and keeping the rest of the parameters constant, we can obtain the Tables 9.4 through 9.15. Figures 9.4 through 9.15 show the behavior of MTTF of the network with respect to different parameters.
- **Sensitivity analysis:** Sensitivity of the network with respect to parameter λ is given as

$$S = \frac{\partial R(t)}{\partial \lambda}$$

TABLE 9.4

Mean Time to Failure (MTTF) w.r.t λ_{25}, λ_{25}, and λ_{1012}

	MTTF w.r.t		
Failure Rate	λ_{25}	λ_{69}	λ_{1012}
0.1	12.20238	10.7881	11.4881
0.2	11.05556	9.735556	10.77778
0.3	10.46875	9.23125	10.36458
0.4	10.04902	8.884314	10.04902
0.5	9.712963	8.612963	9.787037
0.6	9.429825	8.387719	9.561404
0.7	9.184524	8.194524	9.363095
0.8	8.968254	8.025397	9.186508
0.9	8.775253	7.875253	9.027778
1	8.601449	7.74058	8.884058

TABLE 9.5

Mean Time to Failure (MTTF) w.r.t λ_{s6}, λ_{s6}, and λ_{1011}

	MTTF w.r.t		
Failure Rate	λ_{s6}	λ_{s6}	λ_{1011}
0.1	11.4881	12.20238	12.20238
0.2	11.38889	10.83333	12.27778
0.3	11.51042	10.26042	12.34375
0.4	11.66667	9.901961	12.40196
0.5	11.82407	9.638889	12.4537
0.6	11.97368	9.429825	12.5
0.7	12.1131	9.255952	12.54167
0.8	12.24206	9.107143	12.57937
0.9	12.36111	8.977273	12.61364
1	12.47101	8.862319	12.64493

On differentiating $R(t)$ with respect to different parameters, the sensitivity of the network corresponding to that parameter can be obtained. On differentiating $P_{up}(s)$ with respect to λ_{25} and putting the values of failure rates as $\lambda_{35} = 0.4$, $\lambda_{45} = 0.4$, $\lambda_{25} = 0.1$, $\lambda_{1013} = 0.3$, $\lambda_{1113} = 0.4$, $\lambda_{1012} = 0.1$, $\lambda_{89} = 0.4$, $\lambda_{79} = 0.4$, $\lambda_{69} = 0.1$, $\lambda_{s10} = 0.1$, $\lambda_{2t} = 0.2$, $\lambda_{s6} = 0.1$, $\lambda_{s12} = 0.3$, $\lambda_{6t} = 0.3$, $\lambda_{1011} = 0.1$, $\lambda_{67} = 0.2$, $\lambda_{68} = 0.3$, $\lambda_{23} = 0.2$, $\lambda_{24} = 0.3$, $\lambda_{10t} = 0.2$, $\lambda_{s2} = 0.2$, we have

TABLE 9.6

Mean Time to Failure (MTTF) w.r.t λ_{23}

Failure Rate	MTTF w.r.t λ_{23}
0.1	12.11538
0.2	12.20238
0.3	12.27778
0.4	12.34375
0.5	12.40196
0.6	12.4537
0.7	12.5
0.8	12.54167
0.9	12.57937
1	12.61364

TABLE 9.7

Mean Time to Failure (MTTF) w.r.t λ_{67}

Failure Rate	MTTF w.r.t λ_{67}
0.1	12.11538
0.2	12.20238
0.3	12.27778
0.4	12.34375
0.5	12.40196
0.6	12.4537
0.7	12.5
0.8	12.54167
0.9	12.57937
1	12.61364

TABLE 9.8

Mean Time to Failure (MTTF) w.r.t λ_{2t}, λ_{s2}, and λ_{10t}

	Sensitivity w.r.t		
Time	λ_{2t}	λ_{s2}	λ_{10t}
0.1	12.88462	11.60256	13.91026
0.2	12.20238	12.20238	12.20238
0.3	11.83333	12.72222	11.16667
0.4	11.5625	13.17708	10.36458
0.5	11.34314	13.57843	9.696078
0.6	11.15741	13.93519	9.12037
0.7	10.99624	14.25439	8.615288
0.8	10.85417	14.54167	8.166667
0.9	10.72751	14.80159	7.76455
1	10.61364	15.03788	7.401515

TABLE 9.9

Mean Time to Failure (MTTF) w.r.t λ_{6t} and λ_{24}

	MTTF w.r.t	
Failure Rate	λ_{6t}	λ_{24}
0.1	16.18056	15.34722
0.2	13.78205	13.39744
0.3	12.44048	12.20238
0.4	11.44444	11.27778
0.5	10.63542	10.51042
0.6	9.95098	9.852941
0.7	9.358466	9.279101
0.8	8.837719	8.77193
0.9	8.375	8.319444
1	7.960317	7.912698

TABLE 9.10

Mean Time to Failure w.r.t λ_{68}

Failure Rate	MTTF w.r.t λ_{68}
0.1	15.34722
0.2	13.39744
0.3	12.20238
0.4	11.27778
0.5	10.51042
0.6	9.852941
0.7	9.279101
0.8	8.77193
0.9	8.319444
1	7.912698

TABLE 9.11

Mean Time to Failure (MTTF) w.r.t λ_{1013}

Failure Rate	MTTF w.r.t λ_{1013}
0.1	15.34722
0.2	13.39744
0.3	12.20238
0.4	11.27778
0.5	10.51042
0.6	9.852941
0.7	9.279101
0.8	8.77193
0.9	8.319444
1	7.912698

TABLE 9.12

Mean Time to Failure (MTTF) w.r.t λ_{s12}, λ_{89}, and λ_{79}

Failure Rate	MTTF w.r.t. λ_{1113}
0.1	16.21212
0.2	14.44444
0.3	13.20513
0.4	12.20238
0.5	11.35556
0.6	10.625
0.7	9.985994
0.8	9.421296
0.9	8.918129
1	8.466667

TABLE 9.13

Mean Time to Failure (MTTF) w.r.t λ_{1113}

	MTTF w.r.t		
Failure Rate	λ_{s12}	λ_{89}	λ_{79}
0.1	14.79167	16.21212	16.81818
0.2	13.26923	14.44444	15
0.3	12.20238	13.20513	13.71795
0.4	11.33333	12.20238	12.67857
0.5	10.59375	11.35556	11.8
0.6	9.95098	10.625	11.04167
0.7	9.384921	9.985994	10.37815
0.8	8.881579	9.421296	9.791667
0.9	8.430556	8.918129	9.269006
1	8.02381	8.466667	8.8

TABLE 9.14

Mean Time to Failure (MTTF) w.r.t λ_{45}

Failure Rate	MTTF w.r.t. λ_{45}
0.1	16.21212
0.2	14.44444
0.3	13.20513
0.4	12.20238
0.5	11.35556
0.6	10.625
0.7	9.985994
0.8	9.421296
0.9	8.918129
1	8.466667

TABLE 9.15

Mean Time to Failure (MTTF) w.r.t λ_{35}

Failure Rate	MTTF w.r.t. λ_{35}
0.1	16.21212
0.2	14.44444
0.3	13.20513
0.4	12.20238
0.5	11.35556
0.6	10.625
0.7	9.985994
0.8	9.421296
0.9	8.918129
1	8.466667

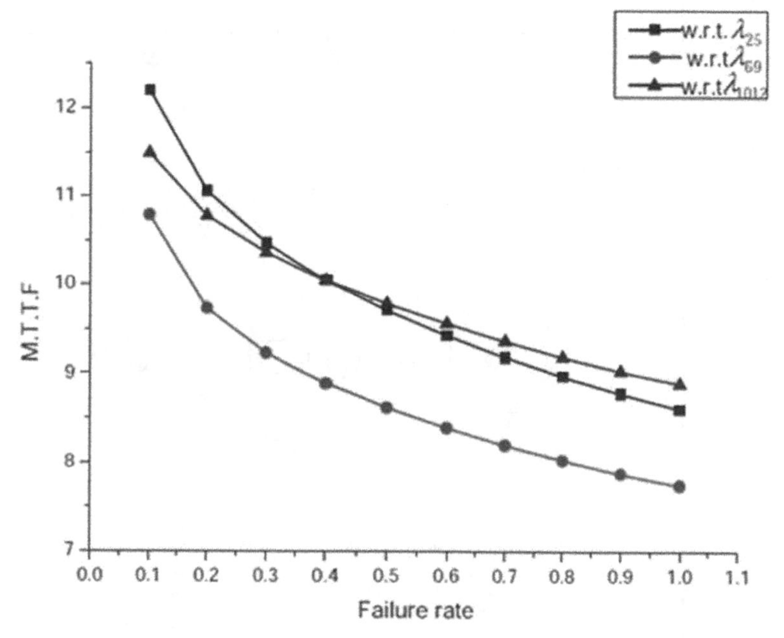

FIGURE 9.4
Mean time to failure (MTTF) vs. λ_{25}, λ_{69}, and λ_{1012}.

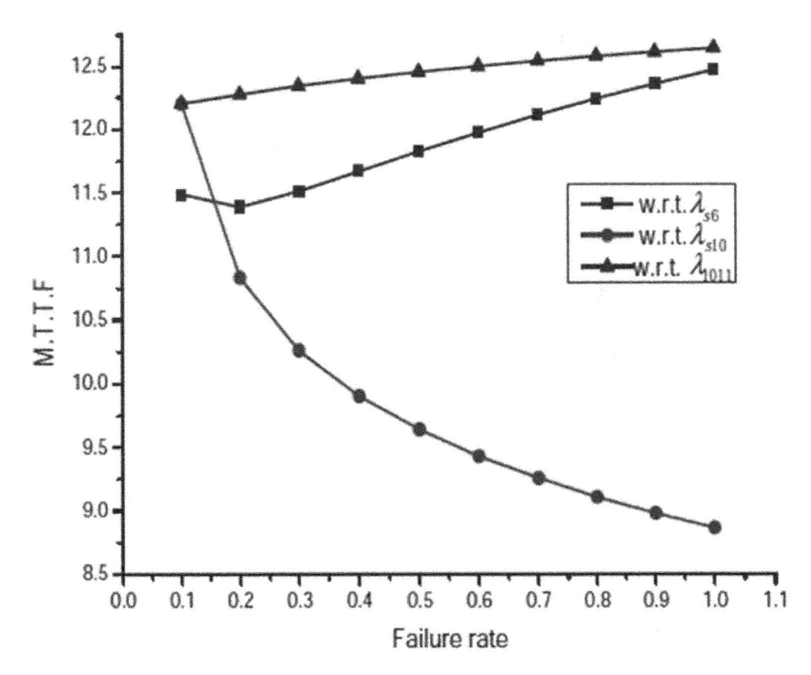

FIGURE 9.5
Mean time to failure (MTTF) vs. λ_{s6}, λ_{s6} and λ_{1011}.

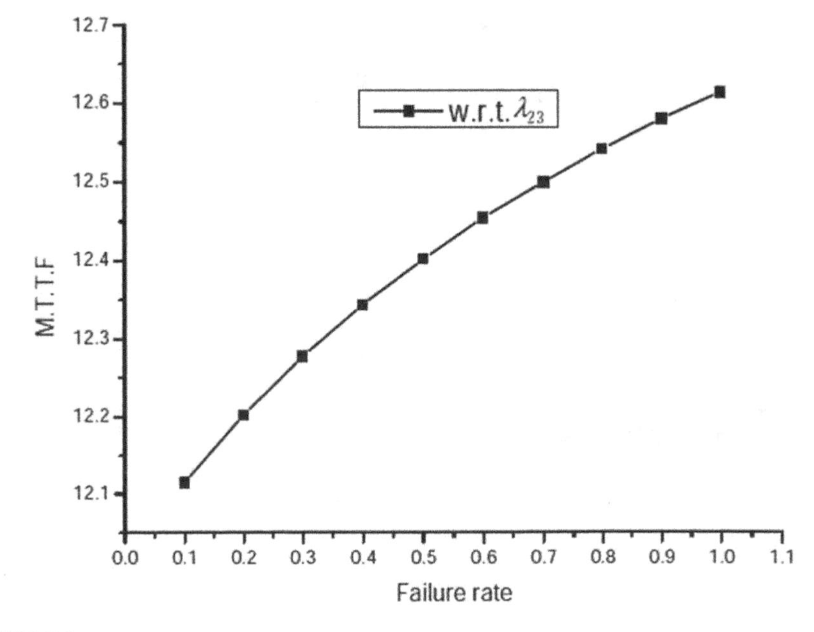

FIGURE 9.6
Mean time to failure (MTTF) vs. λ_{23}.

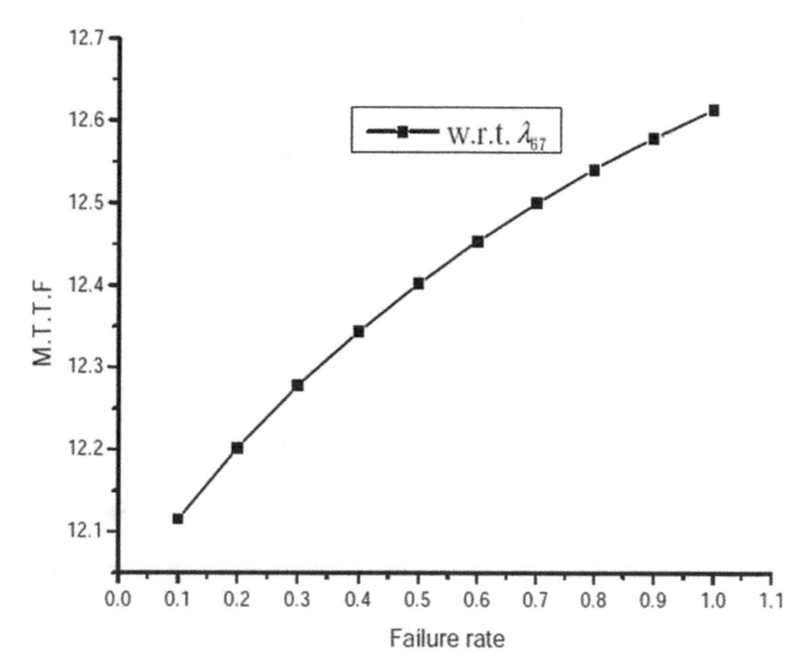

FIGURE 9.7
Mean time to failure (MTTF) vs. λ_{67}.

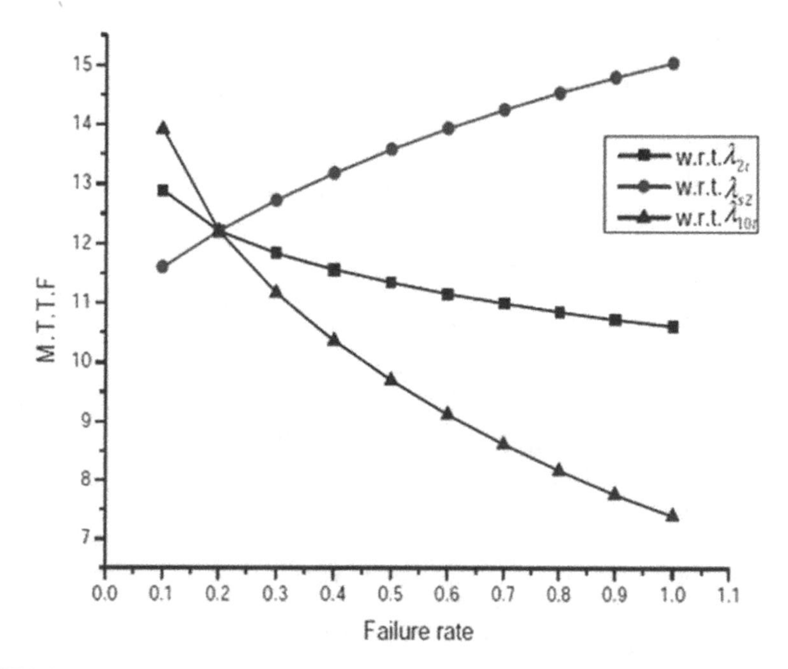

FIGURE 9.8
Mean time to failure (MTTF) vs. λ_{2t}, λ_{s2}, and λ_{10t}.

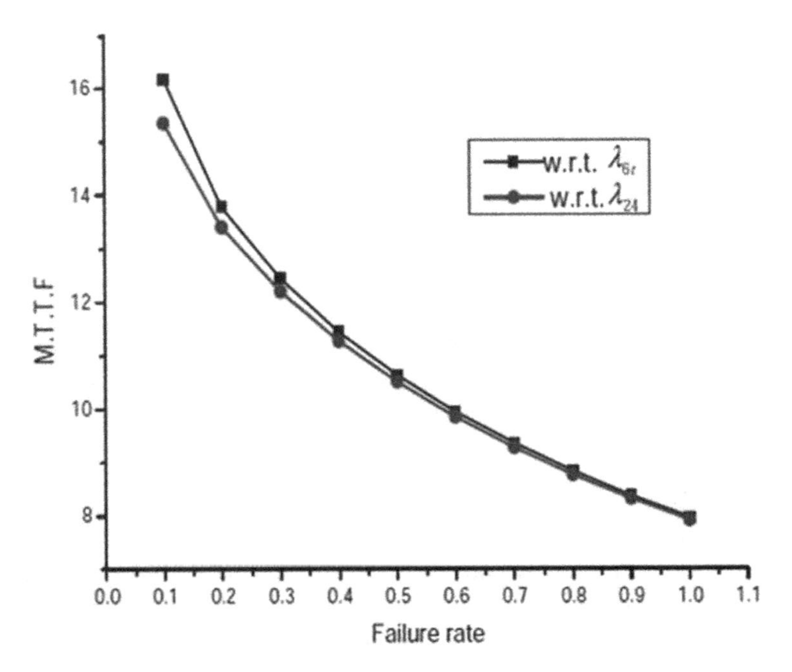

FIGURE 9.9
Mean time to failure (MTTF) vs. λ_{6t} and λ_{24}.

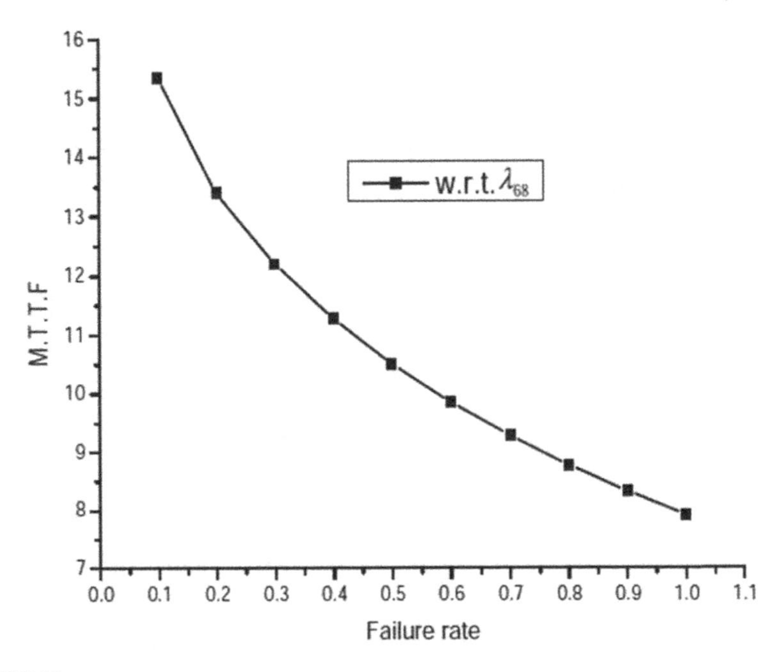

FIGURE 9.10
Mean time to failure (MTTF) vs. λ_{68}.

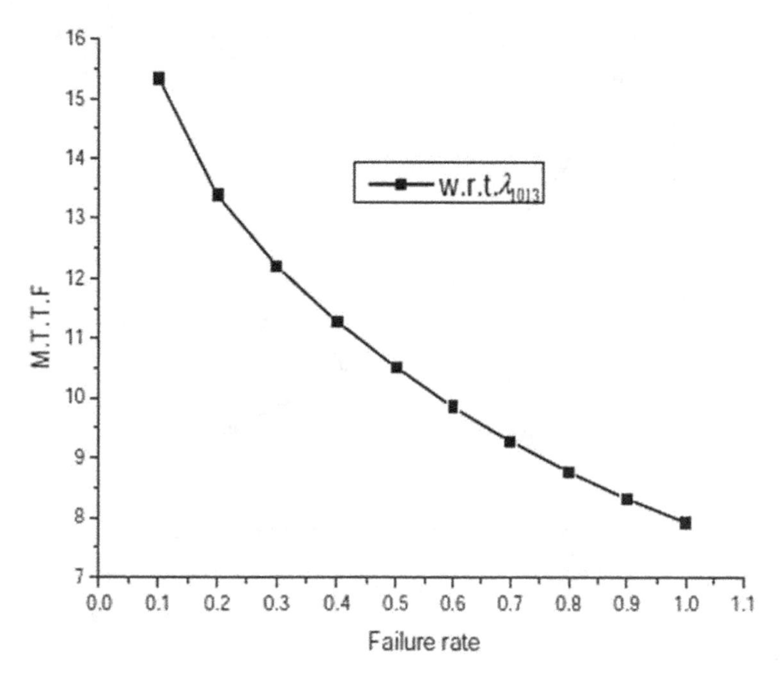

FIGURE 9.11
Mean time to failure (MTTF) vs. λ_{1013}.

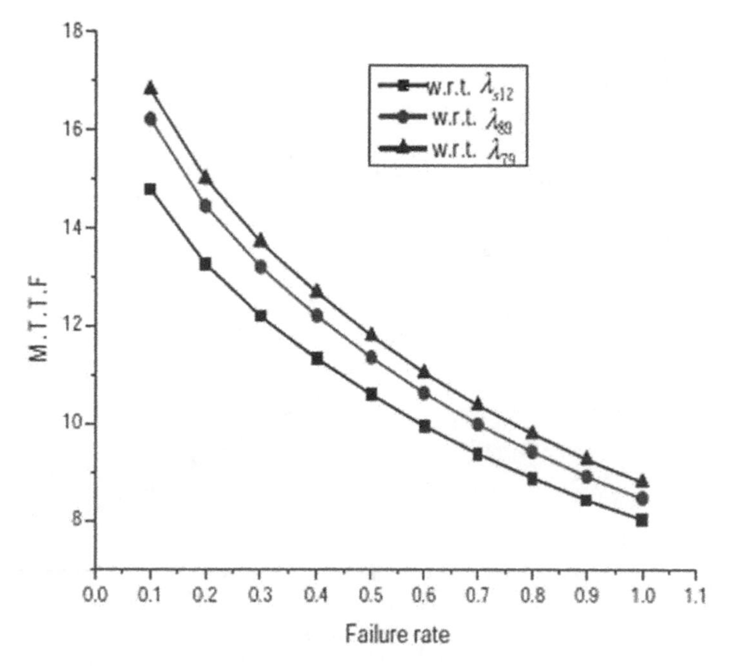

FIGURE 9.12
Mean time to failure (MTTF) vs. λ_{s12}, λ_{89}, and λ_{79}.

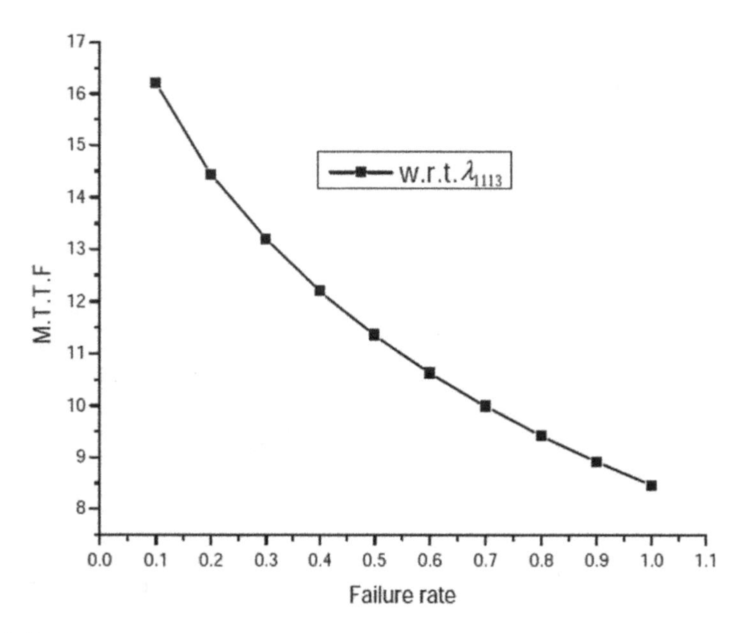

FIGURE 9.13
Mean time to failure (MTTF) vs. λ_{1113}.

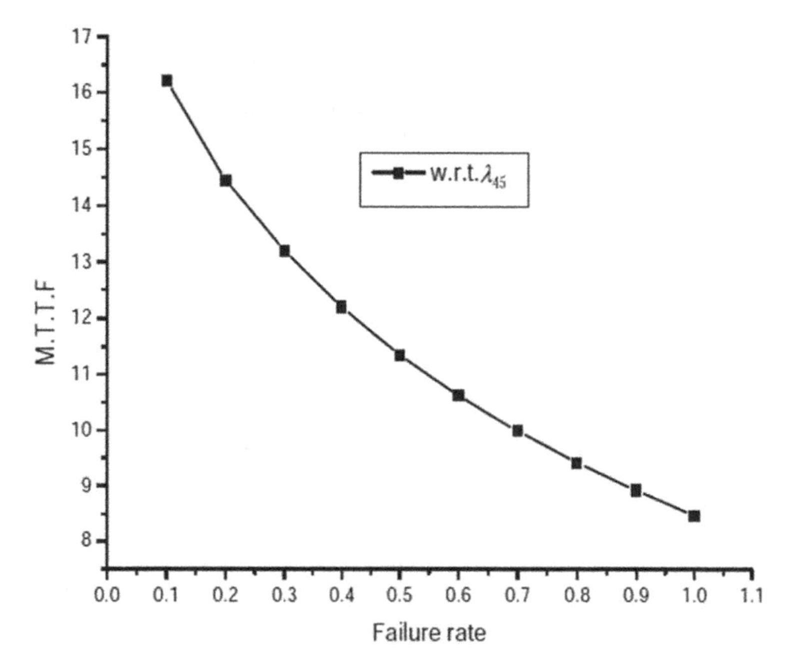

FIGURE 9.14
Mean time to failure (MTTF) vs. λ_{45}.

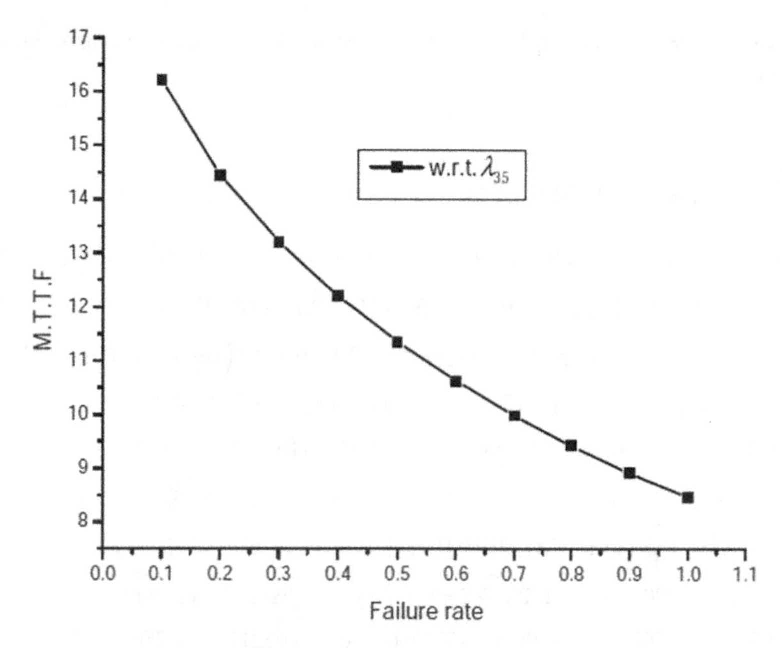

FIGURE 9.15
Mean time to failure (MTTF) vs. λ_{35}.

$$\frac{\partial P_{up}(s)}{\partial \lambda_{25}} = \left[\frac{1}{(s+1.3+\lambda_{25})}\right]\left[\frac{-\lambda_{23}\lambda_{24}}{(s+\lambda_{24})(s+\lambda_{25})^2} + \frac{1}{(s+\lambda_{45})} + \frac{\lambda_{45}}{(s+\lambda_{45})(s+\lambda_{35})}\right]$$
$$+\left[1 + \frac{\lambda_{s2}}{(s+\lambda_{s10})} + \frac{\lambda_{s2}\lambda_{s10}}{(s+\lambda_{s10})(s+\lambda_{s6})} + \frac{\lambda_{2t}}{(s+\lambda_{10t})} + \frac{\lambda_{2t}\lambda_{10t}}{(s+\lambda_{10t})(s+\lambda_{6t})}\right.$$
$$+ \frac{\lambda_{23}}{(s+\lambda_{24})} + \frac{\lambda_{23}\lambda_{24}}{(s+\lambda_{24})(s+\lambda_{25})} + \frac{\lambda_{23}\lambda_{24}}{(s+\lambda_{24})(s+\lambda_{25})} + \frac{\lambda_{67}}{(s+\lambda_{68})}$$
$$+ \frac{\lambda_{67}\lambda_{68}}{(s+\lambda_{68})(s+\lambda_{69})} + \frac{\lambda_{1011}}{(s+\lambda_{1012})} + \frac{\lambda_{1011}\lambda_{1012}}{(s+\lambda_{1012})(s+\lambda_{1013})} + \frac{\lambda_{s10}}{(s+\lambda_{s12})}$$
$$+ \frac{\lambda_{s10}\lambda_{s12}}{(s+\lambda_{s12})(s+\lambda_{6t})} + \frac{\lambda_{s6}}{(s+\lambda_{s10})} + \frac{\lambda_{s6}\lambda_{s10}}{(s+\lambda_{s10})(s+\lambda_{2t})} + \frac{\lambda_{69}}{(s+\lambda_{89})}$$
$$+ \frac{\lambda_{69}\lambda_{89}}{(s+\lambda_{89})(s+\lambda_{79})} + \frac{\lambda_{1012}}{(s+\lambda_{1013})} + \frac{\lambda_{1012}\lambda_{1013}}{(s+\lambda_{1013})(s+\lambda_{1113})}$$
$$\left. + \frac{\lambda_{25}}{(s+\lambda_{45})} + \frac{\lambda_{25}\lambda_{45}}{(s+\lambda_{45})(s+\lambda_{35})}\right]$$

Now, taking its inverse Laplace transformation, the following expression can be obtained:

$$
\begin{aligned}
\frac{\partial P_{up}(t)}{\partial \lambda_{25}} &= \left(\exp\left(-(2\times\lambda_{25})/5\right)\times(5000\times 0.1^2+7500\times 0.1+3500)\right)/\left(5\times(10\times 0.1+9)^3\right) \\
&-\left(3\times\lambda_{25}\times\exp\left(-(3\times\lambda_{25})/10\right)\right)/\left(100\times(0.1+1)^2\right)-\left(\exp(-\lambda_{25}/10)\times(2875\times 0.1+3350)\right) \\
&/\left(20\times(5\times 0.1+6)^3\right)-\left(50\times\exp(-\lambda_{25}/5)\right)/(10\times 0.1+11)^2-\left(\lambda_{25}\times\exp(-\lambda_{25}/10)\right) \\
&/\left(2\times(5\times 0.1+6)^2\right)+\left(\exp(-0.1\times\lambda_{25})\times(600\times 0.1-960)\right)/\left(169\times(10\times 0.1-3)^2\right) \\
&-\left(6\times\lambda_{25}\times\exp(-0.1\times\lambda_{25})\right)/(130\times 0.1-39)+\left(\exp\left(-(3\times\lambda_{25})/10\right)\right. \\
&\times\ (1500\times 0.1^3+1800\times 0.1^2+75\times 0.1+789))/\left(100\times(10\times 0.1-3)^2\times(0.1+1)^3\right) \\
&+\left(\lambda_{25}\times\exp\left(-(2\times\lambda_{25})/5\right)\times(1000\times 0.1+700)\right)/\left(25\times(10\times 0.1+9)^2\right) \\
&-\left(\exp\left(-\left(\lambda_{25}\times(10\times 0.1+13)\right)/10\right)\times\left(218750000000\times 0.1^{10}+1937750000000\times 0.1^9\right.\right. \\
&+7587850000000\times 0.1^8+17228762500000\times 0.1^7+24983705625000\times 0.1^6 \\
&+23987243275000\times 0.1^5+15256895292500\times 0.1^4+6221180388250\times 0.1^3 \\
&+1498480550325\times 0.1^2+176230973295\times 0.1+5485667706))/\left(16900\times(5\times 0.1+6)^3\right. \\
&\left.\times(10\times 0.1+9)^3\times(10\times 0.1+11)^2\times(0.1+1)^3\right)-\left(\lambda_{25}\times\exp\left(-\left(\lambda_{25}\times(10\times 0.1+13)\right)/10\right)\right. \\
&\times\left(2500000\times 0.1^7+14750000\times 0.1^6+35975000\times 0.1^5+46352500\times 0.1^4\right. \\
&+33179000\times 0.1^3+12421250\times 0.1+110.1^2+1840515\times 0.1-29232)) \\
&/\left(100\times(5\times 0.1+6)^2\times(10\times 0.1+9)^2\times(10\times 0.1+11)\times(0.1+1)^2\right)
\end{aligned}
$$

Similarly, sensitivities with respect to the other parameters can also be computed. Obtained sensitivities are $\frac{\partial P_{up}(t)}{\partial\lambda_{35}}$, $\frac{\partial P_{up}(t)}{\partial\lambda_{45}}$, $\frac{\partial P_{up}(t)}{\partial\lambda_{68}}$, $\frac{\partial P_{up}(t)}{\partial\lambda_{1013}}$, $\frac{\partial P_{up}(t)}{\partial\lambda_{1113}}$, $\frac{\partial P_{up}(t)}{\partial\lambda_{1012}}$, $\frac{\partial P_{up}(t)}{\partial\lambda_{89}}$, $\frac{\partial P_{up}(t)}{\partial\lambda_{79}}$, $\frac{\partial P_{up}(t)}{\partial\lambda_{69}}$, $\frac{\partial P_{up}(t)}{\partial\lambda_{s10}}$, $\frac{\partial P_{up}(t)}{\partial\lambda_{2t}}$, $\frac{\partial P_{up}(t)}{\partial\lambda_{s6}}$, $\frac{\partial P_{up}(t)}{\partial\lambda_{s12}}$, $\frac{\partial P_{up}(t)}{\partial\lambda_{6t}}$, $\frac{\partial P_{up}(t)}{\partial\lambda_{1011}}$, $\frac{\partial P_{up}(t)}{\partial\lambda_{67}}$, $\frac{\partial P_{up}(t)}{\partial\lambda_{23}}$, $\frac{\partial P_{up}(t)}{\partial\lambda_{24}}$, $\frac{\partial P_{up}(t)}{\partial\lambda_{10t}}$ and $\frac{\partial P_{up}(t)}{\partial\lambda_{s2}}$, which are shown in Tables 9.16 through 9.22. Figures 9.16 through 9.22 show the behavior of the sensitivity of the network with respect to different parameters.

TABLE 9.16

Sensitivity w.r.t λ_{25}

Time	Sensitivity w.r.t λ_{25}
0	0
1	−0.007573815
2	−0.028004161
3	−0.04293105
4	−0.042493227
5	−0.026009917
6	0.003092006
7	0.040331133
8	0.08154128
9	0.123364058
10	0.163335348

TABLE 9.17

Sensitivity w.r.t λ_{69}, λ_{1012}, and λ_{s6}

	Sensitivity w.r.t		
Time	λ_{69}	λ_{1012}	λ_{s6}
0	0	0	0
1	−0.02190929	−0.0516336	−0.26138
2	−0.10275464	−0.18407208	−0.22912
3	−0.21516064	−0.35346843	−0.21455
4	−0.33139684	−0.52916263	−0.18259
5	−0.43668557	−0.69417266	−0.13555
6	−0.52499904	−0.83997920	−0.09984
7	−0.59499145	−0.96297408	−0.03512
8	−0.64758517	−1.06239475	−0.04515
9	−0.68467305	−1.13910854	−0.0514
10	−0.70844549	−1.19487231	−0.05452

TABLE 9.18

Sensitivity w.r.t λ_{s10}, λ_{1011}, and λ_{67}

	Sensitivity w.r.t		
Time	λ_{s10}	λ_{1011}	λ_{67}
0	0	0	0
1	−0.00757381	−0.0014647	0.0547813
2	−0.02800416	−0.0021401	0.1339398
3	−0.04293105	0.0029072	0.1990924
4	−0.04249322	0.0125815	0.2463291
5	−0.02600991	0.0238776	0.2773404
6	0.00309200	0.0343971	0.2947505
7	0.04033113	0.0428665	0.3012762
8	0.08154128	0.0488472	0.2994483
9	0.12336405	0.0524083	0.2914801
10	0.16333534	0.0538654	0.2792189

TABLE 9.19

Sensitivity w.r.t λ_{23}, λ_{2t}, and λ_{s2}

	Sensitivity w.r.t		
Time	λ_{23}	λ_{2t}	λ_{s2}
0	0	0	0
1	−0.00146457	−0.0056404	−0.00235
2	−0.00214012	−0.0240437	−0.00923
3	0.00290922	−0.0473489	−0.00970
4	0.01258795	−0.0707879	−0.00518
5	0.02387726	−0.0927747	0.000350
6	0.03439711	−0.1127507	0.003776
7	0.04286621	−0.1303769	0.003350
8	0.04884739	−0.1453872	−0.00161
9	0.05240825	−0.1576089	−0.01115
10	0.05386545	−0.1669877	−0.02493

TABLE 9.20

Sensitivity w.r.t λ_{1113}, λ_{89}, and λ_{s12}

	Sensitivity w.r.t		
Time	λ_{1113}	λ_{89}	λ_{s12}
0	0	0	0
1	−0.00272	−0.3253	22.04275
2	−0.01231	−0.4469	16.22714
3	−0.02411	−0.45119	11.94676
4	−0.03394	−0.40365	8.79545
5	−0.04010	−0.3383	6.475219
6	−0.04257	−0.27214	4.766896
7	−0.04208	−0.21283	3.509136
8	−0.03952	−0.16305	2.583137
9	−0.03573	−0.12296	1.901416
10	−0.03138	−0.09158	1.399551

TABLE 9.21

Sensitivity w.r.t λ_{79}, λ_{6t}, and λ_{68}

	Sensitivity w.r.t		
Time	λ_{79}	λ_{6t}	λ_{68}
0	0	0	0
1	0.34502	−1.10042	−0.00202
2	0.45176	0.16346	−0.01023
3	0.45239	0.832562	−0.02265
4	0.40394	1.131026	−0.03622
5	0.33837	1.207928	−0.04886
6	0.27216	1.159943	−0.05944
7	0.21283	1.048257	−0.06752
8	0.16304	0.910453	−0.07310
9	0.12295	0.768676	−0.07639
10	0.09157	0.635146	−0.07771

TABLE 9.22
Sensitivity w.r.t λ_{24} and λ_{10t}

	Sensitivity w.r.t	
Time	λ_{24}	λ_{10t}
0	0	0
1	−0.01065	0.019407
2	−0.02024	0.023441
3	−0.03911	−0.00123
4	−0.06239	−0.04247
5	−0.08665	−0.08829
6	−0.10987	−0.13091
7	−0.13096	−0.16622
8	−0.14947	−0.19263
9	−0.16532	−0.21008
10	−0.17864	−0.21938

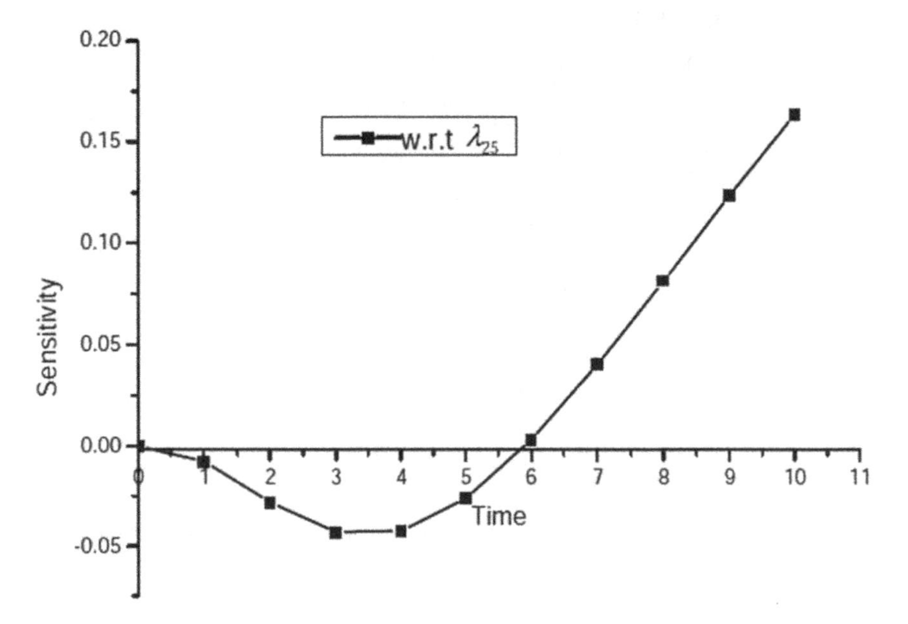

FIGURE 9.16
Sensitivity vs. λ_{25}.

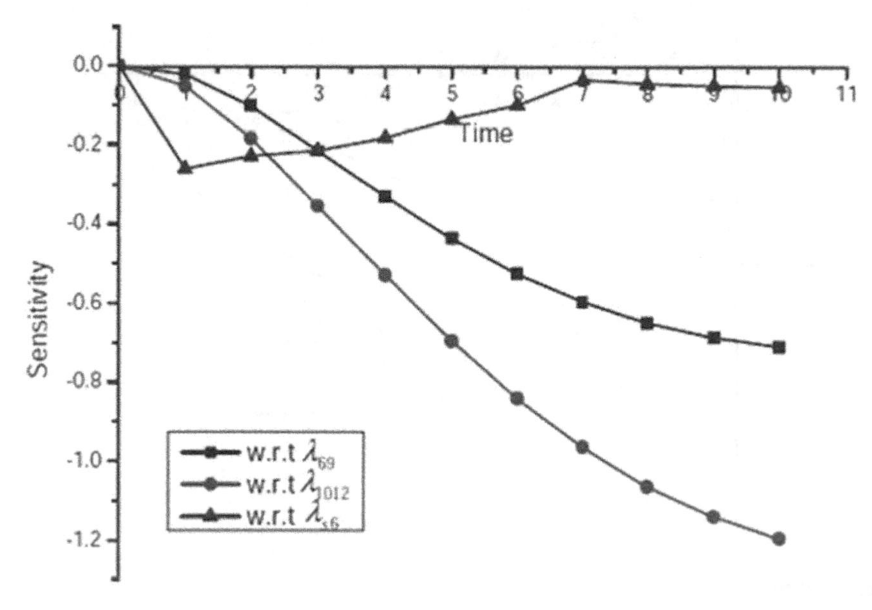

FIGURE 9.17
Sensitivity vs. λ_{69}, λ_{1012}, and λ_{s6}.

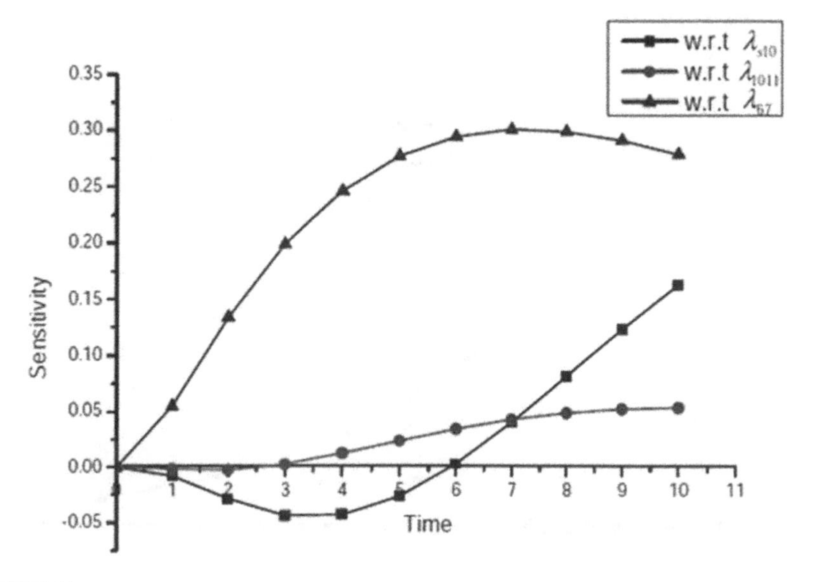

FIGURE 9.18
Sensitivity vs. λ_{s10}, λ_{1011}, and λ_{67}.

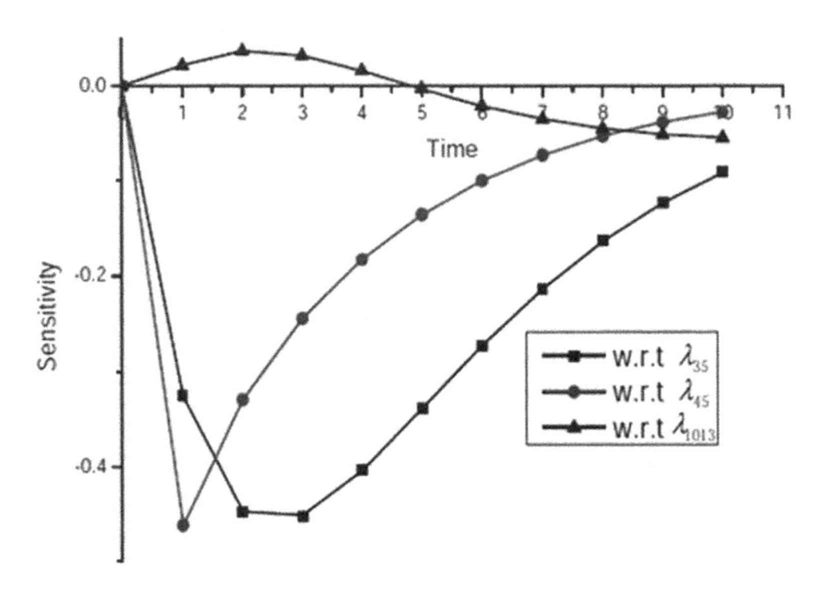

FIGURE 9.19
Sensitivity vs. λ_{35}, λ_{45}, and λ_{1013}.

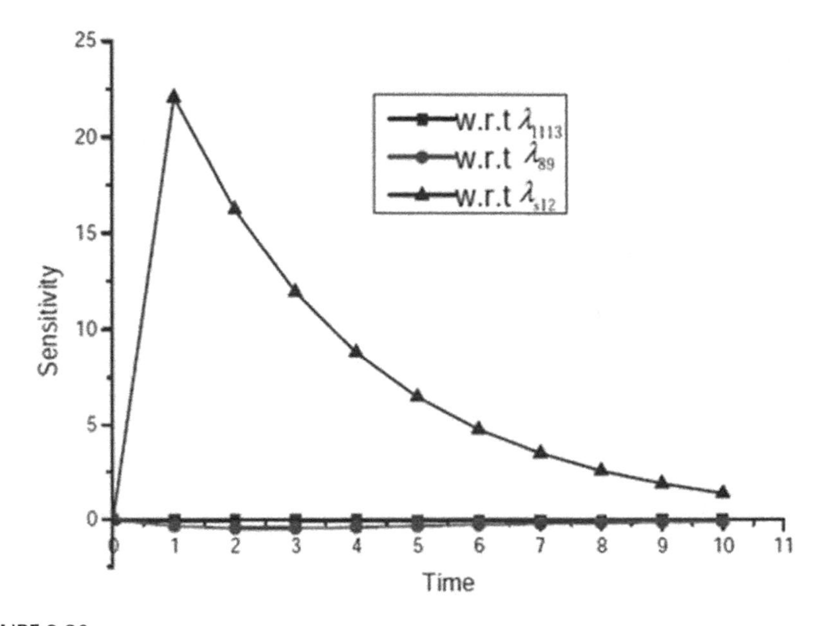

FIGURE 9.20
Sensitivity vs. λ_{1113}, λ_{89}, and λ_{s12}.

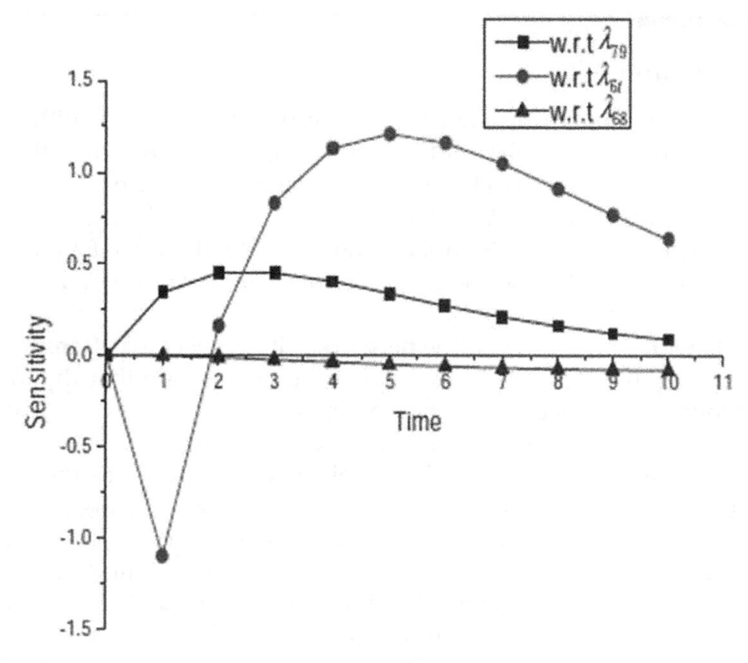

FIGURE 9.21
Sensitivity vs. λ_{79}, λ_{6t}, and λ_{68}.

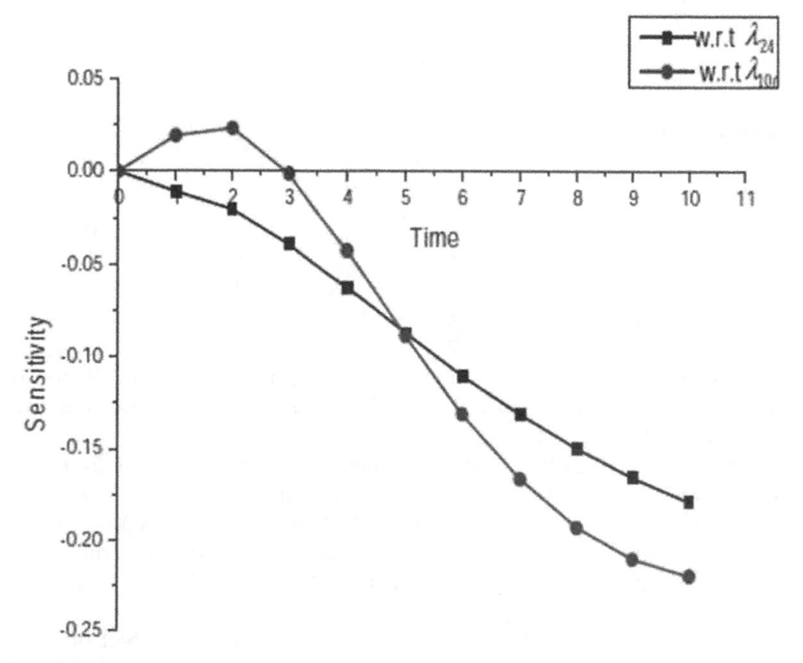

FIGURE 9.22
Sensitivity vs. λ_{24} and λ_{10t}.

9.4 Conclusion

Various reliability measures have been computed for MANET with the help of MCs in this chapter. Different reliability measures like transition state probabilities, reliability, MTTF, and sensitivity of the network with respect to different failure rates have been obtained using MCs, the supplementary variable technique, and Gumbel-Hougaard. Evident from Table 9.3 and Figure 9.3, the reliability of the proposed network decreases with increase of time.

The behavior of the MTTF of the network with respect to different parameters is shown in Figures 9.4 through 9.15; these conclude that the MTTF of the network with respect to the parameter λ_{79} is the highest (16.81818) and with respect to the parameter λ_{69} is the lowest (10.7881).

The sensitivities of the network reliability with respect to the different parameters λ_{35}, $\lambda_{45}\lambda_{25}$, λ_{1013}, λ_{1113}, λ_{1012}, λ_{89}, λ_{79}, λ_{69}, λ_{s10}, λ_{2t}, λ_{s6}, λ_{s12}, λ_{6t}, λ_{1011}, λ_{67}, λ_{68}, λ_{23}, λ_{24}, λ_{10t}, and λ_{s2} are shown in Figures 9.16 through 9.22. The sensitivity of the network with respect to the parameters λ_{1012}, λ_{s6}, λ_{2t}, and λ_{68} decreases as the time increases, with respect to the parameters λ_{69}, λ_{1011}, λ_{67} and λ_{23} it increases as the time increases, and with respect to the parameters λ_{25}, λ_{s10}, λ_{45}, λ_{1013}, and λ_{6t} first decreases and then increases as the time increases. From the aforementioned figures, the sensitivity with respect to the parameter λ_{s12} is the highest and with respect to the parameter λ_{6t} is the lowest.

References

Ahmad, M., and Mishra, D. K. 2012. A reliability calculations model for large-scale MANETs. *International Journal of Computer Applications*, 59(9): 17–21.

Amrutkar, K. P., and Kamalja, K. K. 2017. An overview of various importance measures of reliability system. *International Journal of Mathematical, Engineering and Management Sciences*, 2(3): 150–171.

Bakshi, A., Sharma, A. K., and Mishra, A. 2013. Significance of mobile ad-hoc networks (MANETS). *International Journal of Innovative Technology and Exploring Engineering* (IJITEE), 2(4): 1–5.

Bisht, S., and Singh, S. B. 2018. Signature reliability of binary state node in complex bridge network using universal generating function. *International Journal of Quality and Reliability Management* (accepted). doi:10.1108/IJQRM-08-2017-166.

Chaturvedi, S. K., and Padmavathy, N. 2013. The influence of scenario metrics on network reliability of mobile ad-hoc network. *International Journal of Performability Engineering*, 9(1).

Choudhary, A., Roy, O. P., and Tuithung, T. 2014. Node failure effect on reliability of mobile ad-hoc networks. *Fourth International Conference on Communication Systems and Network Technologies* (CSNT), *IEEE Xplore*, pp. 207–211.

Chowdhury, C., and Neogy, S. 2011. Reliability estimate of mobile agent system for QoS MANET applications. *Proceedings of Annual Reliability and Maintainability Symposium (RAMS), IEEE Xplore*, pp. 1–6.

Dimitar, T., Sonja, F., Bekim, C., and Aksenti, G. 2004. Link reliability analysis in ad hoc networks. *Proceedings of XII telekomunikacioni forum TELFOR*, pp. 23–25.

Egeland, G., and Engelstad, P. E. 2009. The availability and reliability of wireless multi-hop networks with stochastic link failures. *IEEE Journal on Selected Areas in Communications*, 27(7): 1132–1146.

Ekpenyong, M. E., and Isabona, J. 2009. Probabilistic link reliability model for wireless communication networks. *International Journal of Signal System Control and Engineering Application*, 2(1): 22–29.

Kaur, G. 2014. Review paper on reliability of wireless sensor networks. *International Journal of Advanced Research in Computer Engineering & Technology* (*IJARCET*), 3.

Kumar, A., and Ram, M. 2019. Computation interval-valued reliability of sliding window system. *International Journal of Mathematical, Engineering and Management Sciences*, 4(1): 108–115.

Kumar, A., and Singh, S. B. 2019. Signature reliability of A-within-B-from-D/G sliding window system. *International Journal of Mathematical, Engineering and Management Sciences*, 4(1): 95–107.

Kumar, D., and Singh, S. B. 2013. Reliability analysis of embedded system with different modes of failure emphasizing reboot delay. *International Journal of Applied Science and Engineering*, 11(4): 449–470.

Li, J. 2016. Reliability calculation for dormant *k*-out-of-*n* systems with periodic maintenance. *International Journal of Mathematical, Engineering and Management Sciences*, 1(2): 68–76.

Malinowski, J. 2016. A new efficient algorithm generating all minimal *s*-*t* cut-sets in a graph-modeled network. *Proceedings of AIP Conference*, 1738: 480030.

Meena, K. S., and Vasanthi, T. 2016. Reliability analysis of mobile ad hoc networks using universal generating function. *Quality and Reliability Engineering International*, 32(1): 111–122.

Munjal, A., and Singh, S. B. 2014. Reliability analysis of a complex repairable system composed of two 2-out-of-3: G subsystems connected in parallel. *Journal of Reliability and Statistical Studies*, 7: 89–111.

Nailwal, B., and Singh, S. B. 2012. Reliability measures and sensitivity analysis of a complex matrix system including power failure. *International Journal of Engineering-Transactions A: Basics*, 25(2): 115.

Padmavathy, N., and Chaturvedi, S. K. 2015. Reliability evaluation of mobile ad hoc network: with and without mobility considerations. *Proceedings of International Conference on Information and Technologies (ICICT)*, 46: 1126–1139.

Ram, M., and Singh, S. B. 2010. Availability, MTTF and cost analysis of complex system under preemptive-repeat repair discipline using Gumbel-Hougaard family copula. *International Journal of Quality & Reliability Management*, 27(5): 576–595.

Singh, B. 1994. Enumeration of node cutsets for an *s*-*t* network. *Microelectronics Reliability*, 34(3): 559–561.

Singh, S. B., Ram, M., and Chaube, S. 2011. Analysis of the reliability of a three-component system with two repairmen. *International Journal of Engineering Transactions, Part: A*, 24(4): 395–401.

Yeh, W. C. 2001. Search for MC in modified networks. *Computers & Operations Research*, 28(2): 177–184.

10

Hydromagnetic Flow of Copper-Water Nanofluid with Different Nanoparticle Shapes toward a Nonlinear Stretchable Plate

Santosh Chaudhary and KM Kanika

CONTENTS

10.1 Introduction

Magnetohydrodynamic (MHD) flow is an electrically conducting fluid flow with the impact of a magnetic field. MHD flow is sustained in the geothermal applications and industrial technology such as liquid metals fluids, MHD power generation, geothermal heat source pumps, solar flares, and MHD submarines. Magnetic field applications achieve a resistive force, the Lorentz force in the field of flow, which acts as a stabilizing handler to the flow and postpones the separation of the boundary layer. Shin and Kang [1] considered the magnetic-field influence on the bubble shape in a uniaxial straining flow. However, some researchers, like as Singh and Bajaj [2], Jat and Chaudhary [3], Zhang et al. [4], and Chaudhary and Kumar [5], analyzed magnetic-field flow problems under the different conditions. Recently, some excellent illustrations of the effects of MHD flow can be found in Imtiaz et al. [6], Chaudhary and Choudhary [7], and Bourantas et al. [8].

Some heat-transfer fluids such as ethylene glycol, oil, and water have lower thermal conductivity. These fluids are not able to remove the enormous heat in different industrial procedures. Also, a higher rate of heat transfer is required to the development of technology, whereas conventional fluids are not appropriate anymore. Some engineers and scientists are searching for a technique to increase the rate of heat transfer. To enhance heat-transfer properties, the nano solid particles are added into the conventional fluid; this kind of mixture is called nanofluid. Nanoparticles have a higher thermal conductivity, so nanofluids have better thermal characteristics. Several models are introduced to define the nanofluids conductivity, specifically Maxwell [9] and Wang et al. [10], which fully depend on the shape and size of the solid nanoparticles. Furthermore, several authors such as Wen and Ding [11], Nguyen et al. [12], Turkyilmazoglu [13], Khoshvaght-Aliabadi [14], Rokni et al. [15], Saha and Paul [16], and Bezaatpour and Goharkhah [17] have illustrated a significant number of practical and theoretical models of nanofluids to improve the thermal conductivity of fluid.

In manufacturing procedures, a stretching plate has a definite effect on the quality of finished products. So, various processes in engineering and industrial areas take place with several stretching velocities, particularly in a flow created in hot rolling, fiber spinning, glass fiber, polymer extrusion by a dye, rubber plates, annealing wires drawing, and huge metallic plate cooling. Abo-Eldahab and Ghonaim [18] studied the micropolar fluid flow toward a stretching sheet with convective heat transfer. Jat and Chaudhary [19] discuss an analysis of the MHD flow over a stretchable plate near a stagnation region. Later, a comprehensive literature survey of numerous and analytical explorations on the stretching surface problem is presented by Tamizharasi and Kumaran [20], Hsiao [21], Aziz and Mahomed [22], Chaudhary et al. [23], and Ahmad et al. [24].

In the present analysis, the main purpose is to extend the published work of Hamad and Ferdows [25] with the impact of the magnetic field. To enhance the thermal conductivity and the rate of heat transfer, four different shapes, sphere, hexahedron, colum,n and lamina of copper nanoparticles are induced into the base fluid water. The considered physical formulation is of importance in different advanced thermal material processing and nanotechnology fabrication.

10.2 Mathematical Modeling

Figure 10.1 represents a schematic of two-dimensional, steady, viscous copper-water nanofluid flow over a stretched surface along to the four different shapes of nanoparticles. A coordinate system (x, y) is assumed such that x-axis is taken along to the plate and y-axis is taken perpendicular to the plate; it is enough to consider the flow problem in the upper-half plane reason. The velocity and the temperature at the surface is $U_w = cx^n$ and $T_w = T_\infty + bx^{2n}$, respectively, where b and c are the positive constants,

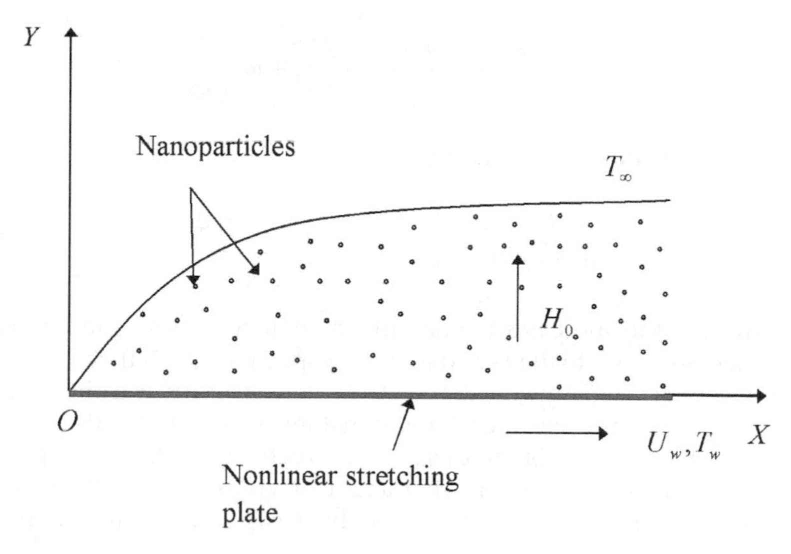

FIGURE 10.1
Flow geometry and coordinate system.

n is the nonlinear stretching parameter, and T_∞ is the temperature of ambient fluid. There is no thermal equilibrium and no slip condition between the conventional fluid and the solid nanoparticles. A uniform transverse magnetic field is applied normal to the sheet and has a constant intensity, H_0, which is thought unchanged by considering small magnetic Reynolds number. Furthermore, the values of thermophysical characteristics of solids and base fluid are depicted in Table 10.1 [26]. Thus, the corresponding basic equations used conventional considerations of the boundary layer are determined in the form given by

$$\frac{\partial u}{\partial x}+\frac{\partial v}{\partial y}=0 \tag{10.1}$$

$$\rho_{nf}\left(u\frac{\partial u}{\partial x}+v\frac{\partial u}{\partial y}\right)=\mu_{nf}\frac{\partial^2 u}{\partial y^2}-(\sigma_e)_{nf}\mu_e^{\,2}H_0^{\,2}u \tag{10.2}$$

TABLE 10.1
Thermophysical Properties of Used Materials

Materials	κ (Wm^{-1}K^{-1})	ρ (Kgm^{-3})	C_p (JKg^{-1}K^{-1})	σ_e (Sm^{-1})
Copper	400	8933	385	5.96×10^7
Water	0.613	997.1	4179	0.05

$$(\rho C_p)_{nf}\left(u\frac{\partial T}{\partial x}+v\frac{\partial T}{\partial y}\right)=\kappa_{nf}\frac{\partial^2 T}{\partial y^2}+\mu_{nf}\left(\frac{\partial u}{\partial y}\right)^2 \tag{10.3}$$

with the relevant boundary conditions as

$$\begin{aligned} &u=U_w, \quad v=0, \quad T=T_w \quad \text{at} \quad y=0 \\ &u\to 0, \quad T\to T_\infty \quad \text{as} \quad y\to\infty \end{aligned} \tag{10.4}$$

where subscript nf indicates the nanofluid characteristics, u and v are the velocity factors along to the x- and y-axes, respectively, ρ is the density, μ is the coefficient of viscosity, σ_e is the electrical conductivity, μ_e is the magnetic permeability, C_p is the specific heat at constant pressure, T is the temperature of nanofluid, and κ is the thermal conductivity. Further, the physical properties of nanofluid namely coefficient of viscosity, density, electrical conductivity, thermal conductivity, and heat capacitance are defined by Garmroodi et al. [27] as:

$$\frac{\mu_{nf}}{\mu_f}=\frac{1}{(1-\phi)^{5/2}} \tag{10.5}$$

$$\frac{\rho_{nf}}{\rho_f}=(1-\phi)+\phi\frac{\rho_s}{\rho_f} \tag{10.6}$$

$$\frac{(\sigma_e)_{nf}}{(\sigma_e)_f}=\frac{(\sigma_e)_s+2(\sigma_e)_f+2\phi[(\sigma_e)_s-(\sigma_e)_f]}{(\sigma_e)_s+2(\sigma_e)_f-\phi[(\sigma_e)_s-(\sigma_e)_f]} \tag{10.7}$$

$$\frac{\kappa_{nf}}{\kappa_f}=\frac{\kappa_s+(m-1)\kappa_f+(m-1)\phi(\kappa_s-\kappa_f)}{\kappa_s+(m-1)\kappa_f-\phi(\kappa_s-\kappa_f)} \tag{10.8}$$

$$\frac{(\rho C_p)_{nf}}{(\rho C_p)_f}=(1-\phi)+\phi\frac{(\rho C_p)_s}{(\rho C_p)_f} \tag{10.9}$$

where subscripts f and s denote the physical properties for ordinary fluid and nanoparticles, respectively; ϕ is the solid volume fraction, and m is the empirical shape factor. Moreover, the values of m are given in Table 10.2 [28] for the pertinent shapes of nanoparticles.

TABLE 10.2

Numerical Values of the Empirical Shape Factor m for Various Copper Nanoparticle Shapes

	Sphere	Hexahedron	Column	Lamina
m	3	3.7221	6.3698	16.1576

10.3 Similarity Analysis

To express the Equations 10.1 through 10.3 in ordinary differential equations, the following similarity variables continued as Hamad and Ferdows [25] are introduced

$$\psi = \sqrt{\frac{2c\nu_f x^{n+1}}{n+1}} f(\eta) \tag{10.10}$$

$$\eta = \sqrt{\frac{c(n+1)x^{n-1}}{2\nu_f}}\, y \tag{10.11}$$

$$\theta(\eta) = \frac{T - T_\infty}{T_w - T_\infty} \tag{10.12}$$

where $\psi(x, y)$ is the stream function, which is exhibited such that $u = \frac{\partial\psi}{\partial y}$ and $v = -\frac{\partial\psi}{\partial x}$ identically satisfies the continuity Equation 10.1, η is the similarity variable, and $\theta(\eta)$ is the non-dimensional temperature. After inserting the similarity transformations equations 10.10 through 10.12, the governing boundary layer Equations 10.2 and 10.3, corresponding to the boundary conditions Equation 10.4 are reduced as follows

$$f''' + E_1 E_2\left(ff'' - \frac{2n}{n+1} f'^2\right) - 2E_1 \frac{(\sigma_e)_{nf}}{(\sigma_e)_f} \frac{M}{(n+1)} f' = 0 \tag{10.13}$$

$$\frac{\kappa_{nf}}{\kappa_f}\theta'' + E_3 Pr\left(f\theta' - \frac{4n}{n+1} f'\theta\right) = 0 \tag{10.14}$$

Subject to the corresponding boundary conditions readily obtained as

$$\begin{aligned} &f = 0, f' = 1, \theta = 1 \quad \text{at} \quad \eta = 0 \\ &f' \to 0, \theta \to 0 \quad \text{as} \quad \eta \to \infty \end{aligned} \tag{10.15}$$

where prime ($'$) shows the differentiation with respect to η, $E_1 = (1-\phi)^{\frac{5}{2}}$, $E_2 = 1 - \phi + \phi\frac{\rho_s}{\rho_f}$, $E_3 = 1 - \phi + \phi\frac{(\rho C_p)_s}{(\rho C_p)_f}$, $M = \frac{(\sigma_e)_f \mu_e{}^2 H_0{}^2}{\rho_f}\frac{Re_x \nu_f}{U_w{}^2}$ is the magnetic parameter, $Re_x = \frac{U_w x}{\nu_f}$ is the local Reynolds number, and $Pr = \frac{\mu_f (C_p)_f}{\kappa_f}$ is the Prandtl number.

10.4 Declaration of Curiosity

As consequential physical quantities are the local skin friction coefficient, C_f, and the local Nusselt number, Nu_x, which can be defined as

$$C_f = \frac{\mu_{nf}\left(\frac{\partial u}{\partial y}\right)_{y=0}}{\frac{\rho_f U_w{}^2}{2}} \tag{10.16}$$

$$Nu_x = -\frac{\kappa_{nf} x\left(\frac{\partial T}{\partial y}\right)_{y=0}}{\kappa_f (T_w - T_\infty)} \tag{10.17}$$

Implementing the similarity variable Equations 10.10 through 10.12, the physical quantities of interest Equations 10.16 through 10.17 are given by

$$C_f = \frac{1}{E_1}\sqrt{\frac{2(n+1)}{Re_x}}f''(0) \tag{10.18}$$

$$Nu_x = -\frac{\kappa_{nf}}{\kappa_f}\sqrt{\frac{(n+1)Re_x}{2}}\theta'(0) \tag{10.19}$$

10.5 Numerical Method for Solution

Convert the boundary value problem in Equations 10.13 and 10.14 subjected by the boundary conditions Equation 10.15 into the initial value problem by applying the shooting technique. For the far field boundary condition, convenient finite value as $\eta \to \infty = 6$ is surmised. A new system of dependent variables w_1, w_2, w_3, p_1, and p_2 are introduced as

$$w_1' = w_2 \tag{10.20}$$

$$w_2' = w_3 \tag{10.21}$$

$$w_3' = -E_1 E_2\left(w_1 w_3 - \frac{2n}{n+1}w_2^2\right) + 2E_1\frac{(\sigma_e)_{nf}}{(\sigma_e)_f}\frac{M}{n+1}w_2 \tag{10.22}$$

and

$$p_1' = p_2 \tag{10.23}$$

$$p_2' = -E_3 \frac{\kappa_f}{\kappa_{nf}} Pr\left(w_1 p_2 - \frac{4n}{n+1} w_2 p_1 \right) \tag{10.24}$$

subject to the boundary conditions

$$\begin{aligned} &w_1 = 0, w_2 = 1, p_1 = 1 \quad \text{at} \quad \eta = 0 \\ &w_2 \to 0, p_1 \to 0 \quad \text{as} \quad \eta \to \infty \end{aligned} \tag{10.25}$$

where $w_1 = f$ and $p_1 = \theta$.

To solve the initial value problem Equations 10.22 and 10.24 along to the boundary conditions Equation 10.25, the values of $w_3(0)$ and $p_2(0)$ are required but no such values are disposed. So to find the solution, the initial guess values of $w_3(0)$ and $p_2(0)$ are selected and employed the fourth-order Runge-Kutta scheme with step size 0.001. Comparing the determined values of w_2 and p_1 for various values of pertinent parameters at the far field boundary condition $\eta \to \infty = 6$ with the mentioned boundary condition $w_2(6) \to 0$ and $p_1(6) \to 0$. To give a better approximation for the solution, the values of $w_3(0)$ and $p_2(0)$ are adjusted from the predicted values. This procedure is replicated again and again until the results are correct up to the appropriate accuracy of 10^{-7} level are achieved.

10.6 Validation of Results

Computational values of the surface shear stress, $f''(0)$, and the surface heat flux, $\theta'(0)$, in the case of various values of the solid volume fraction, ϕ, are compared with Hamad and Ferdows [25], as a test of accuracy of numerical solutions. As shown in Table 10.3, the comparison is found to be in good agreement.

TABLE 10.3

Comparision of the Present Results for $f''(0)$ and $\theta'(0)$ Along to Several Values of ϕ at $n = 10.0$, $M = 0.0$, $m = 3.0$, and $Pr = 10.0$ in Copper-Water Nanofluid

	$-f''(0)$		$-\theta'(0)$	
ϕ	**Hamad and Ferdows [25]**	**Present Results**	**Hamad and Ferdows [25]**	**Present Results**
0.05	1.40049	1.40037	5.62189	5.61195
0.10	1.47769	1.47632	5.17237	5.17102
0.15	1.51794	1.51525	4.77257	4.77071
0.20	1.52880	1.52467	4.41306	4.41067

Source: Hamad, M.A.A. and Ferdows, M., *Appl. Math. Mech.-Engl. Ed.*, 33, 923–930, 2012.

10.7 Discussion of Numerical Results

To draw the clean insight of the physical model, results are discussed via the help of graphical demonstration for the impacts of the controlling parameters like as the solid volume fraction, ϕ, the nonlinear stretching parameter, n, the magnetic parameter, M, and the empirical shape factor, m, on the velocity, $f'(\eta)$, and the temperature, $\theta(\eta)$ profiles. However, impacts of the specified parameters on the surface shear stress, $f''(0)$, and the rate of heat transfer, $\theta'(0)$, are mentioned in the tabular form (Table 10.4).

Figures 10.2 and 10.3 illustrate the velocity, $f'(\eta)$, and the temperature, $\theta(\eta)$, fields, respectively, along to the changeable value of the solid volume fraction, ϕ, while remaining physical parameters are taken as constant. These figures show that the momentum boundary layer thickness decreases and the thermal boundary layer thickness increase with the booming value of the solid volume fraction, ϕ. This happens because the solid nanosized particles have evolving resistance flow that falls down the fluid flow. Furthermore, nanofluid thermal conductivity enhances with the raising value of the solid volume fraction of particles, which implies the development in the temperature field.

The influences of the nonlinear stretching parameter, n, on the dimensionless velocity, $f'(\eta)$, and the temperature, $\theta(\eta)$, can be seen in Figures 10.4 and 10.5, respectively, with all other relevant parameters fixed. From these figures enlargement in the nonlinear stretching parameter, n, leads the

TABLE 10.4

Values of $f''(0)$ and $\theta'(0)$ for Influences of Different Values of Pertinent Parameters in Copper-Water When $Pr = 6.8$

ϕ	n	M	m	$-f''(0)$	$-\theta'(0)$
0.01	0.5	0.01	3.0000	0.92035	3.28019
0.04				0.97770	3.10184
0.07				1.02057	2.93808
0.10				1.05158	2.78688
0.07	1.0			1.14433	3.42198
	5.0			1.36273	4.23537
	10.0			1.40815	4.40005
	0.5	1.00		1.54285	2.75682
		2.00		1.93416	2.61571
		3.00		2.25898	2.49642
		0.01	3.7221	1.02057	2.87006
			6.3698		2.65428
			16.1576		2.13239

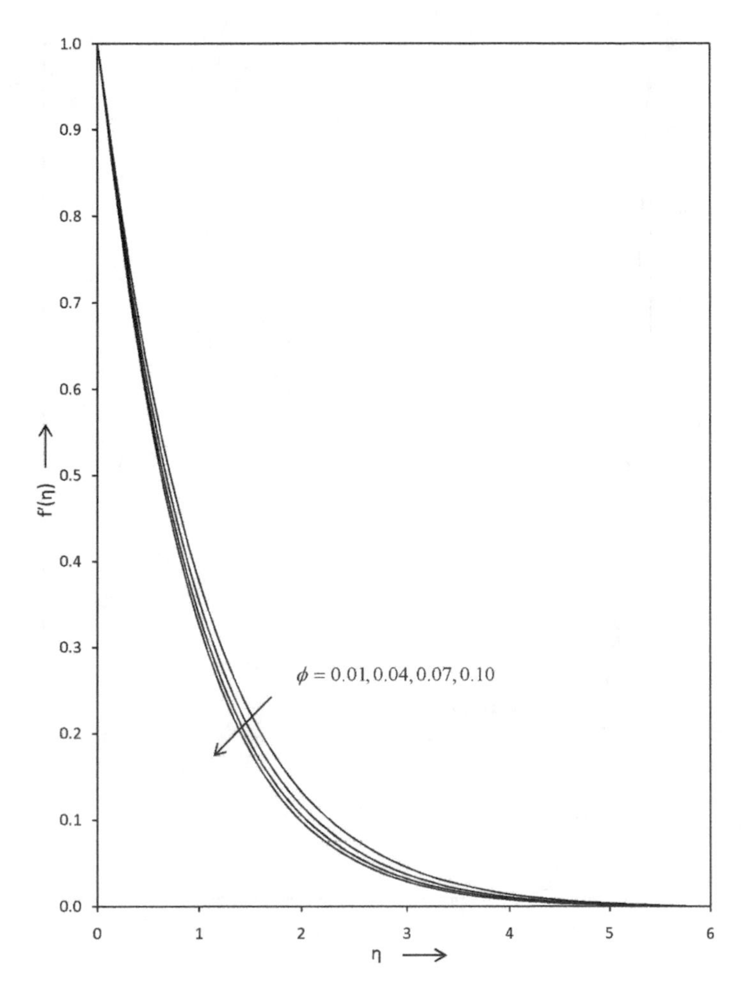

FIGURE 10.2
Effect of ϕ on velocity distribution when $n = 0.5$ and $M = 0.01$.

falls-down in the fluid flow and the fluid temperature. Physically it occurs because the stretching velocity controls the free-stream velocity along to the rising value of the stretching parameter, which implies the velocity of fluid and the temperature of fluid diminish.

Figures 10.6 and 10.7 depict the influence of the magnetic parameter, M, on the velocity, $f'(\eta)$, and the temperature, $\theta(\eta)$, distribution,s respectively, whereas other parameters are considered a constant value. It can be observed from these figures that the velocity profile decreases as the magnetic parameter, M, rises until the reverse happens in the temperature profile. From the physical reason, the higher value of the magnetic parameter yields a Lorentz force, which controls the fluid motion and generates more heat resulting with the increment in the fluid temperature.

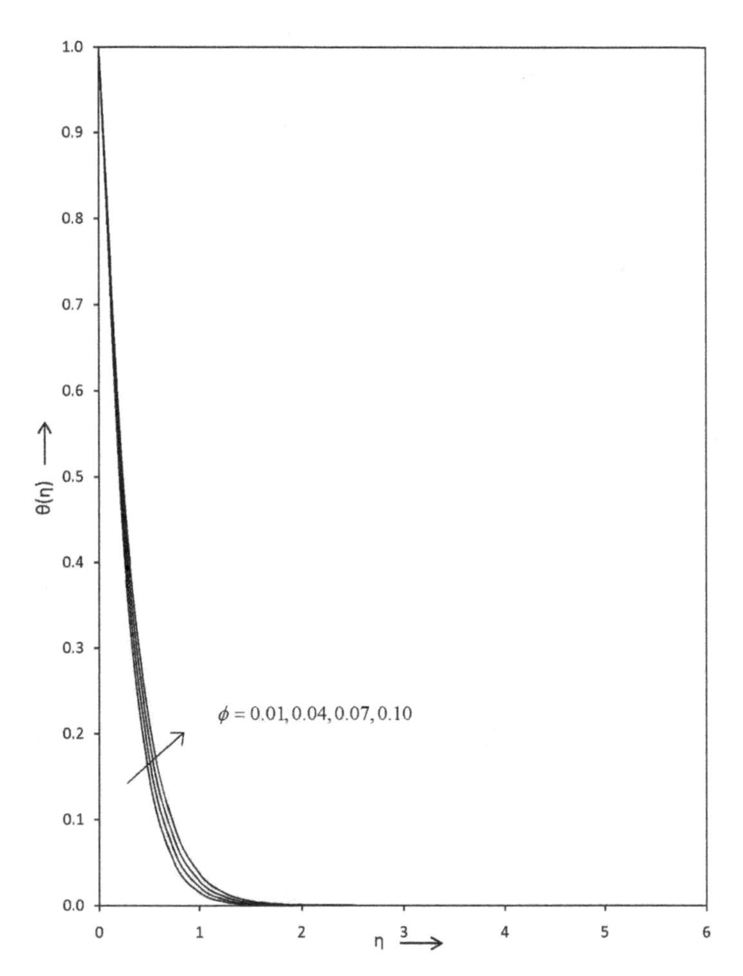

FIGURE 10.3
Effect of ϕ on temperature distribution when $n = 0.5$, $M = 0.01$, $m = 3.0$ and $Pr = 6.8$.

The impact of the empirical shape factor, m, for the shapes of nanoparticles—sphere, hexahedron, column, and lamina—on the dimensionless temperature $\theta(\eta)$ is displayed in Figure 10.8, which has other controlling parameters fixed. From this figure, it is observed that the fluid temperature increases with the shape of copper nanoparticles in an increasing manner such that sphere, hexahedron, column, and lamina.

Natures of the surface shear stress, $f''(0)$, and the surface heat flux, $\theta'(0)$, for the effects of the considering parameters such as the solid volume fraction, ϕ; the nonlinear stretching parameter, n; the magnetic parameter, M; and the empirical shape factor, m, are indicated in Table 10.3, whereas value of the remaining parameters are fixed. It is interesting to be note by Equations 10.18 and 10.19 that the surface shear stress, $f''(0)$, and the rate

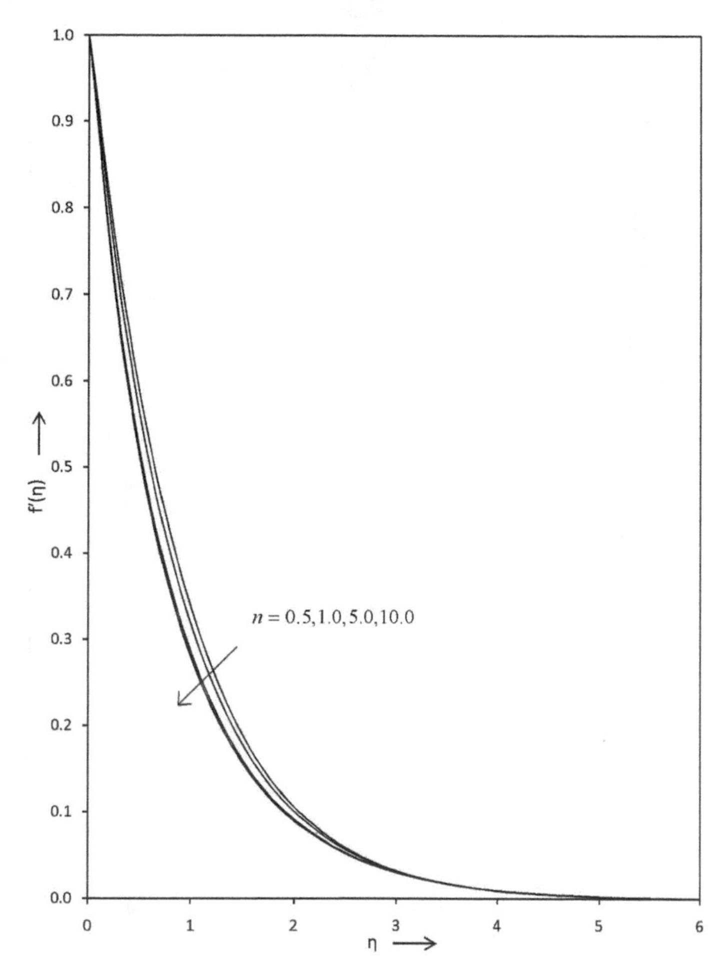

FIGURE 10.4
Effect of n on velocity distribution when $\phi = 0.07$ and $M = 0.01$.

of heat transfer, $\theta'(0)$, are proportional to the local skin friction coefficient, C_f, and the local Nusselt number, Nu_x, respectively. This table indicates that an enlargement in the solid volume fraction, ϕ; the nonlinear stretching parameter, n; and the magnetic parameter, M, tends to the decreasing nature of the local skin friction coefficient and the local Nusselt number, whereas reverse impact is found in the heat transfer rate for the effects of the solid volume fraction, ϕ, and the magnetic parameter, M. Subsequently, as an enhancement in the empirical shape factor, m, the surface heat flux rises. Moreover, all negative values of the surface shear stress and the surface heat flux denote that fluid utilizes a drag like force from the sheet, and there is a heat flow on the sheet, respectively.

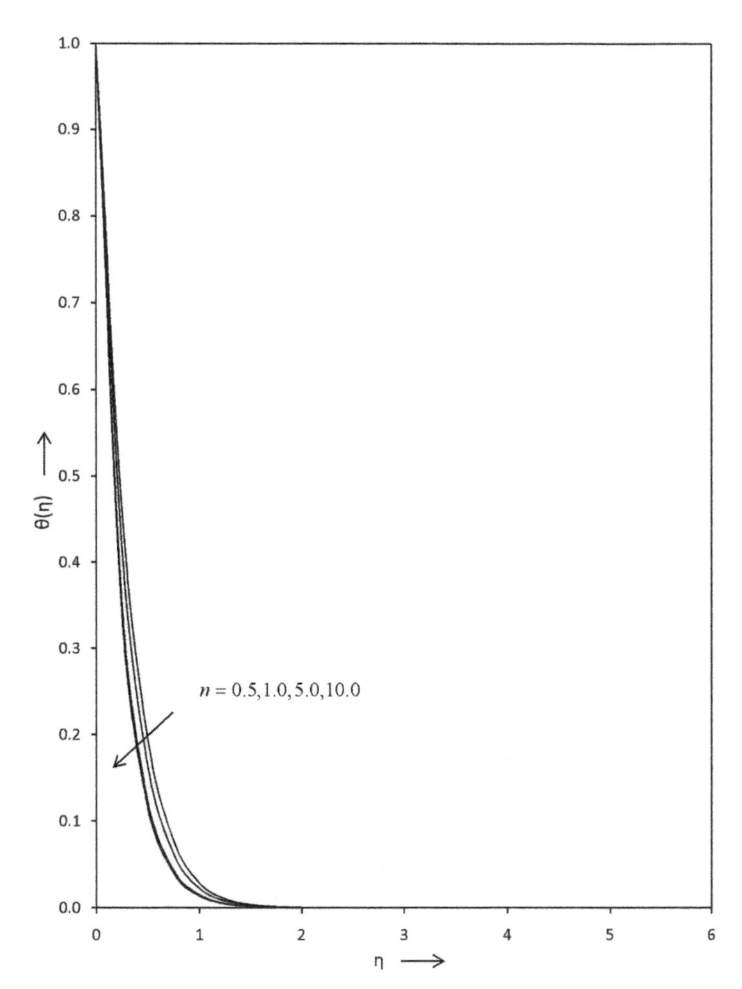

FIGURE 10.5
Effect of n on temperature distribution when $\phi = 0.07$, $M = 0.01$, $m = 3.0$ and $Pr = 6.8$.

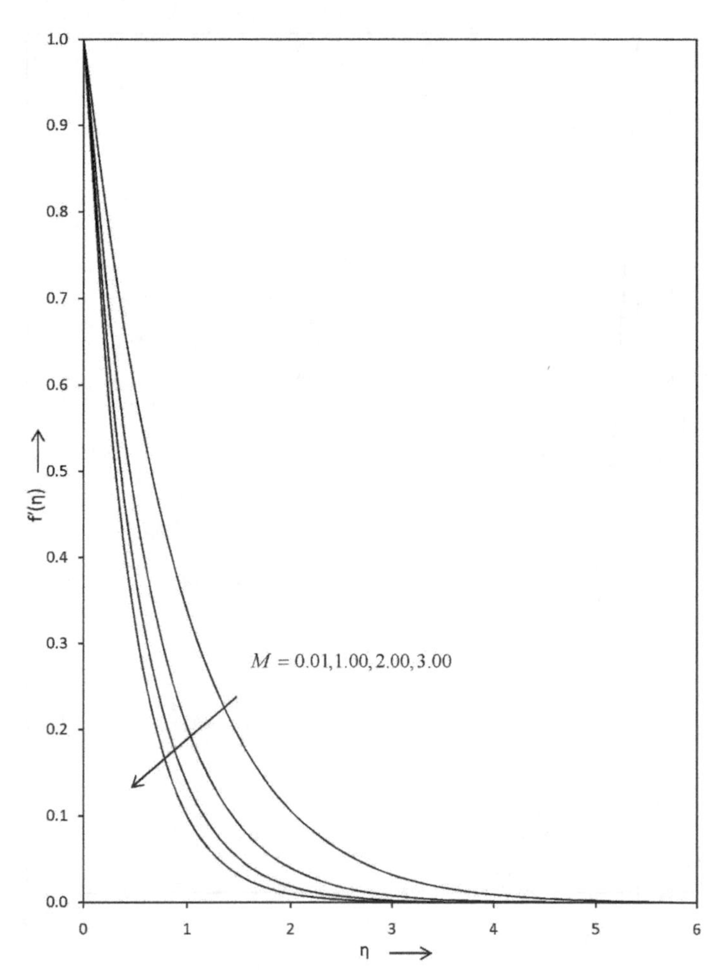

FIGURE 10.6
Effect of M on velocity distribution when $\phi = 0.07$ and $n = 0.5$.

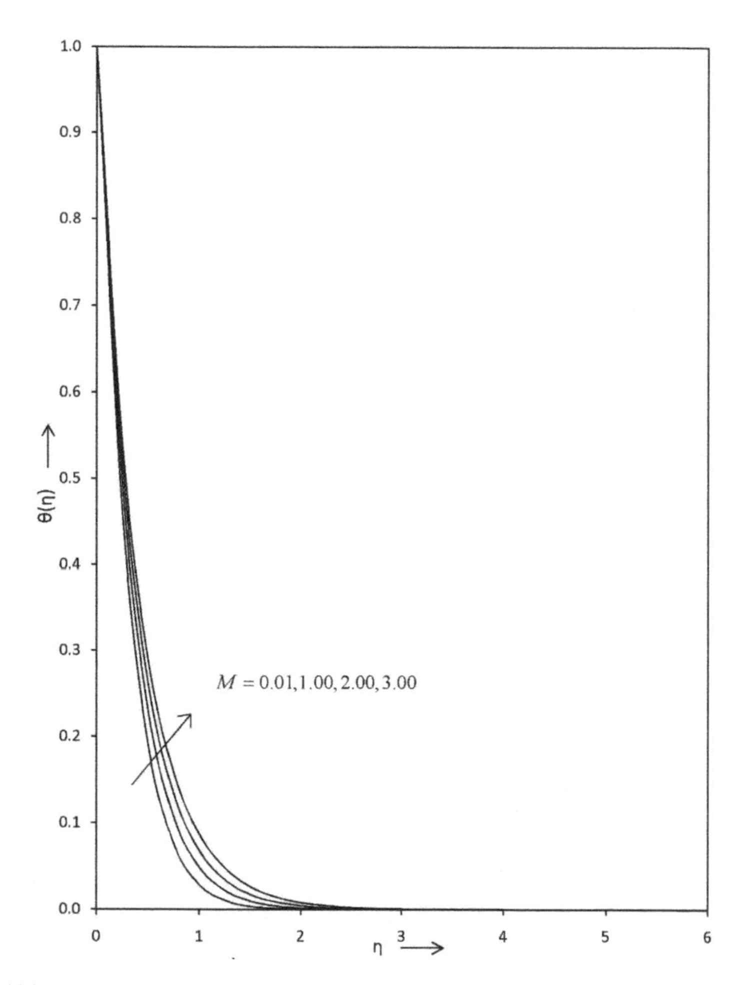

FIGURE 10.7
Effect of M on temperature distribution when $\phi = 0.07$, $n = 0.5$, $m = 3.0$ and $Pr = 6.8$.

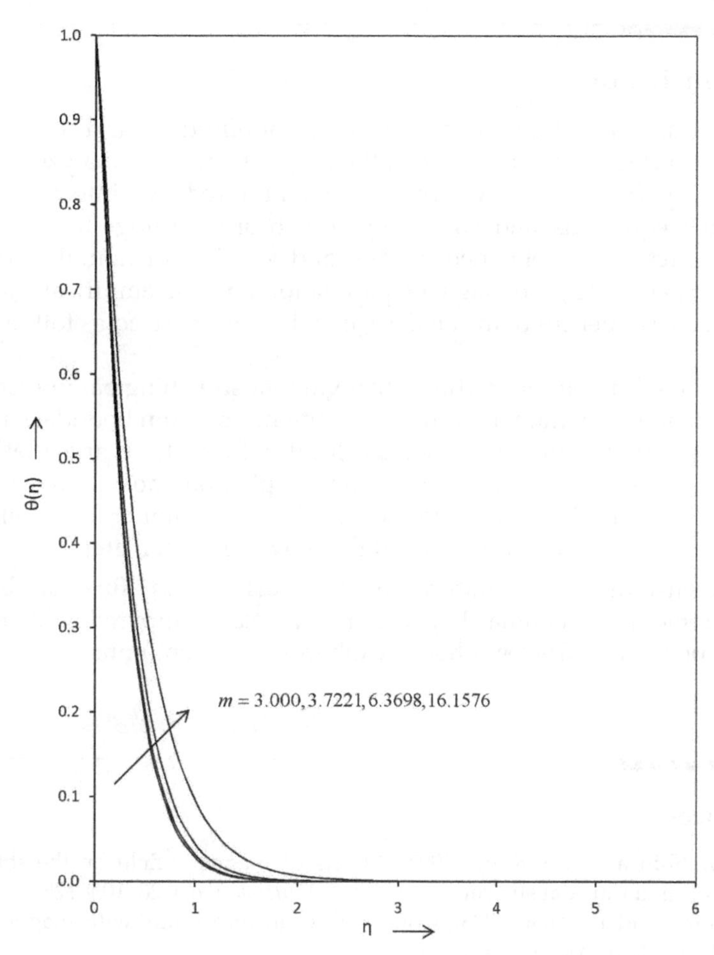

FIGURE 10.8
Effect of m on temperature distribution when $\phi = 0.07$, $n = 0.5$, $M = 0.01$ and $Pr = 6.8$.

10.8 Conclusions

The boundary layer flow of copper-water nanofluid toward a nonlinearly stretching surface in the presence of the magnetic field is analyzed numerically. Basic partial boundary layer equations are reduced into the ordinary differential equations and solved by fourth-order Runge-Kutta method. Numerical results are obtained for the solid volume fraction, the nonlinear stretching parameter, the magnetic parameter, and the empirical shape factor. Main consequences of the analysis may be summarized as follows

1. As the solid volume fraction, the nonlinear stretching parameter and the magnetic parameter grow, all of the momentum boundary layer, the thermal boundary layer, the local skin friction, and the local Nusselt number reduce. The opposite phenomenon occurs in the thermal boundary layer and the local Nusselt number for the effects of the solid volume fraction and the magnetic parameter.
2. Dimensionless temperature and the surface heat flux has better increment for lamina-shaped nanoparticles compared with other shaped nanoparticles sphere, hexahedron, and column.

References

1. S. M. Shin and I. S. Kang (2002), Effects of magnetic field on the shape of a bubble in a uniaxial straining flow, *Int. J. Multiph. Flow*, 28, 105–125.
2. J. Singh and R. Bajaj (2005), Couette flow in ferrofluids with magnetic field, *J. Magn. Magn. Mater.*, 294, 53–62.
3. R. N. Jat and S. Chaudhary (2010), Radiation effects on the MHD flow near the stagnation point of a stretching sheet, *Z. Angew. Math. Phys.*, 61, 1151–1154.
4. X. Zhang, C. Pan and Z. Xu (2013), Effect of contact resistance on liquid metal MHD flows through circular pipes, *Fusion Eng. Des.*, 88, 2228–2234.
5. S. Chaudhary and P. Kumar (2014), MHD forced convection boundary layer flow with a flat plate and porous substrate, *Meccanica*, 49, 69–77.
6. M. Imtiaz, T. Hayat, A. Alsaedi and A. Hobiny (2016), Homogeneous-heterogeneous reactions in MHD flow due to an unsteady curved stretching surface, *J. Mol. Liq.*, 221, 245–253.
7. S. Chaudhary and M. K. Choudhary (2018), Finite element analysis of magnetohydrodynamic flow over flat surface moving in parallel free stream with viscous dissipation and Joule heating, *Eng. Comput.*, 35, 1675–1693.
8. G. C. Bourantas, V. C. Loukopoulos, G. R. Joldes, A. Wittek and K. Miller (2019), An explicit meshless point collocation method for electrically driven magnetohydrodynamics (MHD) flow, *Appl. Math. Comput.*, 348, 215–233.

9. J. C. Maxwell (1904), *Treatise on Electricity and Magnetism*, Oxford University Press, London, UK.
10. B. X. Wang, L. P. Zhou and X. F. Peng (2003), A fractal model for predicting the effective thermal conductivity of liquid with suspension of nanoparticles, *Int. J. Heat Mass Transf.*, 46, 2665–2672.
11. D. Wen and Y. Ding (2004), Experimental investigation into convective heat transfer of nanofluids at the entrance region under laminar flow conditions, *Int. J. Heat Mass Transf.*, 47, 5181–5188.
12. C. T. Nguyen, G. Roy, C. Gauthier and N. Galanis (2007), Heat transfer enhancement using Al_2O_3–water nanofluid for an electronic liquid cooling system, *Appl. Therm. Eng.*, 27, 1501–1506.
13. M. Turkyilmazoglu (2012), Exact analytical solutions for heat and mass transfer of MHD slip flow in nanofluids, *Chem. Eng. Sci.*, 84, 182–187.
14. M. Khoshvaght-Aliabadi (2014), Influence of different design parameters and Al2O3-water nanofluid flow on heat transfer and flow characteristics of sinusoidal-corrugated channels, *Energy Convers. Manag.*, 88, 96–105.
15. H. B. Rokni, D. M. Alsaad and P. Valipour (2016), Electrohydrodynamic nanofluid flow and heat transfer between two plates, *J. Mol. Liq.*, 216, 583–589.
16. G. Saha and M. C. Paul (2017), Transition of nanofluids flow in an inclined heated pipe, *Int. Commun. Heat Mass Transf.*, 82, 49–62.
17. M. Bezaatpour and M. Goharkhah (2019), Three dimensional simulation of hydrodynamic and heat transfer behavior of magnetite nanofluid flow in circular and rectangular channel heat sinks filled with porous media, *Powder Technol.*, 344, 68–78.
18. E. M. Abo-Eldahab and A. F. Ghonaim (2003), Convective heat transfer in an electrically conducting micropolar fluid at a stretching surface with uniform free stream, *Appl. Math. Comput.*, 137, 323–336.
19. R. N. Jat and S. Chaudhary (2008), Magnetohydrodynamic boundary layer flow near the stagnation point of a stretching sheet, *Il Nuovo Cimento*, 123B, 555–566.
20. R. Tamizharasi and V. Kumaran (2011), Pressure in MHD/Brinkman flow past a stretching sheet, *Commun. Nonlinear Sci. Numer. Simul.*, 16, 4671–4681.
21. K. L. Hsiao (2013), Energy conversion conjugate conduction-convection and radiation over non-linearly extrusion stretching sheet with physical multimedia effects, *Energy*, 59, 494–502.
22. T. Aziz and F. M. Mahomed (2016), Remark on classical Crane's solution of viscous flow past a stretching plate, *Appl. Math. Lett.*, 52, 205–211.
23. S. Chaudhary, K. M. Kanika and M. K. Choudhary (2018), Newtonian heating and convective boundary condition on MHD stagnation point flow past a stretching sheet with viscous dissipation and Joule heating, *Indian J. Pure Appl. Phys.*, 56, 931–940.
24. I. Ahmad, H. Zafar, M. Z. Kiyani and S. Farooq (2019), Zero mass flux characteristics in Jeffery nanoliquid flow by a non-linear stretchable surface with variable thickness, *Int. J. Heat Mass Transf.*, 132, 1166–1175.
25. M. A. A. Hamad and M. Ferdows (2012), Similarity solutions to viscous flow and heat transfer of nanofluid over nonlinearly stretching sheet, *Appl. Math. Mech.-Engl. Ed.*, 33, 923–930.

26. A. S. Dogonchia, K. Divsalara and D. D. Ganji (2016), Flow and heat transfer of MHD nanofluid between parallel plates in the presence of thermal radiation, *Comput. Methods Appl. Mech. Eng.*, 310, 58–76.
27. M. R. Daneshvar Garmroodi, A. Ahmadpour and F. Talati (2019), MHD mixed convection of nanofluids in the presence of multiple rotating cylinders in different configurations: A two-phase numerical study, *Int. J. Mech. Sci.*, 150, 247–264.
28. Y. Lin, B. Li, L. Zheng and G. Chen (2016), Particle shape and radiation effects on Marangoni boundary layer flow and heat transfer of copper-water nanofluid driven by an exponential temperature, *Powder Technol.*, 301, 379–386.

11

Reliability and Sensitivity Assessment of a Sugar Mill through Mathematical Modeling

Amit Kumar and Monika Manglik

CONTENTS

11.1 Introduction

As development paced up the reliability and maintainability of engineering and production systems have been become a significant issue during their functioning and design. Some examples of these systems are computers, aircraft, space satellites, nuclear power-generating reactors, automobiles, thermal power plant, sugar mill, and railways. The specific factors that play, directly or indirectly, an instrumental role in increasing the importance of reliability in designed systems include high acquisition cost, maintainability, complexity, productivity, and safety [1,2]. These factors clearly indicate a definite need for maintainability and maintenance for reliability professionals to work closely during operation and designing phases. Weber and Jouffe [3] present a methodology that helps dynamic object-oriented Bayesian networks to formalize complex dynamic models for evaluating the reliability of a manufacturing process. Process industries such as sugar mills, the chemical industry, oil refineries, thermal power plant, fertilizer industry, and paper plants are of major importance in real-life situations because these industries fulfill our numerous daily-living needs. The demand for a quality product and reliable system is of great importance. Failure is a random phenomenon associated with the operating state of any physical system, and one of the consequences of system failure is deterioration of its components. Therefore, various performance measures, such as reliability, availability, expected profit, and productivity, need to be maintained in regard to the failure of the system [4,5]. These measures are the most significant factors associated with repairable and nonrepairable systems. Industries are becoming complex in structure because of technology advancement; therefore, it is quite difficult for a system analyst to formulate proper maintenance policy.

The importance of sugar industry is increasing day by day because of the increase in sugar consumption. In the rapidly growing technology, reliability plays a crucial role in each and every industry from production to operation of various systems/components. So, it is a herculean task for the management of the system to maintain the reliability and quality of the products.

This chapter covers the analysis of various reliability measures of a sugar mill plant, incorporating various types of possible failures. There are various techniques that applied to maintain or enhance the reliability of complex systems in different environments. Dhillon [6] presented a reliability and availability analysis of a two-unit parallel system with warm standby and common-cause failure with the consideration that a failed system's repair time is taken to be arbitrary distributed. He found various reliability measures for the same, but he did not consider human error in the system. Dhillon and Yang [7] consider a system in which two units are in parallel, and one is in standby with critical and noncritical human errors. In this approach, the authors developed a general expression for the system's steady-state availability, but they did not talk about some of the important measures of the system such as mean time to system failure, mean time to repair, and sensitivity analysis. Dhillon and Yang [8] developed a stochastic model for performing different reliability measures for a repairable standby human-machine system with increasing human error and arbitrary repair rates for failed system. They also developed expressions for steady-state availability, time dependent availability, mean time to failure (MTTF), and reliability, but they did not perform sensitivity analysis for the system, which plays an important role. Sharma and Vishwakarma [9] analyzed the reliability and availability of feeding system of sugar industry; incorporating reduced states, the authors used the time-homogeneous Markov process. Kadyan and Kumar [10] analyzed the B-Pan crystallization system in the sugar industry with three subsystems, crystallizer, centrifugal machine, and melter. Also, the availability of the system has been analyzed as a particular case. Sinha and Mukhapadhyay [11] emphasized through the case study that the reliability of cane crusher should be improved. They considered crusher-component failure details for one year. Kumar et al. [12] analyzed the availability of the crushing system in the sugar industry. Lagrange's method for a partial differential equation has been used by the authors for the solution of the considered model. Zhou et al. [13] proposed a reliability-centered predictive maintenance for a system subject to degradation because of the imperfect maintenance. Li and Yang [14] discussed the sugarcane agriculture and sugar industry in China. Authors have also discussed the new challenges for sugar industry in China. Kumar et al. [15] developed an analytic model of the crystallization system in the sugar industry. Availability of the system under common-cause failure has also been analyzed. Jovanovic [16] discussed the background and the needs of industry and showed the relationship of risk-based inspection (RBI)/risk-based life management (RBLM) and other approaches to maintenance. Bertolini et al. [17] developed a risk-based inspection and maintenance procedures of an oil refinery. The manure system of the sugar industry has been analyzed by Kadyan and Kumar [18] with the aid of supplementary variable technique to evaluate its various performance measures. The time to failure of the subsystems follows negative exponential distribution, whereas repair time is taken as arbitrary. Sulphate juice pump in a sugar industry with the assumption that the failure may occur during the maintenance is analyzed by Kaker et al. [19] by using regenerative point technique. Here author analyzed the mean time to system failure (MTSF) for the same.

11.2 Assumptions

Assumption 1: The working stints of each unit of the sugar mill plant are supposed to be independent and identically exponentially distributed.

Assumption 2: Once a working unit breaks down, it will be repaired by the maintenance team.

Assumption 3: The sugar mill plant goes in down or degraded state as soon as the unit failure(s) occurs.

Assumption 4: Average failure and repair rates of each component of the system are taken to be constant.

11.3 Nomenclature and State Narratives

Throughout this study of sugar mill plant, the nomenclatures can be found in Table 11.1 and state narratives can be found in Table 11.2.

TABLE 11.1

Nomenclature

t	Time scale variable
s	Laplace Transforms variable
$P_i(t);\ i=0,1,3$	Probability of the system being in state S_i at instant t
$\overline{P}_i(s)$	Laplace transform of $P_i(t)$
$P_j(x,t);\ j=2,4,5,6,7;$	Probability density function of system being in completely failed state at instant t with elapsed repair time, x
$\overline{P}_i(x,s);$	Laplace transform of $P_i(x,t)$
$\lambda_{CP}/\lambda_{JE}/\lambda_C/\lambda_E/\lambda_J$	The failure rate of cane preparation/juice extraction/cooling/evaporator/juice clarification
$\mu_{CP}/\mu_{JE}/\mu_C/\mu_E/\mu_J$	The recovery rate of cane preparation/juice extraction/cooling/evaporator/juice clarification
K_1/K_2	Revenue/service cost of the sugar mill plant

TABLE 11.2

State Narration

S_0	Good state: All the components of the sugar mill working perfectively
S_1	Degraded state: State occurs due to degradation of cane preparation
S_2	Failed state: State occurs due to complete failure of cane preparation
S_3	Degraded state: State occurs due to degradation of juice extraction
S_4	Failed state: State occurs due to complete failure of juice extraction
S_5	Failed state: State occurs due to complete failure of crystallization process
S_6	Failed state: State occurs due to complete failure of cooling process
S_7	Failed state: State occurs due to complete failure of evaporator

11.4 State Transition Diagram

The sugar mill may work in different states based on working and failure of its components. By critically analyzing these failures and repairs, various states of sugar mill are identified and drawn in Figure 11.1.

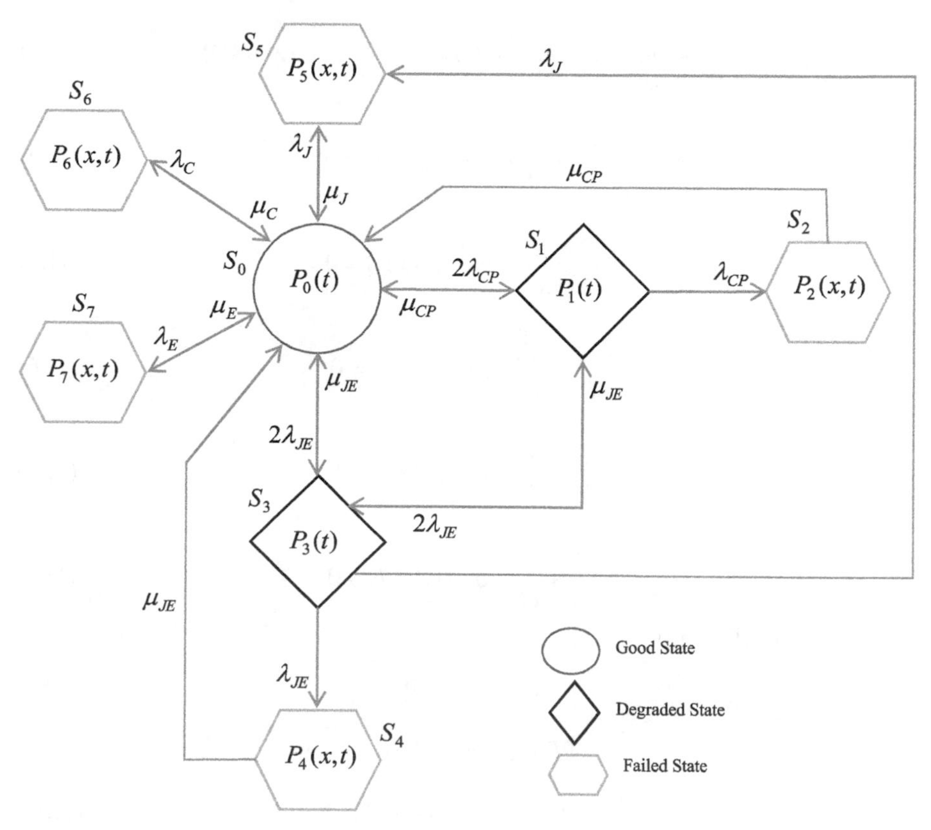

FIGURE 11.1
Transition state diagram of sugar mill.

11.5 Mathematical Formulation and Solution of the Problem

By considering transitions of the Markov process, during time $(t, t+\Delta t)$ and letting $\Delta t \to 0$, we obtained the following set of intro-differential equations.

$$\left(\frac{\partial}{\partial t} + 2\lambda_{CP} + 2\lambda_{JE} + \lambda_E + \lambda_C + \lambda_J\right)P_0(t) = \mu_{CP}P_1(t) + \mu_{JE}P_3(t) + \sum_{k,l}\mu_k\, P_l(x,t)dx \tag{11.1}$$

where $k = CP, JE, J, C, E$

$l = 2,4,5,6,7$ respectively

$$\left(\frac{\partial}{\partial t} + \mu_{CP} + \lambda_{CP} + 2\lambda_{JE}\right)P_1(t) = 2\lambda_{CP}\, P_0(t) + \mu_{JE}\, P_3(t) \tag{11.2}$$

$$\left(\frac{\partial}{\partial t}+2\mu_{JE}+\lambda_{JE}+\lambda_{J}\right)P_3(t)=2\lambda_{JE}\,P_0(t)+2\lambda_{JE}\,P_1(t) \tag{11.3}$$

$$\left(\frac{\partial}{\partial x}+\frac{\partial}{\partial t}+\mu_{CP}\right)P_2(x,t)=0 \tag{11.4}$$

$$\left(\frac{\partial}{\partial x}+\frac{\partial}{\partial t}+\mu_{JE}\right)P_4(x,t)=0 \tag{11.5}$$

$$\left(\frac{\partial}{\partial x}+\frac{\partial}{\partial t}+\mu_{J}\right)P_5(x,t)=0 \tag{11.6}$$

$$\left(\frac{\partial}{\partial x}+\frac{\partial}{\partial t}+\mu_{C}\right)P_6(x,t)=0 \tag{11.7}$$

$$\left(\frac{\partial}{\partial x}+\frac{\partial}{\partial t}+\mu_{E}\right)P_7(x,t)=0 \tag{11.8}$$

Boundary condition

$$P_2(0,t)=\lambda_{CP}\,P_1(t) \tag{11.9}$$

$$P_4(0,t)=\lambda_{JE}\,P_3(t) \tag{11.10}$$

$$P_5(0,t)=\lambda_{J}\,P_0(t)+\lambda_{J}\,P_3(t) \tag{11.11}$$

$$P_6(0,t)=\lambda_{C}\,P_0(t) \tag{11.12}$$

$$P_7(0,t)=\lambda_{E}\,P_0(t) \tag{11.13}$$

Initial condition

$$P_i(t)=\begin{cases}1, & t=0, i=0\\ 0, & otherwise\end{cases} \tag{11.14}$$

The system of intro-differential equations (11.1) through (11.8) together with boundary condition (11.9) through (11.13) and initial condition (11.14) is known as Chapman-Kolmogorov differential equations. To find the various performance indicator of the sugar mill plant, the authors solved the set of equations with the help of Laplace transform and find the various state probabilities as following.

$$\overline{P}_0(s)=\frac{1}{\left\{\left(s+2\lambda_{CP}+\lambda_{J}+\lambda_{C}+\lambda_{E}+2\lambda_{JE}\right)-H_1-H_2-H_3\right\}} \tag{11.15}$$

$$\overline{P}_1(s) = \left\{ \frac{\dfrac{2\lambda_{CP}}{(s+\mu_{CP}+\lambda_{CP}+2\mu_{JE})} + \dfrac{2\lambda_{JE}\mu_{JE}}{(s+2\mu_{JE}+\lambda_{JE})}}{1-\dfrac{2\lambda_{JE}\mu_{JE}}{(s+2\mu_{JE}+\lambda_{JE})}} \right\} \overline{P}_0(s) \qquad (11.16)$$

$$\overline{P}_3(s) = \left\{ \frac{\dfrac{2\lambda_{JE}}{(s+2\mu_{JE}+\lambda_{JE})} \left(1+\dfrac{2\lambda_{CP}}{(s+\mu_{CP}+\lambda_{CP}+2\mu_{JE})} + \dfrac{2\lambda_{JE}\mu_{JE}}{(s+2\mu_{JE}+\lambda_{JE})} \right)}{1-\dfrac{2\lambda_{JE}\mu_{JE}}{(s+2\mu_{JE}+\lambda_{JE})}} \right\} \overline{P}_0(s) \qquad (11.17)$$

where

$$H_1 = \frac{2\lambda_{CP}G_1 + 2\lambda_{JE}\mu_{JE}}{G_1+G_2-2\lambda_{JE}\mu_{JE}} \left(\lambda_{CP} + \frac{\mu_{CP}\lambda_{CP}}{s+\lambda_{CP}} \right)$$

$$H_2 = \frac{2\lambda_{JE}\left\{2\lambda_{CP}G_1 + 2\lambda_{JE}\mu_{JE} + 2\lambda_{JE}\left(G_1+G_2-2\lambda_{JE}\mu_{JE}\right)\right\}}{\left(G_1+G_2-2\lambda_{JE}\mu_{JE}\right)\left(s+2\lambda_{JE}+\mu_{JE}\right)} H_4$$

$$H_3 = \left(\frac{\lambda_C\mu_C}{s+\mu_C} + \frac{\lambda_E\mu_E}{s+\mu_E} + \frac{\lambda_J\mu_J}{s+\mu_J} \right)$$

$$H_4 = \left(\mu_{JE} + \frac{\mu_{JE}\lambda_{JE}}{s+\lambda_{JE}} + \frac{\mu_J\lambda_J}{s+\lambda_J} \right)$$

$$G_1 = (s+2\mu_{JE}+\lambda_{JE}+\lambda_J); \quad G_2 = (s+2\mu_{CP}+\lambda_{JE}+\lambda_{CP});$$

The state probability for the various states of the sugar mill can be obtained by the aid of inverse Laplace transform of Equations 11.15 through 11.17. The numerical computation is carried out by taking various failures and repairs as constant. The availability of the sugar mill has been computed as (from Figure 11.1)

$$A(t) = \sum_{i=0}^{1} P_i(t) + P_3(t) \qquad (11.18)$$

11.6 Computation of Various Performance Indicators for Sugar Mill Plant

11.6.1 Availability

It is the probability that a system or equipment is available for operation when used in understated conditions in an ideal support environment. For evaluating the availability for the sugar mill plant, we take the value of various failures rates as $\lambda_{CP} = 0.006$, $\lambda_{JE} = 0.003$, $\lambda_J = 0.003$, $\lambda_E = 0.013$, and $\lambda_C = 0.011$, and all the repairs as one in Equation 11.18. The availability of the sugar mill plant is obtained as:

$$A(t) = \begin{Bmatrix} 0.990240\,10^{-6}\, e^{-2.006\,t} + 0.007606\, e^{-2.0014\,t} \\ + 0.000760\, e^{-2.00145\,t} + 0.243802\, e^{-1.03609\,t} \\ + 0.00178788\, e^{-1.00599\,t} + 0.973070\, e^{-0.00144\,t} \end{Bmatrix} \tag{11.19}$$

Now varying time unit t in Equation 11.19, we get Figure 11.2, which depicts the behavior of the availability of the sugar mill plant.

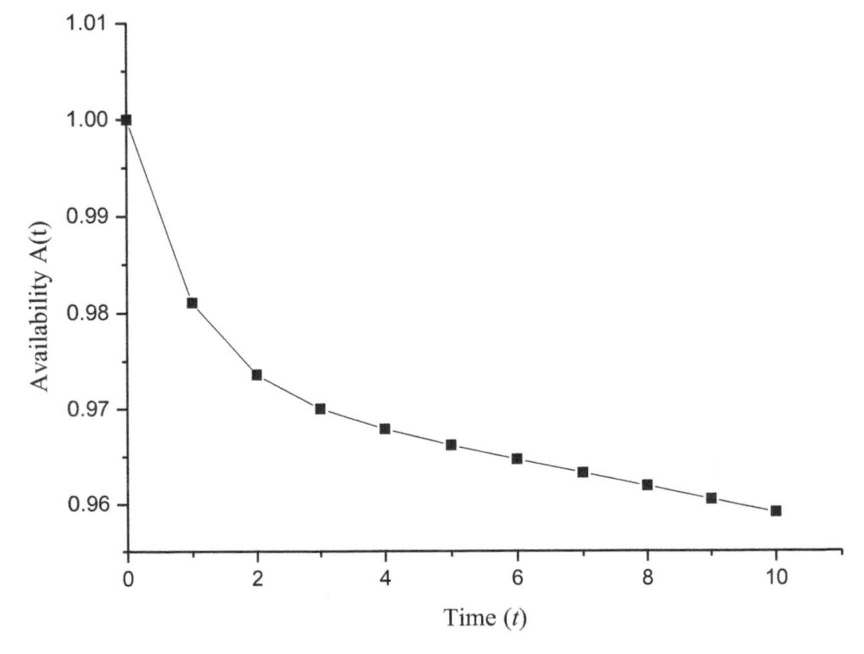

FIGURE 11.2
Behavior of availability vs. time unit (t).

11.6.2 Reliability

"Reliability is the probability of a system performing its purpose adequately for the period intended under the given operating conditions." The reliability of the considered system is calculated by putting the value of various failure rates as $\lambda_{JE} = 0.003$, $\lambda_{CP} = 0.006$, $\lambda_E = 0.013$, $\lambda_J = 0.003$, and $\lambda_C = 0.011$, and all repairs as zero in Equation 11.18, the reliability of the sugar mill plant is obtained as:

$$R(t) = \left\{ \begin{array}{l} 1.0199 e^{-0.036\,t} + 0.4445\, e^{-0.0225\,t} \sinh(0.0135\,t) - 0.1111\, e^{-0.009\,t} \\ + 0.091181\ e^{0.003\,t} \end{array} \right\} \tag{11.20}$$

Now varying time unit t in Equation 11.20, we get Figure 11.3, which depicts the behavior of the reliability of the sugar mill plant.

11.6.3 Mean Time to Failure (MTTF)

Mathematically, the MTTF of a system is calculated as

$$\text{MTTF} = \int_0^t R(t)\,dt = \lim_{s \to 0} \overline{R}(s) \tag{11.21}$$

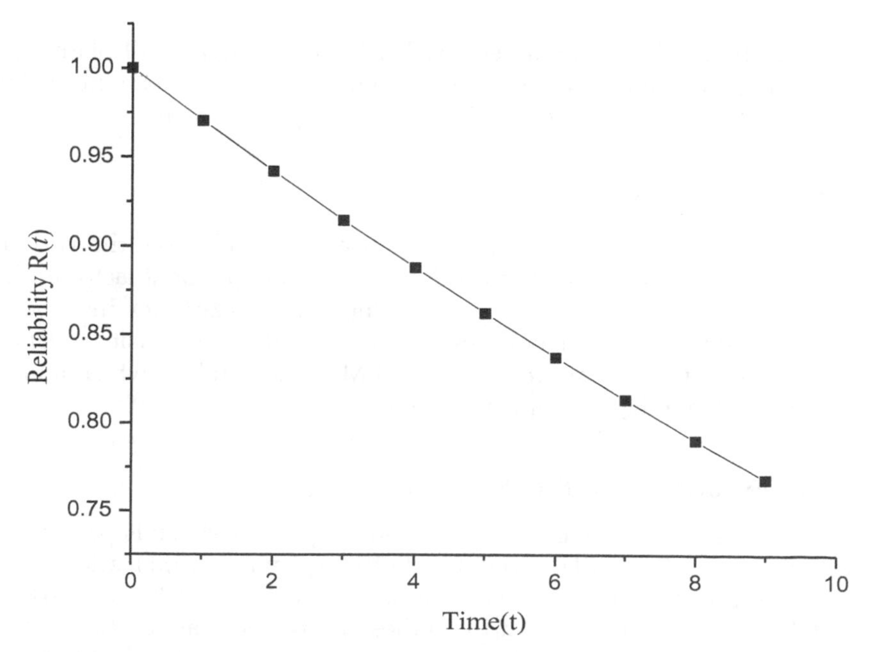

FIGURE 11.3
Reliability vs. time (t).

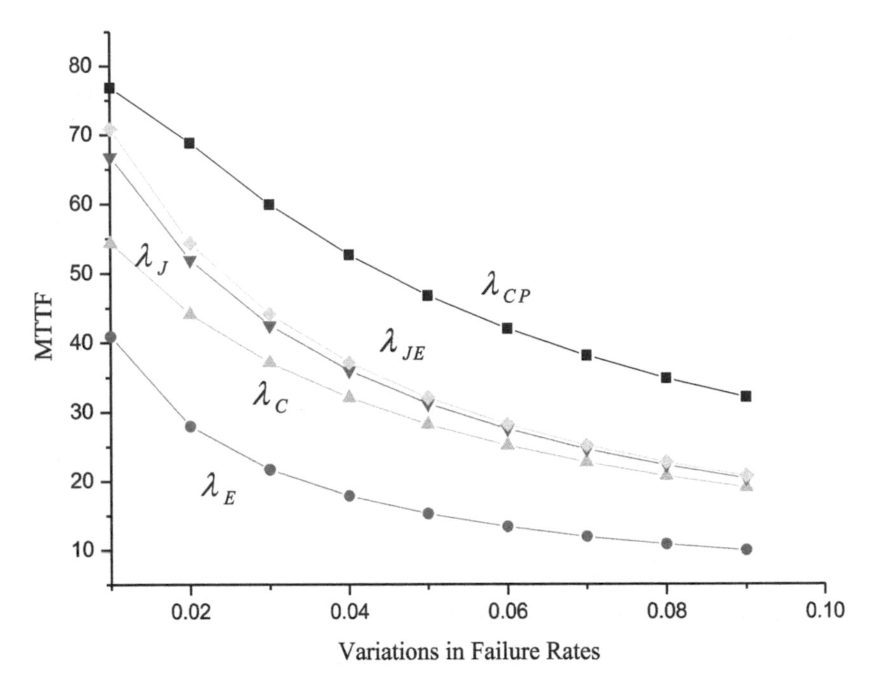

FIGURE 11.4
Mean time to failure (MTTF) vs. failure rates.

Using Equation 11.21, we obtained the MTTF of the sugar mill plant, and then varying the failure rates one by one from 0.01 to 0.09 with an interval of 0.01, Figure 11.4 was obtained for MTTF of the sugar mill plant

11.6.4 Sensitivity Analysis

Sensitivity analysis is a technique that is used to decide how the various values of an independent factor impact a particular dependent factor under some constraints, or it is a factor by which one can analyze that which factor affects the system's performance most. In the present study, sensitivity analysis is performed for system reliability and MTTF for finding which failure affects the system reliability and MTTF most.

11.6.4.1 Sensitivity of MTTF for Sugar Mill

Sensitivity analysis of sugar mill plant with respect to MTTF is performed by differentiating the MTTF expression with respect to various failure rates and then placed the values of various failure rates as $\lambda_{CP} = 0.006$, $\lambda_{JE} = 0.003$, $\lambda_J = 0.003$, $\lambda_C = 0.011$, and $\lambda_E = 0.013$, in these partial derivatives. Now varying the failure rates one by one, respectively, in these partial derivatives, one can obtain the Figure 11.5 for sensitivity of MTTF for the sugar mill plant.

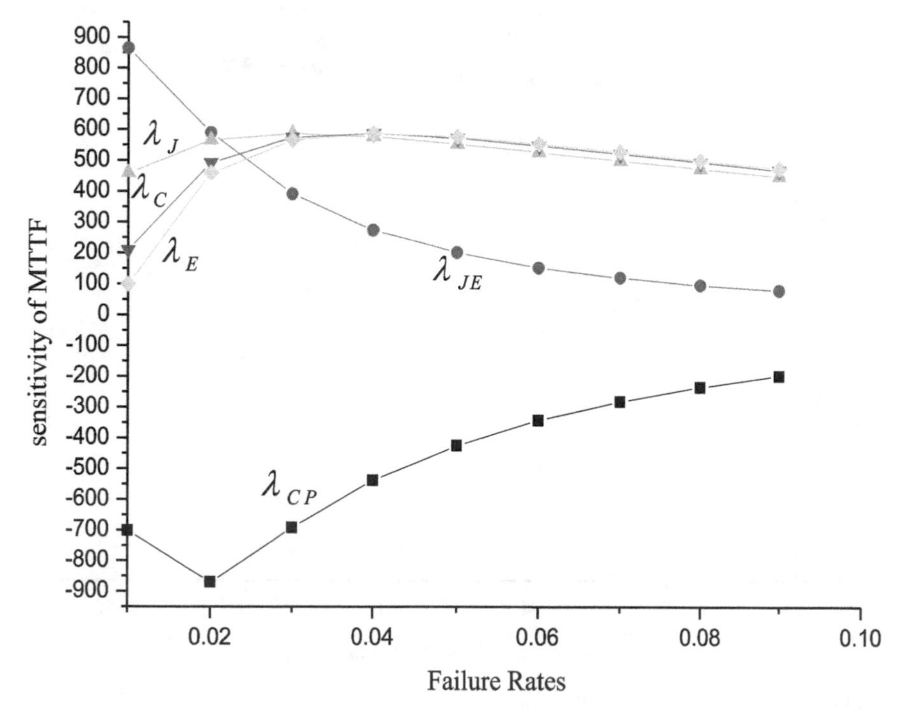

FIGURE 11.5
Sensitivity of mean time to failure (MTTF) vs. failure rates.

11.6.4.2 Sensitivity of Reliability for Sugar Mill

In the same manner as the authors calculated sensitivity of the considered system for MTTF, sensitivity of reliability has been calculated [4–6] by differentiating the reliability expression with respect to the failure rates and then placing the values of various failure rates as $\lambda_{CP} = 0.006$, $\lambda_{JE} = 0.003$, $\lambda_J = 0.003$, $\lambda_C = 0.011$, and $\lambda_E = 0.013$, in these partial derivatives. Now varying the time unit, *t*, one can obtain Figure 11.6 for sensitivity of reliability for sugar mill plant.

11.6.5 Expected Profit Analysis

The expected profit function for the sugar mill in the time interval (0, *t*) is estimated as [4,5]

$$E_P(t) = K_1 \int_0^t A(t)dt - tK_2 \tag{11.22}$$

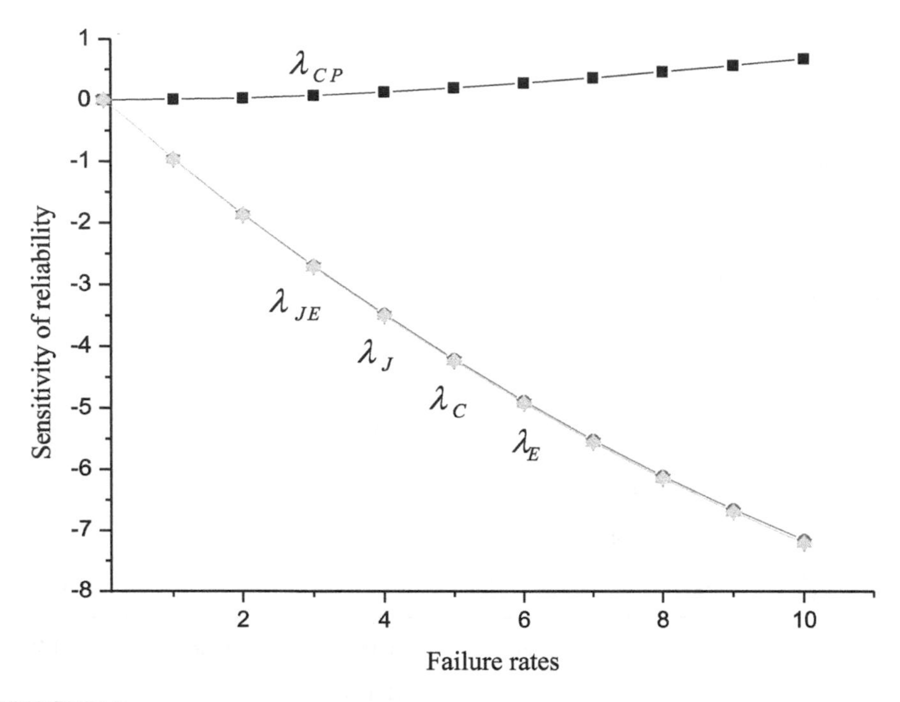

FIGURE 11.6
Sensitivity of reliability vs. time unit, *t*.

Using Equation 11.19 in Equation 11.22, the expected profit function for the same set of failure and repair rates is obtained as

$$E_P(t) = K_1 \begin{Bmatrix} -0.493637614410^{-6}\ e^{(-2.006005967\,t)} - 0.00038 \\ e^{(-2.001457589\,t)} - 0.0235\, e^{(-1.036096131\,t)} - 0.0017 \\ e^{(-1.005993685\,t)} - 672.64726\ e^{(-.001446627760\,t)} \\ +672.67295 \end{Bmatrix} - tK_2 \quad (11.23)$$

Now taking revenue as one and varying service cost K_2 as 0.1, 0.2, 0.3, 0.4, 0.5, respectively, and then varying time unit t in Equation 11.23, Figure 11.7 for expected profit from sugar mill is obtained.

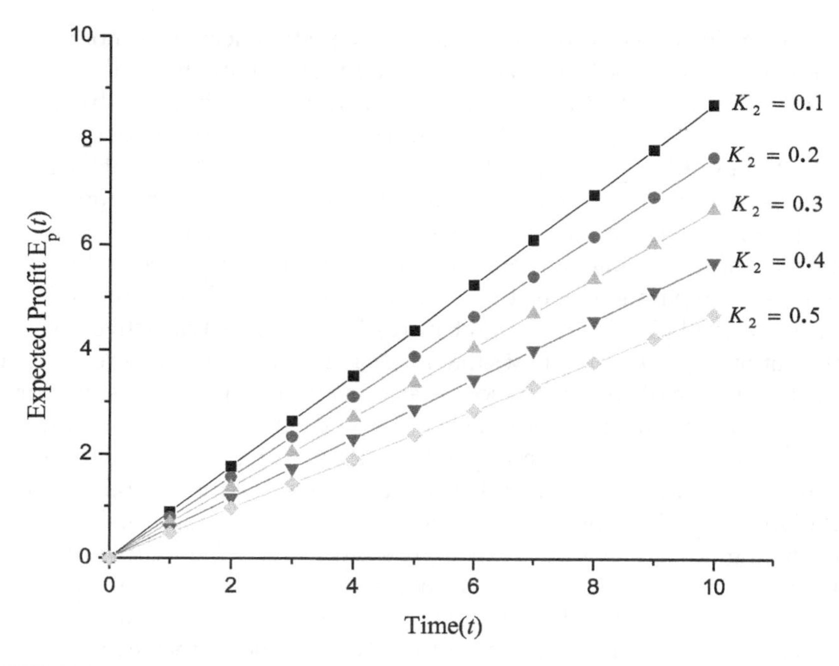

FIGURE 11.7
Expected profit vs. service cost and time unit, t.

11.7 Result Discussion and Conclusion

Sugar production can be optimized by maintaining the reliability, availability, and the MTTF of the sugar mill plant. With the help of supplementary variable technique, Markov process, and Laplace transformations, we are able to explore the various reliability measures of the sugar mill by taking the possible failures into consideration. According to the mathematical illustrations and graphical representation of the various reliability measures, we conclude the impact of the various types of failure on reliability, availability, MTTF, expected profit, and sensitivity of these measures.

Figure 11.2 gives an idea about the availability of the sugar mill plant with respect to time. It can be easily concluded that the availability of the system decreases approximately in even manner with an increase in time.

Figure 11.3 shows the reliability of the system with respect to time by taking some fixed values of failure and repair rates. From the graph (Figure 11.3), it can be easily observed that reliability of the system decreases more sharply in comparison to availability with increase in time. This shows that the reliability of the sugar mill plant can be improved by controlling the failures. Figure 11.4 is the analysis of MTTF of the system with respect to various failure rates. After examining Figure 11.4, it can be observed that MTTF of the system decreases with increase in failure of cane preparation, juice extraction, cooling, evaporator, and juice clarification. Also the MTTF of sugar mill is least with respect to the failure rate of evaporator and highest with respect to the failure rate of cane preparation. Sensitivity analysis of the MTTF and the reliability of the sugar mill plant are given in Figures 11.5 and 11.6, respectively. Critical examination of Figure 11.5 shows that the MTTF is most sensitive with respect to the failure rate of juice extraction. Also it is almost equally sensitive for the failure rate of cooling process and juice clarification. One can say that to optimize the MTTF of the sugar mill plant, the failure of juice extraction must be minimized (as much as possible). Figure 11.6 shows that the reliability of sugar mill plant is equally sensitive with respect to the failure rates of juice extraction, cooling process, evaporator, and juice clarification. Hence to improve the reliability of the sugar mill plant, the maintenance team must pay more attention on these failure rates. Figure 11.7 gives the expected profit from sugar mill versus time. From Figure 11.7, it can be easily seen that increasing service cost results in a decrease in expected profit. The study shows that the minimum service cost leads to maximum expected profit.

This study shows that with the increasing demand of process industries, the proper maintenance of the related plant must be ensured time to time. Also, it is necessary to avoid or control all the related failures to maintain various reliability measures. The future scope related to this study is that applied numerical methods and the matrix method can be used to solve the equations in place of Laplace transformation. Also fuzzy theory can be applied, and other failures can be considered (e.g., human error, common cause failure, etc.).

References

1. B. S. Dhillon, *Maintainability, Maintenance, and Reliability for Engineers*, CRC Press, Boca Raton, FL, 2006.
2. L. R. Higgins, R. K. Mobley, R. Smith, *Maintenance Engineering Handbook*, McGraw-Hill, New York, 2002.
3. P. Weber, L. Jouffe, Complex system reliability modelling with dynamic object oriented Bayesian networks (DOOBN), *Reliability Engineering & System Safety* 91 (2) (2006) 149–162.
4. A. Kumar, M. Ram, R. S. Rawat, Optimization of casting process through reliability approach, *International Journal of Quality & Reliability Management* 34 (6) (2017) 833–848.

5. M. Ram, A. Kumar, Performability analysis of a system under 1-out-of-2: G scheme with perfect reworking, *Journal of the Brazilian Society of Mechanical Sciences and Engineering* 37 (3) (2015) 1029–1038.
6. B. S. Dhillon, Reliability and availability analysis of a system with warm standby and common cause failure, *Microelectronics Reliability* 33 (9) (1993) 1343–1349.
7. B. S. Dhillon, N. Yang, Availability of a man-machine system with critical and non-critical human error, *Microelectronics Reliability* 33 (10) (1993) 1511–1521.
8. B. S. Dhillon, N. Yang, Probabilistic analysis of a maintainable system with human error, *Journal of Quality in Maintenance Engineering* 1 (2) (1995) 50–59.
9. S. P. Sharma, Y. Vishwakarma, Application of Markov process in performance analysis of feeding system of sugar industry, *Journal of Industrial Mathematics* 2014 (2014) 1–9.
10. M. S. Kadyan, R. Kumar, Availability based operational behavior of B-Pan crystallization system in the sugar industry, *International Journal of System Assurance Engineering and Management* 8 (2) (2017) 1450–1460.
11. R. S. Sinha, A. K. Mukhapadhyay, Reliability centered maintenance of cone crusher: a case study, *International Journal of System Assurance Engineering and Management* 6 (1) (2015) 32–35.
12. D. Kumar, J. Singh, I. P. Singh, Availability of the feeding system in sugar industry, *Microelectron Reliability* 28 (6) (1988) 867–871.
13. X. Zhou, L. XI, J. Lee, Reliability-centered predictive maintenance scheduling for a continuously monitored system subject to degradation, *Reliability Engineering & System Safety* 92 (4) (2007) 530–534.
14. Y. R. Lee, L.-T. Yang, Sugarcane agriculture and sugar industry in China, *Sugar Tech* 17 (1) (2015) 1–8.
15. D. Kumar, J. Singh, P. C. Pandey, Availability of the crystallization system in sugar industry under common cause failure, *IEEE transactions on Reliability* 4 (1) (1992) 85–91.
16. A. Jovanovic, Risk based inspection and maintenance in power and process plants in Europe, *Nuclear Energy and Design* 226 (2003) 165–182.
17. M. Bertolini, M. Bevilacqua, F. E. Ciarapica, G. Giacchetta, Development of risk based inspection and maintenance procedures of an oil refinery, *Journal of Loss Prevention in the Process Industries* 22 (2009) 244–253.
18. M. S. Kadyan, R. Kumar, Availability and profit analysis of a feeding system in sugar industry, *International Journal of System Assurance Engineering and Management* 8 (2017) 301–316.
19. M. Kakkar, A. K. Chitkara, J. Bhatti, Probability analysis of a complex system working in a sugar mill with repair equipment failure and correlated life time, *Mathematical Journal of Interdisciplinary Sciences* 1 (1) (2012) 57–66.

Index

M

N

O

P

R

S

T

V

W